D0825454

GEORGE GAYLORD SIMPSON

The Major Features of Evolution

A Clarion Book

PUBLISHED BY SIMON AND SCHUSTER

A CLARION BOOK
PUBLISHED BY SIMON AND SCHUSTER
ROCKEFELLER CENTER, 630 FIFTH AVENUE
NEW YORK, NEW YORK 10020

THIS VOLUME INCLUDES MATERIAL FROM *Tempo and Mode in Evolution*
PUBLISHED BY COLUMBIA UNIVERSITY PRESS, 1944

REPRINTED BY ARRANGEMENT WITH COLUMBIA UNIVERSITY PRESS

FIRST PAPERBACK PRINTING

LIBRARY OF CONGRESS CATALOG CARD NUMBER: 53-10263
MANUFACTURED IN THE UNITED STATES OF AMERICA
PRINTED BY MURRAY PRINTING CO., FORGE VILLAGE, MASS.
BOUND BY ELECTRONIC PERFECT BINDERS, INC., BROOKLYN, N.Y.

Gratefully dedicated to colleagues in the study of the major features of evolution:

EDWIN H. COLBERT
NORMAN D. NEWELL
OTTO H. HAAS
BOBB SCHAEFFER

Preface

When a revision of *Tempo and Mode in Evolution* was called for, my feeling was that the book had served its purpose and should be allowed to fossilize quietly. That work was written at intervals between the spring of 1938 and the summer of 1942. At that time it was to me a new and exciting idea to try to apply population genetics to interpretation of the fossil record and conversely to check the broader validity of genetical theory and to extend its field by means of the fossil record. That idea is now a commonplace. I like to think that *Tempo and Mode in Evolution* helped to produce its own rapid obsolescence, although of course many others had equal and greater parts in the movement that has brought paleontology, genetics, systematics, and other biological disciplines into a synthesis of evolutionary theory.

Much more work in the field of that book has been done since it was written than in all the years before. The last ten years have produced a veritable revolution in knowledge of evolution and, what is even more important, in the breadth and nature of approach to evolutionary theory. The Society for the Study of Evolution has been founded and has published five volumes of its international journal *Evolution*, every issue of which has contained articles pertinent to the themes of *Tempo and Mode in Evolution*. Innumerable other short studies and monographs in what may be called the new spirit of evolutionary theory have appeared in other journals. The National Research Council's Committee on Common Problems of Genetics, Paleontology, and Systematics completed its deliberations, held a conference in cooperation with Princeton University, and prepared a symposial volume (Jepsen, Mayr, and Simpson, 1949). Other conferences in this general field were held, not only in the United States but also in Great Britain, France, Italy, Australia, and elsewhere.

A number of important summarizing and reviewing volumes in the new spirit of evolutionary theory have also appeared in the last ten years, outstanding among them those by Huxley (1942),[1] Mayr (1942),[1] Heberer (1943),[1] Rensch (1947), Schmalhausen (1949), Stebbins (1950), Carter (1951), and Dobzhansky (1951).[2] Even more numerous are works less immediately of this school or in this field, but with a strong bearing on it. Included are important general works on evolutionary theory by such dissenters as Schindewolf (1950) and Cuénot (1951). In annectant fields, merely by way of example, there are the volumes by Mather (1949) and by Darlington and Mather (1950) on some aspects of genetics, by Florkin (1949) and by Blum (1951) on biochemical and biophysical factors in evolution, by de Beer (1951) on embryology and evolution, by White (1945) on cytology and evolution, by Zeuner (1946) on geological time, by Romer (1945) on vertebrate paleontology, by Gregory (1951) on vertebrate phylogeny, and by Emerson (in Allee, Emerson, Park, Park, and Schmidt, 1949) on ecology and evolution. Added to all this is the relevant serial literature, running to some thousands of titles in this period.

When I suggested that *Tempo and Mode in Evolution* could now fade away without great loss, the publishers assured me that it must still have some usefulness because it continued in steady demand. Obviously it was too out of date to permit mere reprinting indefinitely, so I reluctantly undertook revision. The plan was to go through tear sheets and simply delete or reword statements that now seem incorrect or in need of modification. This immediately proved to be completely impractical. So much has been learned and said since 1942 that revision in any such limited sense would leave the book almost as badly out of date as before. Moreover, it is not even the *sort* of book to publish in 1953. To revise it now would be like installing central heating and air conditioning in a pioneer's log cabin. What is needed is replacement by a modern structure, to fit the city that has grown up around the cabin. So, as revision proceeded, not only were practically all the original sentences discarded or rewritten but also the selection of topics was changed and expanded and the whole sequence and structure radically modified.

This, then, is another book, in the field of *Tempo and Mode in Evo-*

[1] Although *Tempo and Mode in Evolution* was not published until 1944, circumstances prevented my making any changes or additions after the summer of 1942. No note could be taken of these three volumes or of any other publications of latter 1942 and thereafter.

[2] This revision, essentially a new work, was received after completion of the present manuscript. Some references to it have been inserted.

lution and therefore drawing heavily on that earlier work, but still quite a different book. Its predecessor was a series of technical theses, speculative in large part, developed in a rather personal way, and necessarily thin in background and support. This book, although not an attempt at exhaustive review of a whole field from all points of view, is more rounded and complete within its scope. Then there was embarrassingly little in the way of directly pertinent previous evidence and discussion; now the embarrassment is one of choice among riches. The change is partly due to the fact that the author personally has grown a little less ignorant in the ten years, but it mainly results from the movement and publications referred to above. As just one example among many, the theses now draw some support from plant evolution, which was not even mentioned in the earlier book. This is possible not because I have overcome my botanical incompetence but because Stebbins has summarized the botanical evidence from just this point of view. As regards the original theses, it is gratifying that some quite speculative suggestions can now be firmly supported. Others of course require modification in the light of later knowledge. Whatever measure of originality the preceding book had is now past, but its loss may be compensated by increased solidity.

The purpose of the present book is to consider and to try to explain some of the major features of the evolution of life, as far as this can be done between two covers. These major features are most directly seen in the fossil record, but their explanation must bring in every branch of biological research. In addition to paleontology, population genetics and systematics are the sciences most directly contributing to evolutionary theory today. The fact that the latter two are by practical necessity concerned mainly with minor features of evolution only makes their synthesis in application to major features the more necessary and useful. Perhaps no one is fully competent to treat all the pertinent sciences at once. I, at least, am not; but as I wrote ten years ago: "The effort to achieve such a synthesis is so manifestly desirable that no apology is in order; the intention will hardly be criticized, whatever is said about its execution."

The quotation brings to mind a final prefatory remark. The critics were almost uniformly kind to *Tempo and Mode in Evolution.* Such critical suggestions as were made were almost all constructive and have been followed in the present book. Two reviewers did criticize the intention, however. The most adverse criticism was also the longest. It ran serially for three weeks in a British Sunday school paper and can be fully paraphrased as saying that I should never have written the book because evolution is all a lie anyway. More important were remarks by a colleague who seemed to imply that it is not quite fair for someone labeled "paleontologist" to base interpretations and conclusions on anything but fossils. I should think that this would make me a better paleontologist when I am being one, but the point here is that for the purposes of this book, at least, I am not a paleontologist. I am trying to pursue a science that is beginning to have a good many practitioners but that has no name: the science of four-dimensional biology or of time and life. Fossils are pertinent to this field when they are treated as historical records (paleontologists do not always treat them so), but *Drosophila* is equally pertinent when it exemplifies changes of populations in time.

Acknowledgments should be made first of all to the many colleagues who wrote critical reviews of *Tempo and Mode in Evolution,* who have considered its theses in the course of their own work, and who have discussed those and related topics with me. It is unfortunately impossible to make explicit mention of all who have helped in those ways. I am again indebted to Dr. Anne Roe for essential encouragement and for reading the whole manuscript for style and comprehensibility. The manuscript was edited by Miss Gladys Fornell for Columbia University Press, and I am further grateful to the Press for care in design and manufacture of the book. The index is by Publications Indexers.

Some of the illustrations are taken without special acknowledgment from *Tempo and Mode in Evolution.* Most of them are new, drawn by the Staff Illustrators Corps of The American Museum of Natural History from my sketches and data. Mr. David Kitts assisted in compilation and some other details. Mrs. Rachel H. Nichols checked some of the bibliographic data.

The final draft of the manuscript was completed in December, 1951, and only minor verbal changes have since been made. A few later publications already known to me in manuscript in 1951 were taken into consideration, but no later revision has been practicable.

G.G.S.

The American Museum of Natural History
and Columbia University
New York

Contents

Figures

Tables

The Major Features of Evolution

CHAPTER I

Rates of Evolution; Morphological Rates

HOW FAST, as a matter of fact, do animals evolve in nature? That is one of the fundamental questions regarding evolution. Some attempt to answer it is a necessary preliminary for the whole consideration of the history of life. It is therefore the first topic for discussion here, even though it plunges us at once into numerous complications and some uncertainties.

The older literature is full of subjective statements about fast or slow rates of evolution, without giving sufficient objective data to evaluate or compare rates, still less to reduce them to numerical terms, and without specifying what sort of rate is meant. Recently many excellent quantitative observations have been gathered by numerous students. These still cover only a small part of the organic realm and, in each case, only limited aspects of evolutionary rate. They do, nevertheless, provide instructive and probably representative examples of different sorts of rates in a wide variety of organisms.

Examination of these data soon convinces any thoughtful student that the expression "rate of evolution," without further qualification, is extremely ambiguous. There are fundamentally different sorts of rates. Much confusion, even in recent work, has arisen from the fact that different students were talking about different sorts of rates without clear realization or statement of the sort of rate really under consideration. The first requirement here is, therefore, some attempt to define and classify rates of evolution.

KINDS OF EVOLUTIONARY RATES

An evolutionary rate may be defined as a measure of change in organisms relative to elapsed time or to some other independent variable correlated with time. The sort of measurement used, the respect in which change has occurred, the nature of the group of organisms involved, and the scale against which change is measured may each be of several different kinds. Evolution involves changes of so many sorts that measurements of its rates must necessarily be complicated if they are to cover important aspects of the subject in an unambiguous and instructive way.

In many respects the ideal measurement of evolutionary rate would be the amount of genetic change in continuous (ancestral and descendant) populations per year or other unit of absolute time. Unfortunately such genetic rates are not now available or obtainable for the vast majority of cases in which evolutionary rates are to be studied. Quite recently a significant, but still extremely limited, start has been made, for instance by Dobzhansky (e.g., 1947) and his associates who have studied temporal changes in frequencies of chromosome arrangements in some wild populations of *Drosophila.* Such measurements have been made only for a handful of the many millions of populations now living; they do not include the genetic system as a whole; and they cover such short periods of time that the changes observed are mostly in short cycles and still cast little direct light on the secular changes involved even in such relatively small scale evolutionary phenomena as the rise of subspecies and species in nature. For periods of time such as are normally involved in the differentiation of a subspecies or species, not to mention the origin and deployment of larger taxonomic groups, it is necessary to use paleontological data. At present little or nothing can be learned directly from fossils about the genotypes of the organisms represented. There is little hope that fossils can ever be used for any extensive study of genetic evolution directly as such.

The practical use of genetic rates in any broad or long-range consideration of evolution is thus excluded. The changes that can be usefully measured for consideration of such problems are in almost all cases either morphological or taxonomic. In practice, taxonomic data represent inferences and constructs based mainly on morphological data, but they sum up and categorize the morphological changes in special ways. Moreover, they do involve concepts and usually also data that are not morphological. The distinction between morphological and taxonomic rates of evolution is important and of practical use.

Morphological and taxonomic rates have a decided, even though indirect, relationship to genetic rates. If this were not so, their bearing on evolutionary theory would be quite different. It has become a commonplace that changes in morphology, or phenotype, may be induced by factors other than changes in genotype and therefore may not reflect the latter accurately. More recently it has been recognized that changes in genotype may not be accompanied proportionately, or at all, by changes in the phenotype (see, e.g., Stern, 1949). Nevertheless, there can seldom be any doubt that well-defined morphological changes in phenotypes of successive populations, particularly as these occur over considerable periods of time in the fossil record, run parallel to genetic changes in those populations. It is therefore a proper assumption in such cases that morphological rates do reflect genetic rates, even though they are probably not exactly proportional to the latter. The assumption is even more reliable for taxonomic rates, because the concepts and usages of modern taxonomy are in part genetical even when the observed data are morphological.

The morphological and taxonomic approaches to study of tempo and mode in evolution are, moreover, valid on their own grounds and not merely as an indirect reflection of genetics. In this field, it is the organisms, themselves, the phenotypes as agents in a changing world, and the populations of these organisms that really concern us. The implicit genetic factors are not important to us for their own sake, but only because they are among the various determinants of phenotypic evolution.

In consideration of rates of evolutionary phenomena there is a fundamental dichotomy, absolutely essential to clarity but not always explicitly recognized. The changes measured may be those involved in descent within a single lineage or phylum [1] or within the multiple lineages of a larger group. The rates of such changes may be specified as "phylogenetic." [2] On the other hand, changes also occur in the numbers, or frequencies, of phyla or larger taxonomic groups existing at a given time. The rates of these changes may be called "taxonomic frequency rates." Phylogenetic rates and taxonomic frequency rates are not comparable with each other and they reflect decidedly different evolutionary phenomena.

[1] "Phylum" here means a line of descent, not a major subdivision of the Linnaean hierarchy. The word is needed in both senses and they are so different as hardly to be confused.

[2] The terminology here proposed is developed from but not quite identical with that of two previous studies (Simpson, 1944a, 1949b). It seems possible now to use a somewhat more consistent and more nearly self-explanatory set of terms.

The morphological rates most frequently studied and those usually considered in this book are phylogenetic rates; that is, they measure changes involved in descent rather than in frequency of units at any given time. They may be rates for a single, unit character, such as a standard dimension of a bone or shell, or they may be rates for character complexes. The character complexes may be linked only by occurrence in the same organism or may be more closely related either functionally or (by inference) genetically.

Among other possible sorts of morphological rates, one has received particular attention: change in one part of an organism relative to change in another or in the organism as a whole—that is, allometry. Allometry in individual development (heterauxesis) or between different contemporaneous organisms (race, species, etc., allomorphosis) may be related to but does not itself represent a rate of evolution. Allometry in successive populations of a lineage does represent an evolutionary rate, although one of a very special sort. Westoll (1950) has proposed that this phenomenon be called "lineage allomorphosis" and its measurement may be called a "lineage allomorphic rate."

There are also different ways in which morphological changes may be measured and their rates quantified. It is simplest, clearest, and usually adequate to stick to the terms of original observation, a primary measurement of some sort or a qualitative datum, or to some likewise simple ratio or frequency count of such observations. In a stimulating discussion of this aspect of rate measurement, Haldane (1949) has suggested that rates of linear and related sorts of change are more significant and more comparable if given in terms not of original measurement but of percentage increase or of change in natural logarithm (both of which are essentially equivalent ratios of the original measurements). If genetic factors for size change tend to act in proportion to existing size rather than in absolute steps—and this certainly seems to be the case—then some such system of proportional rate measurement is desirable in most instances. Of course its use depends on previous calculation of absolute change.

In the same paper, Haldane has also suggested that the length of time involved in a (population mean) change of one standard deviation may be a significant measure of evolutionary rate. The logic is that the variation in a population constitutes the raw material for evolution [insofar as this is controlled by natural selection], and that the standard deviation measures the [absolute, not relative] variation, so that a rate in terms of standard deviations measures utilization of the raw material. Rate of change in standard deviations also tends to measure rate of reduction of overlap in two related populations. This is an interesting suggestion and it is to be hoped that it may be tried out for various groups, but it seems unlikely that this measure will be widely used. The calculation involved is extensive, although far from prohibitive if justified by the results. The standard deviation is less accurately determinable than the mean (as Haldane notes). Absolute or relative change of the mean is usually the essential point in evolutionary rate problems and the relationship between this and change in units of standard deviations is not simple or constant. With equal absolute or relative changes in the means of two populations (therefore equal rates of evolution in the sense usually considered effective) the rate in standard deviations will be lower for the more variable of the two.

Haldane also suggests the term "darwin" for an increase or decrease of size by a factor *e* per million years, or of $\frac{1}{1000}$ per 1,000 years. He shows, for instance, that changes in horse tooth dimensions exemplified later in this chapter proceeded at rates on the order of 40 millidarwins. At present this seems an unnecessary complication and darwins will not be used as rate units in this book, but the suggestion is repeated here for possible future trial.

Taxonomic rates may, of course, apply to any recognized taxonomic units from subspecies to classes or phyla. They may be either phylogenetic or frequency rates. Phylogenetic taxonomic rates are in terms of the amounts of change evaluated as representing a given taxonomic level. Such a rate may, for instance, be the average duration of a genus in millions of years. It has proved useful to distinguish two different but related rates of this sort, depending essentially on the sort of data available. In some cases a direct, ancestral and descendant phylum is represented by a sequence of fossils and divided into strictly successive taxonomic units. The real duration of each unit can then be determined more or less accurately and used as a measure of what may be called a "phyletic rate." More widely available are data for larger groups, including various separate, related lineages. In these groups the real duration of each unit is not necessarily determinable.

Such data may nevertheless give very suggestive approximations of average durations and serve to measure what may be called a "group rate" of phylogenetic, taxonomic evolution.

Taxonomic frequency rates are based on counts of taxonomic units at given, successive times. Data on taxonomic frequencies have been compiled for many groups and are often given in the form of what Zeuner (1946a and elsewhere) calls "time-frequency curves," which plot total numbers of known, more or less contemporaneous taxonomic units against a time scale of some sort. A time-frequency curve does not directly represent rates of evolution, but the slope of such a curve is a rate which may be calculated and represented as rate of change of total frequency. Still more significant among taxonomic frequency rates may be a "rate of origination," or "diversification," based on the frequency of first appearances of taxonomic groups per million years or other unit, or a "rate of extinction" similarly based on last occurrences.

All of these rates of evolution, with the special exception of the lineage allomorphic rate, involve the use of a time scale or, at least, of some scale that can reasonably be correlated with time. In almost all cases the preferred scale would be that of absolute time, in years or multiples (or fractions) thereof, giving a "temporal rate" in the literal sense of "temporal." Temporal rates will be extensively used and discussed on following pages, but it must be recognized that these are almost always only very rough approximations. Absolute dating against the geological time scale is still subject to very large margins of error (see Knopf, 1949). Yet the approximations do seem to be close enough to give generally consistent and comparable results which, at worst, indicate correct orders of magnitude. Except as they may copy each others' scales, no two authors give precisely the same figures for absolute dates in geological time, but most recent workers use scales sufficiently similar so that results, understood to be approximate in any case, are sufficiently comparable for most studies of evolutionary rates. For examples of such different but comparable scales see Schindewolf (1950a and b), Simpson (1947a, 1949a), and Zeuner (1946a).

Particularly for relatively short spans of geological time, attempted use of an absolute time scale may be quite excessively inaccurate in the present state of knowledge, and for this or other reasons it may be more convenient or significant to express rate of change on some other scale. The measurement of duration then used must be correlated with time, even though the correlation may be loose and irregular and the absolute time equivalence not determinable. Most generally useful of such "correlative temporal rates" is one measured against thickness of strata, a "stratigraphic rate," which may be fairly accurate over relatively short spans and in beds uniform in character and, presumably, in rate of deposition. Cyclic phenomena of sedimentation, such as varves or cyclothems, may give particularly accurate scales which may even (in the case of varves) be convertible to absolute time, but such deposits are very restricted in occurrence.

Finally, rates may be suggested and sometimes graphed in an enlightening way against the sequential (not durational) geological time scale, in eras, periods, epochs, and ages, but these units are known to be of very unequal length and therefore such data do not directly permit usable numerical expression of rates. The usual raw data for rate estimates are nevertheless in terms of the geological scale. For quantification, these data may then be converted to a more uniform scale, usually by means of estimates of the absolute durations of periods and epochs.

Although the usual approach in scaling evolutionary events is to seek sequential, relative, or absolute dating from geological data, a reversed procedure is also rather frequent and will be exemplified below. Evolutionary changes may be used to obtain estimates of the duration of geological events or sequential time subdivisions. Still another procedure, not now in frequent use but perhaps more promising than has yet been realized, is strictly biological rather than biological plus geological. Evolution in one group may be studied by its rate relative to changes in another contemporaneous group, especially if the latter can be presumed to have evolved at a fairly uniform rate. A special method of this sort would be an adaptation of allometry to relative changes in size and proportions between homologous parts of different, contemporaneous phyla.

For fuller discussion of temporal and correlative temporal scales see Zeuner (1946a) and Schindewolf (1950a and b). For summary of some pertinent sedimentational and stratigraphic methods and data see Krumbein and Sloss (1951).

A classification of evolutionary rates along the lines sketched above is given in Table 1. The more important and significant of these rates will be exemplified and further discussed in the rest of this chapter. The list is far from exhaustive. Rates of change in abundance of individuals, in ecological associations, and others are, for instance, also evolutionary rates, but their consideration is less necessary for the purposes of the present book.

TABLE 1

CLASSIFICATION AND TERMINOLOGY OF SOME RATES OF EVOLUTION

I. Genetic rates
II. Morphological rates (phylogenetic)
 A. Rates of unit characters
 B. Rates of character complexes
 C. Lineage allomorphic rates
III. Taxonomic rates
 A. Phylogenetic taxonomic rates
  1. Phyletic rates
  2. Group rates
 B. Taxonomic frequency rates
  1. Rates of change of total frequency
  2. Rates of origination (first appearances)
  3. Rates of extinction (last occurrences)

Most of the above rates may be expressed against the following sorts of scales:

I. Temporal (by absolute time), or
II. Correlative temporal, among which are:
 A. Stratigraphic (by thickness of strata)
 B. Cyclic (by varves, cyclothems, etc.)
 C. Sequential or geological (by periods, ages, etc.)
III. Biological (by relative change in other organs or organisms)

Allometric (specifically, lineage allomorphic, II C, above) rates do not usually involve a temporal or correlative temporal scale but a biological scale (III, here).

MORPHOLOGICAL RATES OF UNIT AND OF RELATED CHARACTERS

The simplest sort of morphological rate of evolution in unit characters may be exemplified by changes in linear dimensions of horse (i.e., broadly equid) teeth. At the same time, the example may be used to show how two such characters, unitary from the descriptive, morphological point of view, may combine functionally to form another character of great importance for the evolution of the animals concerned. The unit characters involved are a vertical and a horizontal dimension of the crowns of cheek teeth. These characters may be studied separately, but they are two dimensions of the same structures, with an obvious anatomical relationship. Together, they are functionally involved in a striking, progressive change in equid teeth, the rise of hypsodonty, or increase in height of cheek tooth crowns relative to their horizontal dimensions.

Both vertical and horizontal dimensions of these teeth are positively correlated with gross size of the animal (and hence with almost all its other linear dimensions) in all three of the possible ways: within populations, between contemporaneous populations, and between successive populations. Hypsodonty, the relationship between vertical and horizontal dimensions, is positively correlated with size and with most linear dimensions among successive populations, but shows no such correlations among individuals or among contemporaneous populations.[3] The successive intergroup correlation is thus spurious, like so many correlations between temporal sequences. Hypsodonty and size both developed progressively but did so independently. The horses became larger and more hypsodont, but the two characters seem to be separately determined in a genetic sense. Any real relationship was evidently indirect and nongenetic, for instance, through natural selection because greater hypsodonty assists the survival of larger animals.

Hypsodonty is one of the most important elements in horse evolution, but there are few good numerical data on it. For the present illustrative purposes five small samples have been selected from the American Museum collections and the essential data gathered and analyzed. Among many possible measures of hypsodonty, the following index was selected as best adapted to the available material: 100 × (paracone height) / (ectoloph length).

Measurements were made on unworn M^3. The samples have the following identifications and specifications:

Hyracotherium borealis: Early Eocene, Willwood Formation, Bighorn Basin, Wyoming

Mesohippus bairdi: Middle Oligocene, lower Brulé Formation, Big Badlands, South Dakota

[3] This statement agrees with, but is not proven by, calculated correlation coefficients. Hypsodonty can only be measured as a ratio, and the statistical correlation of a ratio with one of its elements or with a variate correlated with the latter is frequently spurious. Nevertheless, the stated independence is an evident and, I believe, incontrovertible biological fact.

Merychippus paniensis: Late Miocene, associated in a "Lower Snake Creek" deposit, Nebraska

Neohipparion occidentale: Late early or early middle Pliocene, associated in an "Upper Snake Creek" deposit, Nebraska

Hypohippus osborni: Late Miocene, associated in a "Lower Snake Creek" deposit, Nebraska.

Hyracotherium–Mesohippus–Merychippus–Neohipparion and *Hyracotherium–Mesohippus–Hypohippus* represent approximate genetic phyla. *Neohipparion* and *Hypohippus* are thus typical of divergent phyla of common ancestry (slightly beyond the *Mesohippus* stage).

Some of the pertinent statistics [4] are given in Table 2.

The same data are graphically shown in Figure 1, set up to represent the hypothesis that the (geometric) growth rate of ectoloph length

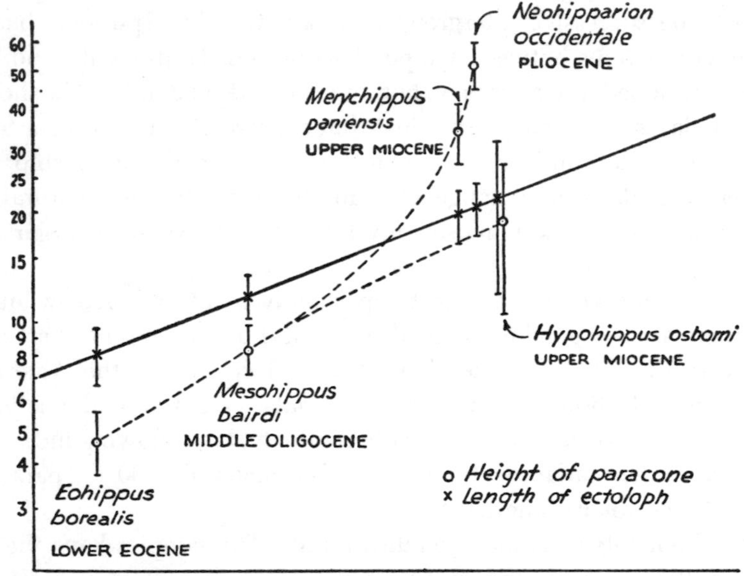

FIGURE 1. RELATIONSHIPS OF STRUCTURAL CHANGES IN TWO CHARACTERS OF FIVE GENERA OF EQUIDAE. Height of unworn paracone and length of ectoloph of M^3, data in text. Ordinate scale logarithmic, no abscissal scale. Arranged on hypothesis of rectilinear mean increase in ectoloph length. Circles and crosses are means of samples. Vertical lines are standard ranges, statistical estimates of variation in a population of 1,000 individuals.

[4] Here and elsewhere, it is assumed that the reader is familiar with elementary statistics. For an introduction to this subject, see Simpson and Roe (1939).

was constant (by placing these values in a straight line on semilog coordinates). The assumption that gross size increase (with which ectoloph length is closely correlated) was approximately constant in rate has been made for this and other so-called orthogenetic series. The diagram shows, in fact, that the hypothesis is false, for if it were true the horizontal distances between species would be proportionate to the geologic ages, whereas *Hypohippus osborni* comes out much too far to the right, the distance from *Neohipparion occidentale* to *Merychippus*

TABLE 2

MEASUREMENTS OF M^3 OF FIVE SAMPLES OF FOSSIL HORSES

	N[a]	*O.R.*	*S.R.* (*S.D.*)	*M*	σ
		A. PARACONE HEIGHT IN MILLIMETERS			
Hyracotherium borealis	11	4.2– 5.1	1.9	4.67 ± 0.09	0.29 ± 0.06
Mesohippus bairdi	14	7.8– 9.4	2.6	8.36 ± 0.11	0.40 ± 0.08
Merychippus paniensis	13	29.6–37.6	13.0	34.08 ± 0.56	2.01 ± 0.39
Neohipparion occidentale	5	49–55	15.6	52.40 ± 1.08	2.41 ± 0.76
Hypohippus osborni	4	16.7–22.4	17.0	18.75 ± 1.31	2.62 ± 0.93
		B. ECTOLOPH LENGTH IN MILLIMETERS			
Hyracotherium borealis	11	7.6– 8.9	3.0	8.21 ± 0.14	0.46 ± 0.10
Mesohippus bairdi	14	11.0–13.0	3.6	11.89 ± 0.15	0.55 ± 0.10
Merychippus paniensis	13	17.7–21.7	6.9	19.96 ± 0.29	1.06 ± 0.21
Neohipparion occidentale	5	19–22	7.1	20.80 ± 0.49	1.10 ± 0.35
Hypohippus osborni	4	19.4–26.4	20.5	22.03 ± 1.59	3.17 ± 1.12
		C. 100 × A/B			
Hyracotherium borealis	11	54–60	11.7	57.0 ± 0.5	1.8 ± 0.4
Mesohippus bairdi	14	64–77	20.7	70.4 ± 0.9	3.2 ± 0.6
Merychippus paniensis	13	155–184	48.0	170.7 ± 2.1	7.4 ± 1.5
Neohipparion occidentale	5	241–262	61.6	252.2 ± 4.3	9.5 ± 3.0
Hypohippus osborni	4	84–88	11.0	85.5 ± 0.9	1.7 ± 0.6

[a] N, size of sample; O.R., observed range (by extreme measurements); S.R. (S.D.), standard range from standard deviation (by span); M, means; σ, standard deviation. The last two with standard errors.

paniensis is surely too small, and that from the latter species to *Mesohippus bairdi* is probably too large, relative to the *Hyracotherium–Mesohippus* distance. The more important conclusion from the diagram is, however, that the rate of evolution of paracone height behaved in a very different way from that of ectoloph length. Plotting these as for relative growth (Fig. 2), it is seen that the rate for paracone height

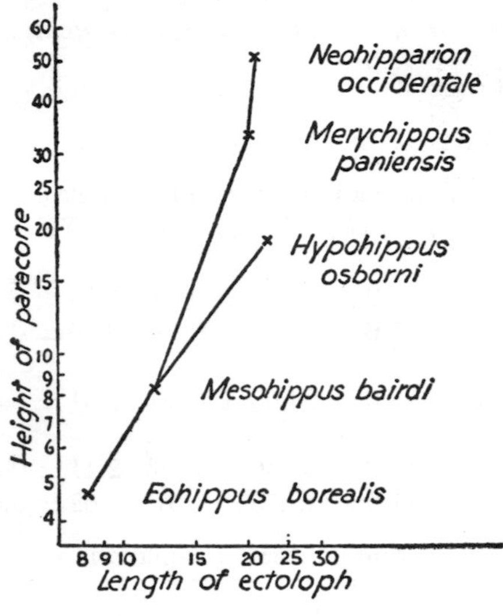

FIGURE 2. RELATIVE RATES OF EVOLUTION IN EQUID MOLARS. Same data as for Figure 1 (means of samples only) with paracone height plotted against ectoloph length on double logarithmic coordinates.

was higher than for ectoloph length throughout, that the ratio of these rates was approximately constant in the *Hyracotherium–Hypohippus* line, but that the rate for paracone height showed differential acceleration in the *Mesohippus–Neohipparion* line.

In Figure 3 an attempt is made to show the true temporal trends in these rates. The plot is semilog, with time on the arithmetic scale. The precise absolute or relative lengths of the Tertiary epochs are unknown, but the relative lengths were probably more or less as shown. It is seen that neither rate was constant in the two phyletic series involved. Both show acceleration toward the end of the Oligocene. The acceleration in increase of paracone height was greater and continued until about the middle Miocene. Both show deceleration in the late Miocene; this is more pronounced in ectoloph length, so that increase of this dimension almost ceases in the Pliocene. The rate is more nearly constant for ectoloph length than for paracone height. For ectoloph length the rates in the two phyla diverge very little—the data are insufficient to prove

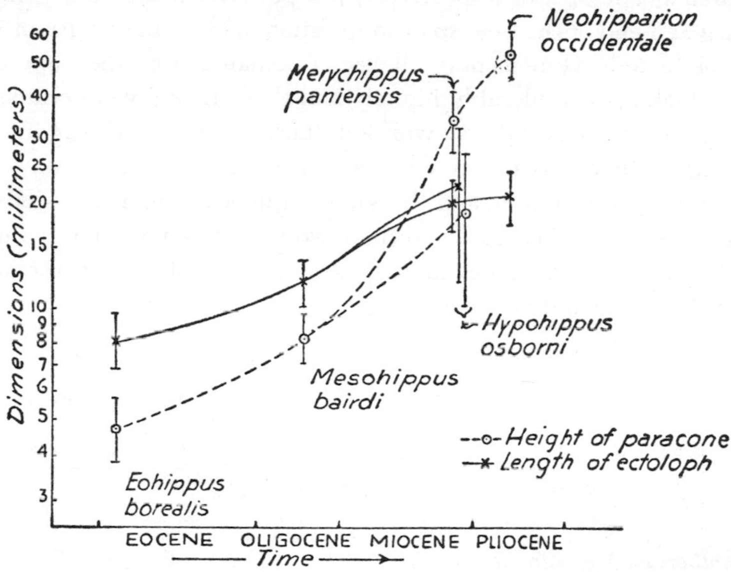

FIGURE 3. PHYLOGENY OF FIVE GENERA OF EQUIDAE AND EVOLUTION OF TWO RELATED TOOTH CHARACTERS. Data as for Figure 1. Ordinate scale, logarithmic, absolute dimensions. Abscissal scale, arithmetic, approximation of absolute lapse of time (probable relative lengths of epochs represented in arithmetic proportion). Slopes of lines are proportionate to rates of evolution of the unit characters.

that the small apparent divergence is not due to chance. For paracone height, however, the rates in the two phyla diverge markedly and significantly. In the Miocene this dimension increased much more rapidly in the line leading to *Neohipparion* than in that leading to *Hypohippus*. The result is that *Hypohippus*, although decidedly more hypsodont than the common ancestry of the two lines, is much less hypsodont than the contemporary *Merychippus*.

Morphological rates for paracone height, ectoloph length, and the hypsodonty index, considered as three unit characters, are given in Table 3. Each rate is given in two ways, in terms of mean absolute change of the population means per million years and in terms of mean percentage increase per million years. The figures for mean percentage increase are calculated by the formula of Haldane (1949), i.e.,

$$\frac{\log_e x_2 - \log_e x_1}{t}$$

in which x_1 and x_2 are, respectively, the population means at the beginning and end of the time span in question and t is the length of that span. Haldane gave percentage figures calculated from the same data all of which are considerably higher than those here given because he used figures for elapsed time lower than those I now use. The figures are rough approximations in any case, and Haldane's and mine have the same orders of magnitude and the same sequence of relative values, so that general conclusions based on them would not be importantly different. The various differences in rates are clearly reflected in the same way in this table and in Figures 2–3.

TABLE 3

EVOLUTIONARY RATES IN DIMENSIONS OF M^3 OF FOSSIL HORSES

LINEAGE STAGES	MEAN CHANGE PER MILLION YEARS					
	A. Paracone height		*B. Ectoloph length*		*C. 100 A/B*	
	Mm.	*Percent*	*Mm.*	*Percent*	*Absolute*	*Percent*
Hyracotherium-Mesohippus	.15	2.4	.15	1.5	.56	.9
Mesohippus-Hypohippus	.58	4.5	.56	3.4	.89	1.1
Mesohippus-Merychippus	1.43	7.9	.45	2.9	5.58	4.9
Merychippus-Neohipparion	1.61	5.9	.12	.6	11.65	5.5

Hypsodonty is a single character from a physiological point of view, and it is unquestionably a unit in its interaction with the pressure of natural selection; but it here appears as a resultant of two other and simpler characters evolving with considerable, not complete, independence from each other. The probable reasons for the slow advance of hypsodonty throughout and its marked acceleration in one line during the Miocene will be discussed in a later section.

Study of the other essential differences between *Merychippus* and *Hypohippus* shows that most of them arose because of differences in rates of evolution, both the rates and the differences between them being distinctive for each character. This is a widespread evolutionary phenomenon in diverging phyla of abundant animals evolving at moderate rates.

This example from the Equidae illustrates three basic theorems concerning rates of evolution:

1. The rate of evolution of any character or combination of characters may change markedly at any time in phyletic evolution, even though the direction of evolution remain the same.

2. The rates of evolution of two or more characters within a single phylum may change independently.

3. Two phyla of common ancestry may become differentiated by differences in rates of evolution.

MORPHOLOGICAL RATES ON A CORRELATIVE TEMPORAL SCALE

A correlative temporal rate can be determined when the values of a morphological variate are correlated with some variate external to the animals and the latter variate is, in turn, correlated with time. A frequently obtainable correlative scale may be based on thickness of strata, and rates relative to such thicknesses have frequently been used, especially for marine invertebrates. That the stratigraphic succession is temporal, when correctly determined, is a proven fact, but the validity of the method also demands a constant relationship between not only succession but also thickness of strata and time. It demands that the rate of deposition shall have been approximately constant. This is difficult to prove and frequently is not true, so that the usefulness of the method is sharply limited. Moreover, the usual proof of sedimentary continuity is faunal continuity, and then the determination of rates of evolution relative to sedimentary units is logically questionable. Nevertheless, the method can give some information about fluctuations in rates of evolution, and it is a valid way of comparing the rates of different characters or of different phyla during the same span of time (identical sequences of strata).

Demonstration that a character has significantly evolved in a given series of strata may be made, among other ways, by calculation of a correlation coefficient or by analysis of variance into intersample and intrasample, the samples being successive. These figures do not measure any rate of evolution, however, and of several possible measures of rate probably the most useful is the regression coefficient in cases of nearly rectilinear regression against thickness of strata.

One of the most extensive and completely analyzed sets of data from which correlative temporal rates can be obtained was given by Brinkmann (1929) for the ammonite genus *Kosmoceras* through about 13 meters of predominantly fine-grained sediments near Peterborough, England. Table 4 presents typical data, rearranged and somewhat revaluated, from statistics calculated by Brinkmann.

Only for the groups of strata in which the correlation coefficient is significant do the data demonstrate that any change in the character took place (i.e., that either the correlation coefficient or the regression coefficient is likely to exceed zero). The value of such subdivided data is that they reveal that rates have changed, not simply that evolution did occur or what its average value is over a long span. In dealing with a single character, comparisons are valid only if no hiatus is present within any one class of the stratigraphic sequence. This is not apparent from the table itself, but other information in Brinkmann's paper shows that there is no significant hiatus within this table.

TABLE 4

CORRELATION AND REGRESSION WITH THICKNESS OF STRATA FOR TWO CHARACTERS OF KOSMOCERAS (ZUGOKOSMOCERAS)

Character	*Strata Distance in cm. from bottom of section*	*N* [a]	*r*	*b*
I Terminal diameter	b. 26–28	23	.32	3.0 ± 1.8
	c. 29–39	32	.34	0.55 ± 0.26
	d. 40–50	25	.24	0.58 ± 0.46
	e. 56–78	19	.45 [b]	1.6 ± 0.6
	f. 79–134	32	.47 [c]	0.22 ± 0.06
II Diameter at disappearance of outer nodes	a. 7–20	35	.08	0.10 ± 0.22
	b. 26–28	67	.07	0.40 ± 0.70
	c. 29–39	80	.34 [c]	0.55 ± 0.16
	d. 40–50	74	.14	0.33 ± 0.28
	e. 56–78	147	.52 [c]	0.96 ± 0.11
	f. 79–134	96	.23 [b]	0.11 ± 0.05

[a] N, size of sample; r, correlation coefficient; b, regression coefficient, change of character in millimeters per 1 cm. of strata.
[b] Significant.
[c] Highly significant.

For the terminal diameter (character I), the table shows that change was certainly occurring while strata groups *e* and *f* were laid down and that the regression was considerably faster in *e* than in *f*. It does not show that any change occurred in *b–d,* inclusive, or whether the regression was then faster or slower than in *e* and *f*. For the diameter at disappearance of outer nodes (character II), change is shown to have occurred in *c, e,* and *f,* and regression was faster in *e* than in *f*. In *c* the rate was probably intermediate, but it is not certain that it was either slower than in *e* or faster than in *f*. The rate was probably slower in *a* than in *e*. Both characters I and II were evolving at about the same rate and accelerated and decelerated together.

Since these rates are relative to thicknesses of strata, the demonstrated changes in them may mean either that rate of evolution changed or that rate of sedimentation did. If the two characters were not correlated with each other, the tendency for rates to vary together would suggest (but not prove) that the variation was mainly in sedimentation. In fact, they are highly correlated with each other ($r = +.66 \pm .15$ for beds 65–70, where the greatest change in regression occurs, and $r = +.85 \pm .02$ for the whole sequence). Other data show that there are, for some characters at least, real changes in rate of evolution in the lines studied by Brinkmann. For instance, in the subgenus *Kosmoceras* (*Anakosmoceras*) regression of the "bundling index" (*Bündelungsziffer*) is negative in strata 1080–1093 and positive thereafter, a change that cannot be caused by rate of sedimentation. In other cases a fairly high positive regression is followed or preceded by one so nearly zero over so long a sequence of strata that explanation by rapidity of sedimentation is incredible. For instance, after strata 26–134 of the table, in which it shows well-marked regression, terminal diameter in *Kosmoceras* (*Zugokosmoceras*) shows no significant regression from 136 to 380, a sequence of strata more than 2½ times as thick.[5]

The ammonites form one of the groups in which great regularity of evolution, as to both direction and rate, have been claimed. Brinkmann's materials afford exceptionally good conditions for the demonstration of this regularity, if it exists, and his data are remarkably complete and objective. Although they cannot rigidly prove irregularity, because alternative explanations cannot be completely ruled out, they strongly suggest it. In any event, they fail to confirm the usual conclusion, based on fewer observations and far more subjective methods of inference.

RATES OF CHARACTER COMPLEXES

A balanced rate summing up total morphological change, as opposed to rates for one or a few selected characters, would be of the greatest

[5] This character, like several others, also shows reversal of trend; but, since it occurs during a hiatus, this may have been caused by local extinction and repeopling by a less progressive stock from elsewhere.

value for the study of evolution. Matthew (1914) wrote that "to select a few of the great number of structural differences would be almost certainly misleading; to average them all would entail many thousands of measurements for each species or genus compared." Although the mere multiplicity of measurements is not an insuperable problem, I was formerly (in the predecessor of this book, Simpson, 1944a) inclined to agree with Matthew that measurement of total morphological change is impracticable and that the fruitful approach to this sort of problem is by taxonomic and not morphological rates. This skepticism need not now be quite so strong. It is still true that the taxonomic approach is more generally useful and that a really global measurement of morphological change has not been devised. Nevertheless, progress is being made in treatment of character complexes that are far from global but that probably can lead to essentially the same conclusions about rates of evolution that would be obtainable from the impossible measurement of total change.

The most important difficulties here are two: first, the difficulty inherent in correlation and weighting of characters and second, that arising from the need to combine observations on characters which cannot all be measured in the same terms and some of which are exceedingly difficult to measure at all. Some morphological characters are highly correlated with each other, some are slightly correlated, and some vary quite independently. Some correlations are produced by genetic factors, some by natural selection for functional harmony, and some apparently in other ways. Evidently the inclusion in a global rate of a large number of closely correlated characters, all affected by some one factor, will give a misleading idea of the real nature and degree of the evolutionary change involved. The problems of choice and weighting are illustrated by equid hypsodonty, previously discussed. By the criteria of physiological function and selection value, hypsodonty is an important unit character, but morphologically it is the resultant of two other characters which are correlated in one way and not in another, and genetically it is probably affected by various different genes some or all of which may simultaneously affect other quite distinct phenotypic characters.

It is improbable that this sort of difficulty can be fully overcome, but a limited compromise seems possible. Statistical methods such as the path coefficients of Wright or the vector analysis of Thurstone may make it practical, first, to evaluate and weigh covariance and relative contributions to this. Second, on this basis it is theoretically possible to designate a limited number of characters for measurement, changes in which are relatively independent, and combination of which is more practicable and significant than use of a very large number of unselected characters. Third, it may be possible to produce a reasonable combined measurement of changes in all these key characters simultaneously by some such method as Fisher's discriminant functions. The amount of work involved is formidable and it is not yet clear that the results would be sufficiently useful in a study of evolutionary rates. To the best of my knowledge, no adequate study of this sort has yet been published and no concrete example can here be given, but work of this sort is under way (e.g., by Miller and Olson in Chicago) and its potentialities may be great.

Studies of the sort just suggested can hardly be expected to solve the problem of incommensurate characters. A combined measure by most conventional methods must be limited to the same sorts of characters, e.g., to linear dimensions, and omit others that may be equally or more important—things such as color, angulation, segmental counts, and, especially, characters of form and pattern that defy any practicable or, at least, simple means of quantitative measurement. Another sort of compromise evades this difficulty and is already in fruitful use for studies of evolutionary rates. This is to select a list of characters that can be designated as present or absent, or as developed in greater or less degree, and to compile a score on this basis for each stage in a phylogeny. Rate of change in this score is then a valid measure of rate of evolution in the whole complex of selected characters, which may be of any sort. Obviously this method, too, has strict limitations, but it has proved useful and seems to have great promise.

Selection of characters for such use and their evaluation, if they are not clearly and simply either present or absent, are necessarily rather subjective. Nevertheless a good anatomist thoroughly familiar with the group can produce consistent and clearly significant results. After all, the personal element in taxonomic rates is even greater and yet these have certainly revealed a great deal about evolution. Characters selected will generally be those with little variation in any one population but with distinct change in the course of evolution. They will generally be of some functional importance. As far as possible, variation should be independent, or nearly so, in the different characters used. For these same reasons, the best characters for arriving at a score of this sort will often be the characters most used in formal taxonomy—to distinguish genera, for instance.

Olson (1944) has scored mammallike reptiles on the basis of morphological approach in various characters to a mammalian condition. He was thus able to show approximately the degree and rate of progression in different groups toward a more advanced, mammallike structure. Westoll (1949) has applied similar methods to the lungfishes. He selected 21 (with certain subdivisions, actually 26) rather broad sorts of characters, such as skull proportions, nature of dermal bones, type of dentition, and nature of paired fins, for each of which he could designate from 3 to 8 grades of structure. The grades were scored according to whether they appear in oldest, most primitive lungfishes, in which case they were given highest value, or in latest and most specialized, in which case they were rated zero. Total scores ranged from 89–96 for the earliest known form (100 for its presumed ancestor) to 0 for two of the three living genera. Plotting these scores against the approximately 300 million years of lungfish history gives a picture of changing degree of average primitiveness of the animals through that span. The slope of the resulting curve is approximately proportional to rate of loss of primitive (or, at least, ancient) characters, a form of morphological, character complex rate of evolution.

Westoll's example is so instructive and he has based on it so important a discussion of topics treated later in this book, that his data are summarized here in altered forms in Tables 5–6 and Figure 4.

Westoll's ratings of characters in these lungfishes are taken as basic data, but for present purposes they are handled somewhat differently. New scores have been calculated by Westoll's method and on his ratings—but giving 0 value to the most primitive condition and 2 to 7 to the most advanced, specialized, or latest. This scoring indicates acquisition of new rather than loss of old characters, making it more comparable with other examples and giving positive rather than negative values for rate of evolution. Scores have also been somewhat simplified so as to give a single figure for each form included. In spite of the subjective nature of the scoring and the uncertainty of the time scale, the result is a smooth and plausible curve, Figure 4B. The only poor fit is for *Conchopoma,* indicated by Westoll on other grounds as

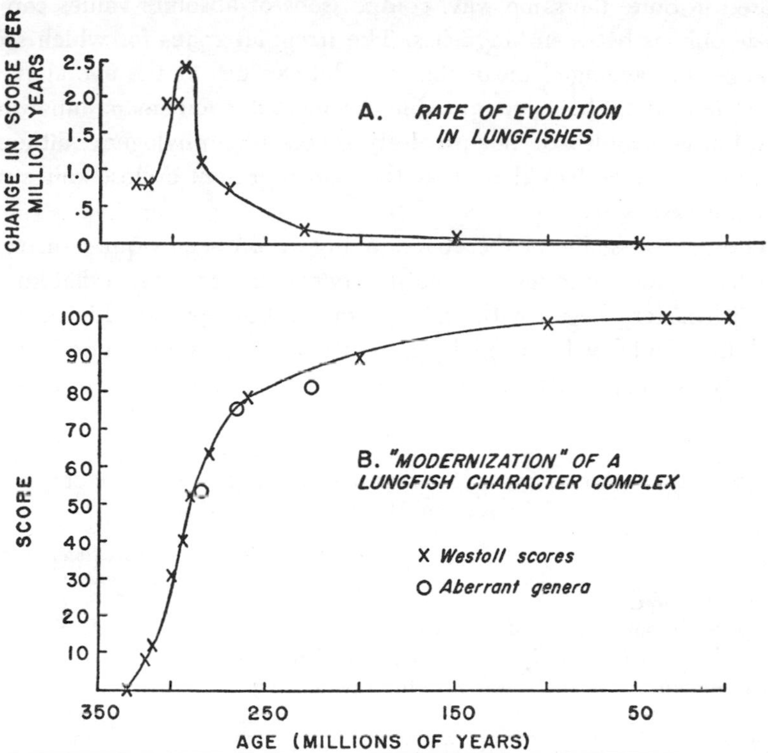

FIGURE 4. EVOLUTION OF A CHARACTER COMPLEX IN LUNGFISHES. A, rate, see Table 6. B, scores, see Table 5. A is essentially a reflection of the slope of B.

being an aberrant form outside the main lines of lungfish evolution. The other two aberrant genera designated by Westoll, *Fleurantia* and *Uronemus*, do fit the curve fairly well. In Table 6, changes in scores are tabulated against rough estimates of elapsed time for each stage, omitting the three forms considered aberrant by Westoll and combining two that were contemporaneous and had the same score, *Ctenodus* and *Sagenodus*. The average change in score per million years for each stage represents a character complex rate of evolution.

The values of this measure through the span of lungfish history are plotted in Figure 4A. In spite of the further approximations and uncertainties here involved, this result is also a consistent and plausible curve. The exact values have little significance. Since no other group can be scored in quite the same way, comparisons of absolute values can be made only as between lungfishes. The irregular spans for which score changes are averaged mean that absolute values of the averages are not strictly and reliably comparable throughout, even among lungfishes. The forms included do not precisely represent a phylogeny, although the curves themselves show that they do represent a close-knit structural sequence.

The general shape of the curve in Figure 4A seems quite surely to represent a factual series of evolutionary events. It is shown that known lungfish history began with an appreciable but not very high rate of evolution, quickly followed by a relatively very short period of extremely fast change. From this high point, the rate fell off rapidly at

TABLE 5

SCORING FOR ADVANCED OR SPECIALIZED CHARACTERS IN LUNGFISHES (Based on Westoll, 1949)

Form	*Age*	*Score*	*Approximate first appearance (millions) of years ago*
Ancestor (hypothetical)	Earliest Devonian?	0	325?
Dipnorhynchus lehmanni	Late early Devonian	8	315
D. sussmilchi	Early middle Devonian	12	310
Dipterus	Late middle Devonian	31	300
Pentlandia	Latest middle Devonian	40½	295
Scaumenacia	Early late Devonian	52½	290
Fleurantia *	Late Devonian	53½	285
Phaneropleuron	Latest Devonian	63½	280
Uronemus *	Middle to late Mississippian	75½	265
Ctenodus	Late Mississippian to late Pennsylvanian	78½	260
Sagenodus	Late Mississippian to early Permian	78½	260
Conchopoma *	Early Permian	81	225
Ceratodus	Triassic to Cretaceous	89½	200
Epiceratodus	Late Cretaceous to Recent	98	100
Protopterus	Oligocene to Recent	100	35
Lepidosiren	Recent	100	(No fossils)

* These are aberrant genera listed as "specialized" by Westoll.

TABLE 6

EVOLUTIONARY RATES OF A CHARACTER COMPLEX IN LUNGFISHES

Step	*Change in score*	*Elapsed time (Millions of years)*	*Ave. change per million years*
Ancestor–*D. lehmanni*	8	10?	.80
D. lehmanni–D. sussmilchi	4	5	.80
D. sussmilchi–Dipterus	9	10	.90
Dipterus–Pentlandia	9½	5	1.9
Pentlandia–Scaumenacia	12	5	2.4
Scaumenacia–Phaneropleuron	11	10	1.1
Phaneropleuron–Ctenodus, Sagenodus	15	20	.75
Ctenodus, Sagenodus–Ceratodus	11	60	.17
Ceratodus–Epiceratodus	8½	100	.08

first, then more and more slowly, approaching zero. The rate has been extremely slow, or virtually nil, for the last 150 million years or more, about half of the whole history of the group.

The theoretical significance of these facts will be discussed in later chapters. Schaeffer (1952) has recently presented a somewhat similar modification of Westoll's data and has compiled similar data for the related coelacanth fishes. The shapes of the curves for coelacanth evolution are closely comparable to those for lungfishes.

LINEAGE ALLOMORPHIC RATES

The general concepts of relative growth, formerly called heterogony and now called allometry, were made familiar by Huxley's book (1932). Huxley and his associates later modified and expanded the concepts and terminology (especially Huxley, Needham, and Lerner, 1941; Reeve and Huxley, 1945). Westoll (1949) has suggested a slight further modification of terminology in connection with evolutionary studies. In a general way, at least, the subject of allometry can now be assumed to be familiar to all biologists. It has given rise to an extensive literature on methodology, significance, and concrete applications. Review of this literature is not called for here, but special attention is directed to the recent symposium led by Zuckerman (1950).

A basic idea of allometry is that with increase in general size of organisms, parts may increase (or decrease) at different rates and that the relative values (or the ratios) of such rates may be approximately constant. The relationship can frequently be described by the allometric equation:

$$y = bx^k$$

in which y is any usable anatomical variate, x is a similar variate for another anatomical part, or some measure of gross size, and b and k are constants, designated by Huxley and others as, respectively, the initial growth constant and the equilibrium constant or growth ratio.[6] There has been, and continues to be, dispute as to the biological significance of these constants and the logical status of the equation as a whole. (See, among many others, Reeve and Huxley, 1945; Richards and Kavanagh, 1945; Sholl, 1950; Waddington, 1950.) The consensus at present seems to be that this equation does not now have a satisfactorily established logical basis—it cannot be taken as a "law of growth" in which the constants have certainly definable biological significance—but that relative growth is a real and very widespread phenomenon in nature and that this equation is often at least an adequate description of the way in which relative growth occurs.

A pioneer study of allometry and evolution was that of Robb (1935a and b) who plotted face length and total skull length in fossil and living horses. He concluded that the allometric relationship has been constant in all these forms, that changes in relative length of face have resulted not from change in growth pattern but only from changes in head (or gross body) size, and that the same sequence of changes in proportions occur in the ontogeny of recent horses. The allometric equation nearly fitting all of Robb's data is:

$$y = .25x^{1.23}$$

Inherent in Robb's conclusion is the belief that in the horses several different allometric relationships all followed the same course. Among such different relationships are:

1. Allometry in the ontogeny of individuals = heterauxesis.

2. Allometry between adults of the same stock or population but of different size = individual allomorphosis.[7]

3. Allometry between adults of different stocks, races, or subspecies of one species = race allomorphosis.

[6] Huxley and his associates now use α (Greek letter alpha) for the latter constant. I continue to use k, as originally proposed by Huxley and widely used for more than twenty-five years. Use of k is typographically simpler and at least as clear.

[7] New term on the analogy of those proposed by Westoll (1950) in extension of the Huxleyan terminology. Westoll omitted this particular sort of allomorphosis. The other terms here are those he gave.

4. Allometry between members of different species of the same genus, not lying in a single line of descent = species-form allomorphosis.

5. Allometry between members of a single phylum in direct (or, at least, approximate) line of descent = lineage allomorphosis.

Reeve and Murray (1942)[8] later pointed out that Robb's method of plotting one dimension against another including the first produces an artificial similarity in regressions that are, in fact, different. They used Robb's data, and some others, to plot face length not against total skull length but against the other component of the latter, cranium length. With this methodological improvement, the following conclusions are suggested (based on, but slightly modified from, Reeve and Murray):

1. From *Hyracotherium* through *Miohippus,* at least, growth in these two skull dimensions apparently did nearly follow simple allometry, with a growth ratio, *k,* of about 1.8.

2. At about the stage of *Merychippus* there may have been a change in growth pattern involving, in terms of the allometric equation, an increase in the initial growth constant, *b,* and a decrease in *k* [but this is extremely uncertain because it is based on only six fossil skulls the dimensions of which do not, in fact, closely approximate a simple allometric sequence and two of which, at least, were far removed from the lineage of *Equus*].

3. Early stages of ontogeny in *Equus* nearly accord with simple allometry with *k* = about 1.5, which seems to be smaller than, but might not prove significantly different from, *k* for the sequence *Hyracotherium–Mesohippus.* The value of *b* is probably larger in *Equus* ontogeny than in the latter sequence. [If conclusion 2, above, is correct, the modern growth pattern may have arisen at about the stage of *Merychippus.*]

4. The early ontogenetic growth pattern does not persist in *Equus.* After cranium length reaches about 150 mm., face and cranium elongate at nearly the same absolute rate, i.e., *k* is about 1.

5. One specimen studied by Robb, a *Hypohippus,* does not seem to lie in any of the allometric lines suggested above. [*Hypohippus* is far outside the *Merychippus–Equus* lineage. It is more nearly in a continuation of the *Hyracotherium–Mesohippus* line and the scanty data

[8] This paper was published after the manuscript of Simpson, 1944a, was in press and no further changes were possible. In that manuscript I accepted Robb's work at face value, although no conclusions essential to the main theses of the book were based on erroneous aspects of Robb's paper.

do not exclude the possibility that simple allometry in that line could lead to *Hypohippus* by extrapolation.]

This is a series of quite uncertain conclusions. Obviously many more data on skull dimensions of fossil horses are needed. (Many appropriate specimens are in collections but for various reasons have not been available for allometric study.) The final picture will certainly be much more complicated than Robb's oversimplification but his data established, and later corrections do not contradict, that relative growth is an essential factor in evolution of the horse skull.

Several other studies on allometry in fossil sequences have been made, among them Hersh (1934) on titanotheres, Robb (1936, 1937) on equid feet, Phleger (1940) on cats and oreodonts, Phleger and Putnam (1942) on oreodonts, Gray (1946) on ceratopsian dinosaurs, Romer (1948) on pelycosaurs, and Lull and Gray (1949) on ceratopsians. As might be expected, these studies show that lineage allomorphosis is complicated and that simple allometry (i.e., persistent following of the relationship $y = bx^k$) in phylogeny may occur but is far from universal.

Regardless of the status of the allometric equation, the work summarized or cited above warrants the following theses regarding lineage allomorphosis:

1. This phenomenon does occur, that is, proportions of parts of evolving organisms do commonly change in some systematic (but not necessarily simple) way with changes in gross size.
2. Allometric relations may vary within populations and may change in the course of phylogeny. They also evolve.
3. But allometry for a given anatomical region may also persist while changes in gross size or other respects occur. Proportions of parts may then change and appear to evolve even though the growth pattern producing them has not changed.

The special case of allometry is important because it provides a possible mechanism for various phenomena of evolution otherwise hardly explicable except by metaphysical speculation. Some of these points will be mentioned later. In a wider sense, these relationships of allometry to evolution are only an aspect of the general fact that it is a growth mechanism (and in a sense—not a teleological one—a growth pattern) that is inherited. The characters developed by this mechanism are not what is really inherited. From this point of view, evolution of the mechanism is more fundamental than that of the characters. Another potential importance of lineage allometry, unsatisfactory and incomplete as is present knowledge of it, is that it seemingly may provide an unusually good and direct means of inference from observable characters to the unobservable genetic mechanism (see Reeve, 1950).

CHAPTER II

Taxonomic Rates of Evolution

PHYLETIC TAXONOMIC RATES

THE POINT OF MATTHEW'S REMARK, quoted above (p. 20), is that the taxonomy of an experienced student is a more reliable indication of the total difference between organisms (better, between populations of organisms) than any collective measure yet derived for differences in objective characters. In spite of progress in evaluation of character complexes, as in the example of the lungfishes, this is still generally true. It will be noticed, too, that the stages of change in character complexes in the lungfishes (and this is almost always the case) are associated with the previously erected taxonomic system; the populations had already been defined as species and genera. The two methods are not, then, entirely alternative but also supplementary.

The weaknesses of taxonomic data for quantitative studies of evolutionary rates are fairly obvious and do not need belaboring. The assumption that genera, say, cover more or less equivalent amounts of total evolutionary change seems sufficiently valid when the genera in question are closely related and have been defined by a single, skillful student or by more or less equally able students using similar criteria. That assumption becomes, however, increasingly uncertain when the groups compared are more distantly related and more dissimilar and when their definition has been done by students with different taxonomic tendencies and using different criteria. Yet even in the less favorable cases, taxonomic data treated with reasonable judgment yield rough but useful approximations. An approximation, recognized as such, is more valuable than a spuriously exact measure, or than no measure at all. Reliability of the general result may also be considerably greater than that of any one taxonomic determination when large bodies of data can be averaged, as in taxonomic group rates.

Species can sometimes be used to advantage in studies of taxonomic rates of extinct groups, but genera are usually the most useful units at present. It is true that species, carefully defined on modern criteria and with fully adequate samples, are the least arbitrary (in this sense, the most "objective") of taxonomic units (Simpson, 1951b). But paleontological samples are seldom adequate for group studies of species. Their inadequacy results from a number of causes, among which the local nature of most paleontological sampling is especially important. Early Paleocene mammals, for instance, are known from only four localities, all in the same general region, and are very inadequately represented at two of these localities. Species usually arise allopatrically and tend in many cases to remain allopatric (Mayr, 1942). Obviously, then, our knowledge of early Paleocene mammals is worthless for study of frequencies and group evolutionary rates of species. Although the same sort of limitation applies also to paleontological genera, it is much less disabling in this case because genera are usually much more widespread than species. Sampling at few, restricted localities certainly reveals a much higher percentage of the genera than of the species that existed at any one time (see Simpson, 1936a).

This inadequacy is alleviated by constant improvement of paleontological sampling, but as regards species it can never be wholly, or perhaps even sufficiently, overcome except in the case of a few groups over relatively short periods of time. For many or most periods of earth history, paleontological samples so distributed over the earth as to reveal all or a reliably large fraction of the existing species of a major taxonomic group simply do not exist in the rocks. A second important difficulty in the use of paleontological species for study of taxonomic rates is eradicable but is nevertheless crippling at the present time. This is the fact that relatively few fossil species have been defined by what can be considered correct and sufficient modern criteria from a biological, evolutionary point of view. Many definitions can only be called capricious. Generic definition has hardly reached perfection, either, but as a rule paleontological genera seem to be adequately defined and fairly comparable with each other. At worst, they are better in definition than are species at present.

Even if species could be more extensively used in taxonomic rate studies, use of genera would also be desirable, because the latter permit clearer inferences and valid simplification in reference to broader and longer range features of evolution. Use of higher categories, along with genera, is similarly helpful, but the genus is certainly now and probably will continue to be basic in such studies. To the extent that they are well defined, genera meet the requirements of taxonomic rate determination that they be essentially monophyletic in origin, that they have a significant extension in time, and that they be horizontally divided from preceding and following units of the same rank (even though such division is usually an artifact of taxonomy).

If all genera were strictly comparable, phyletic rates of evolution would be proportional to the reciprocals of the durations of the genera in question. For a sequence of successive genera, a more reliable value would be obtained by dividing number of genera by total duration. Thus in the line *Hyracotherium–Equus* (but omitting *Equus* because its span is incomplete and indeterminate) there are eight successive genera according to some good modern classifications (e.g., Stirton, 1940). The time covered is about 60 million years,[1] and the average rate can therefore be expressed as .13 genera per million years, or reciprocally as 7.5 million years per genus.

Strictly comparable averages can only be obtained for genera that arise at known times from known ancestors and that disappear not by extinction but by evolution into other genera. The number of such genera now known in any one group is small. Some comparisons are nevertheless possible as suggested by the data in Table 7.

With due allowance for all the uncertainties involved, it appears that the rate of evolution in chalicotheres was about the same as in

[1] Here and throughout the present book, higher figures are used for the durations of the Cenozoic and its epochs than were used in its predecessor (Simpson, 1944a). One of the more reliable radioactivity dates in earth history is based on Colorado minerals about 60 million years old. It was formerly believed that these minerals came from around the beginning of the Cenozoic, and 60 million years was therefore the generally accepted span for the whole of that era. More recently (e.g., Knopf, 1949) it has appeared probable that this date belongs somewhere near the Paleocene–Eocene transition. The Cenozoic is therefore now assigned a length of 60 million years *plus* the Paleocene. The length of the Paleocene is not directly or well determined, but on criteria of amount of evolution, numbers of distinct successive faunas, thickness and complexity of strata (where these represent anything like the full span of the epoch), etc., the Paleocene must have been comparable to other Tertiary epochs in length. I now tentatively and roughly assign it about 15 million years, which results in about 75 million years for the whole length of the Cenozoic.

TABLE 7

RATES OF EVOLUTION IN HORSES, CHALICOTHERES, AND AMMONITES IN TERMS OF NUMBER OF GENERA PER MILLION YEARS

Group or Line	*Number of Genera* [a]	*Average Genera (in one line) per Million Years*	*Phylogeny and Classification Used as Basis*
Hyracotherium–Equus	8	.13	Stirton 1940
Chalicotheriidae	5	.13	Colbert 1935
Triassic and earlier ammonites	8	.05	Swinnerton 1923 and others

[a] With approximately known time of origin and time of transformation into another genus.

horses and that in generic terms it was faster in both groups of perissodactyls than in these early ammonites.[2]

Analogous estimates of changes in rate within a single line are less reliable and require weighting. Matthew (1914) tried this with horse genera, although his purpose was the reverse of the present attempt: he postulated a uniform rate of evolution and attempted to estimate lapses of time by relative amounts of evolution. Matthew's figures were severely criticized by Abel (1929) first because of their subjective and

[2] *"In generic terms"* may be stressed. Obviously no one can say whether one of these ammonite genera really represents the same amount of something as one of these horse genera. Schindewolf (1950b) objected to this figure for rate of evolution in ammonites and suggested that the usual rate was about 1.0 to .33 genera per million years, i.e., much faster than in horses. The reasons for this great discrepancy are: (1) Schindewolf's figures are not based on the same genera as mine; there is no implication here that all ammonites evolved at rates near .05, indeed this certainly was not true. (2) As he states, Schindewolf's figures are based on guide genera ("Leitgattungen"), which are deliberately selected because of their (real or apparent) exceptionally short durations, i.e. (real or apparent) high rates of evolution. (3) If I understand correctly, Schindewolf is referring to average known duration of selected genera in various lineages and not to durations of successive genera in single phyla with determinate beginning and ending; if this is the case, he is comparing a group rate with a phyletic rate. The two are not strictly comparable and group rates may tend to appear somewhat faster (because the *complete* spans of all the genera are not known). (4) Schindewolf apparently uses a more split classification than my sources, so that he is comparing rates of smaller with those of larger units; rates are usually much faster for small than for large taxonomic categories, as would be expected. The broader ammonite genera seem to me to be more comparable with those of perissodactyls, but this is admittedly a highly subjective and uncertain judgment. Schindewolf's rates are of course correct for his data, but they are not correctly comparable with my figure for rate in certain early ammonites and in no way invalidate the latter, on its own stated basis.

approximate nature and, second, because of some differences of opinion, e.g., that the step *Parahippus–Merychippus* should have been relatively longer, *Eohippus–Orohippus* and *Orohippus–Epihippus* relatively shorter. Despite the very rough nature of such approximations, which was freely admitted by Matthew as it is here, they do have considerable interest and an attempt to revise the estimates on more recent data may have some value. Such an attempt is made in Table 8.

TABLE 8

ESTIMATES OF RELATIVE AMOUNTS AND RATES OF EVOLUTION IN HORSES

Genus	*A* [a]	*B*	*C*	*D*
Equus	10	7	7	1
Pliohippus [b]	10	11	10	1
Merychippus	15	18	7½	2
Parahippus	5	5	4½	1
Miohippus	5	5	5	1
Mesohippus	15	16	7	2
Epihippus	10	9	7	1
Orohippus	10	9	7	1
Hyracotherium [c]				

[a] A, Matthew's weighting, on a basis of an average weight of 10 per genus. In each case the weight is understood to be for the approximate total advance to the midpoint of this genus from the midpoint of that preceding. B, similar weights adjusted to subsequent criticism and discovery. C, estimates of approximate time involved, in millions of years. D, rates obtained by dividing the adjusted weights (B) by the approximate time (C).

[b] *Hipparion* in Matthew. *Pliohippus* is now known to be nearer the direct line and its evolutionary stage is roughly comparable.

[c] *Eohippus* in Matthew. *Eohippus* and *Hyracotherium* are now believed to be synonymous.

On the whole, the rates thus obtained are reasonable relative values, at least to the point of showing acceleration in the late Eocene to early Oligocene and early to middle Miocene, with fairly uniform over-all rates at other times in this particular sequence of genera (which of course is not the only lineage in the Equidae).

Reverting to Matthew's original purpose, Table 9 gives estimates of the lengths of the Tertiary epochs, except Paleocene, postulating a total of 60 million years.

The general agreement of all three estimates is striking and is certainly well within the large margin of error for such estimates. Agreement between the adjusted weights and the independent estimates suggests that the average rates of horse evolution in each epoch for this

TABLE 9

ESTIMATES OF DURATIONS OF TERTIARY EPOCHS
(In millions of years)

	A [a]	*B*	*C*
Pliocene	12	10	12
Miocene	15½	17½	17
Oligocene	12	13	11
Eocene	20½	19½	20

[a] A, based on horse genera, Matthew's weighting, on his hypothesis of uniform evolution. B, same, revised weighting (B of preceding table). C, independent estimates based on a balance of all available evidence (sedimentation, radioactivity, general faunal change, etc.).

particular lineage did not differ greatly despite acceleration and deceleration during periods that were less than an epoch or that overlapped epochs. To this extent Matthew's approach to the chronological problem seems justified: it produced results about as good as those from other methods. From the point of view of evolutionary rates, however, the averaging of rates by whole epochs has concealed real and important changes in rate.

AGES OF LIVING TAXONOMIC GROUPS

Another approach to the general problem of phyletic rates of evolution involves estimates of the ages of living species, genera, or other taxonomic groups. Here must be emphasized a point related to the fundamental distinction between frequency and phylogenetic rates of evolution. The time at which two related, contemporaneous species became distinct from each other may (in theory) be a perfectly definite date. The time at which an evolving population became different enough from its ancestry to be called a different species cannot, even in theory, be a precise, naturally defined date unless the new, descendant species arose in a single, abrupt step. The latter point will be discussed later and the conclusion will be that species very seldom, groups of higher taxonomic rank practically never, do arise in a single step. Therefore determination of the time of origin of any living group involves in each case an arbitrary element, although with reasoned judgment the data surely do tell something about evolutionary rate.

Statement as to how long a species, say, has been in existence, i.e., how long its ancestry has been morphologically so similar to the living representatives as to be placed in the same species by an experienced taxonomist, clearly does somehow involve the rate of phyletic evolution, but for two reasons it does so imperfectly and somewhat ambiguously. First, the whole natural span of the species remains indeterminate because there is no terminal date; the species will presumably continue to exist for an unknown length of future time. Second, determination of the span of existence of a population more or less in its present form does not necessarily measure the time necessary for the evolutionary process of achieving that form. Neither ambiguity has much importance if the population has evolved continuously at a fairly steady rate. This condition is often probable but it certainly is not met in many cases. Exceptionally old living species (or other taxonomic groups) clearly owe this status to the fact that evolutionary change practically ceased after a remote time of origin, in which the rate may have been rather rapid.

Such cases of arrested or bradytelic evolution will be specially considered later. At present the striking examples of *Triops cancriformis,* a species around 170 million years old (known from middle late Jurassic), and of *Lingula,* a genus some 400 million years old (known from the Ordovician [3]), are merely mentioned.

On less exceptional cases of recent subspecies, species, and genera a good many data have now been accumulated. Doutt (1942) has recorded the case of *Phoca vitulina,* normally a marine seal, isolated in a Canadian lake some 3,000 to 8,000 years ago, now outside the range of variation of its marine congeners in some respects, and authoritatively given separate subspecific status. Degerbol (1939) also considers subspecifically distinct a stock of field mice (*Apodemus sylvaticus*) introduced in Iceland about a thousand years ago. Ashton and Zuckerman (1950) show that green monkeys (*Cercopithecus aethiops*) introduced on the island of St. Kitts about 300 years ago are now significantly different from the African parental stock.

On a larger scale and dealing with more usual sorts of zoogeographic

[3] Ordovician *Lingula* is specifically different from recent forms. This difference could, of course, be called generic and is so called by some specialists. Nevertheless the Ordovician and recent species are as much alike as are some recent species competently referred to one genus. It has also been argued against this extraordinary record that profound change may have occurred in parts unknown in the fossils. On this point, Schindewolf (1950b) has well remarked that the fossils do include impressions of extensive, complicated soft parts. There is really no reason to assume any important change, and the probabilities are against this.

occurrences, Beirne (1947) has divided British land mammals into a group probably isolated since the first interstadial phase of the last glaciation, 80,000 years ago by Zeuner's (somewhat disputed) dating, or possibly from the second glacial phase, 50,000 years ago; a second group isolated since the second interstadial phase, 25,000 years ago; and a third group postglacial in isolation. In the oldest group, 12 (57 percent) have become specifically distinct and the other 9 (43 percent) subspecifically. In the second, 25,000-year group, none is specifically distinct but 20 of the 22 (91 percent) are subspecifically distinct. None in the postglacial group is subspecifically distinct.

Broader considerations suggest that all the figures cited above are low, i.e., that they represent more than average rates of evolution, even for mammals. (As will later be emphasized from other but equally cogent sorts of evidence, many groups, notably marine molluscs, evolve at slower average rates than mammals.) Many if not most living species of mammals are found well back into the Pleistocene, with spans on the order of 10^5 years and upward. On the other hand, late Tertiary beds, with ages on the order of 10^6 years and upward, contain very few and usually only rather dubiously identified recent species of mammals. Of North American mammalian stocks introduced into South America in latest Pliocene and/or early Pleistocene, that is, perhaps from a million to 500 thousand years ago, probably all have become specifically distinct, many have become generically distinct, but none has become so distinct as to be reasonably placed in a new subfamily or family. As in some of the single examples noted above, this represents an increased average rate of evolution accompanying occupation of new territory. Members of the same groups remaining in North America have usually changed in appreciable but somewhat lesser degree during the same time: most of them have changed sufficiently to be referred to different species but relatively few have progressed to an extent rated as generic. In some more slowly evolving groups, such as the molluscs, most of the living species (apparently about 60 percent of molluscs on an average) have survived since the late Pliocene.

Many other estimates of ages of living species have been made, such as, for instance, a noteworthy example by Zimmerman (1948) for many groups in the Hawaiian Islands. Compilations of such data have been made by, among others, Huxley (1942), Rensch (1947), Schindewolf (1950b), and especially Zeuner (1946a).

Interesting as they undoubtedly are, such examples of the ages of individual living species seem to me to have only limited value for study of rates of evolution in general and to be far less instructive than group rates. They do show that durations and hence, although not necessarily in close proportion, rates of evolution of subspecies, species, and genera may vary enormously from group to group and even in related lines of a single taxonomic group. They also suggest that phyletic evolutionary change sufficient to be recognized as specific in rank usually takes an appreciable and often a geologically long time under natural conditions, as has been emphasized by Zeuner (1946a and b). The evidence is such that it cannot really prove that there is always and for all groups an absolute minimum time for specific progression, or an absolute maximum of possible rates. Even though the time is appreciable in all natural examples known, the evidence hardly warrants Zeuner's general rule of an absolute maximum rate of 500,000 years per species-step.

GROUP RATES AND SURVIVORSHIP

When obtainable, estimates of the average duration of genera within a phylum, like the phyletic rates exemplified above, are on the whole probably the most satisfactory phylogenetic taxonomic rates. They are, however, greatly limited by the small number of genera for which both ancestral and descendant genera are surely recognized. Use of all the genera of a larger taxonomic group introduces other sources of error: (1) most genera certainly have a fossil record shorter than their real duration; (2) the numerous genera that disappeared by extinction cannot, on the average, have undergone evolutionary changes comparable to those of genera that disappeared by transformation; and (3) the fossil record of more slowly evolving genera probably is in general more complete than that of rapidly evolving genera in the same group. The first two sources of error will tend to make estimates of rate too high, and the last to make them too low. The extent of compensation by these opposite tendencies cannot be determined. The errors are, however, systematic and more or less independent of the particular nature of the fossils in question. Thus, they deprive the rate estimates of absolute accuracy, but do not necessarily invalidate the estimates as relative rates in the comparison of different groups. With all their shortcomings, such data have proved to have considerable value and to reveal facts of great importance in evolution, as will be shown.

To explore and illustrate the possibilities, two very different groups may be taken: pelecypod molluscs and carnivorous placental mammals (excluding the pinnipeds, for which the record is wholly inadequate).

TABLE 10

DISTRIBUTION OF GENERA OF PELECYPODA
(Figures are numbers of known genera)

	LAST KNOWN APPEARANCE										TOTALS, FIRST KNOWN APPEARANCES
FIRST KNOWN APPEARANCE	*Ordovician*	*Silurian*	*Devonian*	*Carboniferous*	*Permian*	*Triassic*	*Jurassic*	*Cretaceous*	*Tertiary*	*Recent*	
Ordovician	13	8	6	1	0	4	0	0	0	1	33
Silurian	..	17	13	0	1	2	1	0	0	4	38
Devonian	..	..	30	10	4	5	1	0	0	4	54
Carboniferous	..	..	..	16	3	0	0	1	0	3	23
Permian	..	..	..	..	3	1	0	1	0	3	8
Triassic	..	..	..	..	..	30	6	12	0	20	68
Jurassic	..	..	..	..	..	..	25	12	3	16	56
Cretaceous	..	..	..	..	..	..	..	37	0	24	61
Tertiary	..	..	..	..	..	..	..	..	18	64	82
Totals, last known appearances	13	25	49	27	11	42	33	63	21	139	423

The raw data consist of the geological distributions of all the known genera in each group. These data are summarized in Tables 10 and 11, in which the numbers of genera running through any given sequence of the geological time scale are entered. Data on pelecypods were gathered chiefly from the latest editions of the standard Zittel *Grundzüge*, in German, English, and Russian revisions. These genera are broadly drawn and not exhaustively listed, but the data are sufficiently good for present purposes and prove to be adequately enlightening.

Schindewolf (1950b) has remarked that these data for pelecypods are out of date. This is true, but it is improbable that more recent data will make any real difference in orders of magnitude of rates, in general shapes of curves based on the data, or in such conclusions as are here drawn. This optimistic prognosis is supported in part by Schindewolf's

Table 11

Distribution of Genera of Carnivora (Except Pinnipedia)
(Figures are numbers of known genera)

First known appearance		Paleocene L	Paleocene M	Paleocene U	Eocene L	Eocene M	Eocene U	Oligocene L	Oligocene M	Oligocene U	Miocene L	Miocene M	Miocene U	Pliocene L	Pliocene M	Pliocene U	Pleistocene	Recent	Total first known appearances
Paleocene	L	5	0	0	1	..	..	..	..	..	..	..	..	..	..	..	..	..	6
	M	..	12	0	1	..	..	..	..	..	..	..	..	..	..	..	..	..	13
	U	..	..	5	3	..	..	..	..	..	..	..	..	..	..	..	..	..	8
Eocene	L	..	..	..	8	5	3	..	..	..	..	..	..	..	..	..	..	..	16
	M	..	..	..	..	7	4	..	..	..	..	..	..	..	..	..	..	..	11
	U	..	..	..	..	..	13	6	1	0	0	2	..	..	..	..	..	..	22
Oligocene	L	..	..	..	..	..	..	9	1	4	5	2	0	1	..	..	..	..	22
	M	..	..	..	..	..	..	..	3	..	..	..	..	..	..	..	..	..	3
	U	..	..	..	..	..	..	..	..	0	3	1	..	..	..	..	..	..	4
Miocene	L	..	..	..	..	..	..	..	..	..	20	0	3	2	..	..	..	..	25
	M	..	..	..	..	..	..	..	..	..	..	9	3	7	1	1	1	..	22
	U	..	..	..	..	..	..	..	..	..	..	..	8	8	1	0	0	2	19
Pliocene	L	..	..	..	..	..	..	..	..	..	..	..	..	23	5	4	1	7	40
	M	..	..	..	..	..	..	..	..	..	..	..	..	..	4	2	0	2	8
	U	..	..	..	..	..	..	..	..	..	..	..	..	..	..	4	2	3	9
Pleistocene		..	..	..	..	..	..	..	..	..	..	..	..	..	..	..	15	19	34
Total last known appearances		5	12	5	13	12	20	15	5	4	28	14	14	41	11	11	19	33	262

Column heads: LAST KNOWN APPEARANCE.

own similar tabulations, graphs, and calculations for prosobranch gastropods, brachiopods, and foraminifera on more up-to-date data (although I suspect that some specialists might query the modernity of his brachiopod data). The results are thoroughly consistent with those from my pelecypod data and in part quite surprisingly similar. In view of this agreement, continued use of the pelecypod example (from Simpson, 1944a) is justified. Recalculation would be premature before publication of the exhaustive handbook of invertebrate paleontology now being compiled under the leadership of R. C. Moore. That publication will be a rich source of data for studies of the sort here exemplified.

The tendencies of new discovery are to extend known ranges of genera and to add new genera, often of short range. To some extent these tendencies cancel out and may not greatly change the pertinent curves. The tendency of revision is to multiply the number of known genera by splitting. This, too, may leave the shape of the curves essentially the same but tends to decrease average durations and survivorship. Excessive splitting might obscure real phenomena more clearly seen by use of the broader genera of earlier students.

Data on carnivore genera are more complete and accurate, having been taken from my classification of mammals (Simpson, 1944b), which in turn is based upon almost all the literature of the subject. In both cases the many recent genera that are not known as fossils are omitted. Among the pelecypods, genera unknown before the Pleistocene are also omitted, and the Pleistocene data for the carnivores are less complete than for the Tertiary and not entirely comparable.

Because the lengths of the various periods and epochs differ greatly, these tables do not directly yield estimates of rates of evolution. The geological ages must be translated into terms of relative or absolute time, which introduces another source of error, since this translation cannot as yet be exact. As above, however, use of recent estimates of absolute time produces consistent and plausible results more than likely to be significant in use as frank approximations.

One interesting method of presentation and analysis of these data after statement in terms of years of duration is by modified survivorship curves (as explained, for instance, in their customary form in Pearl, 1940). One method of construction of such curves adapted for the present use is shown by the solid lines in Figure 5. Here only genera now extinct are counted, and the plotted points represent the percentage

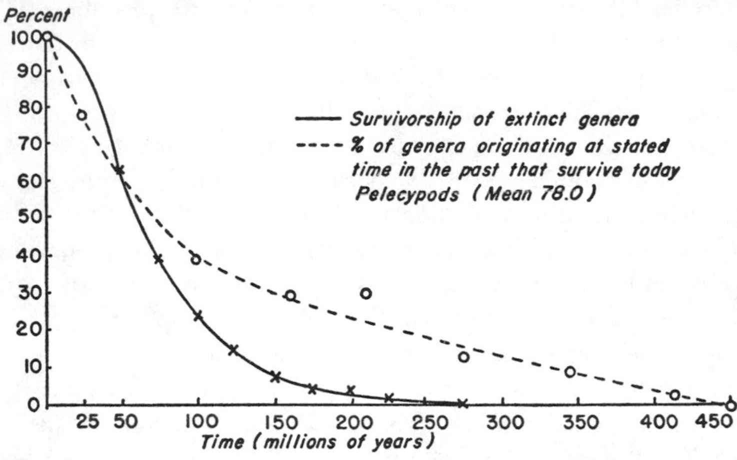

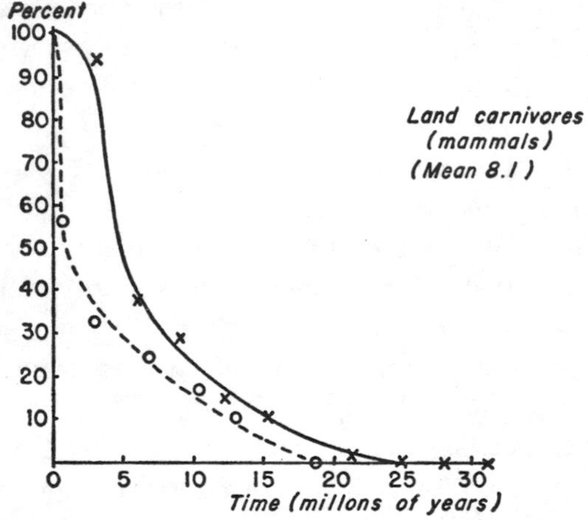

FIGURE 5. SURVIVORSHIP CURVES FOR GENERA OF PELECYPODS AND OF LAND CARNIVORES. Continuous lines, survivorship in genera with completed span (extinct); broken lines, survivorship on basis of ages of genera now living (and known as fossils). Crosses and circles are calculated values to which curves are roughly fitted. Arithmetic coordinates; time scale absolute.

of all these genera with a given known duration equal to or higher than the various stated numbers of years. The actual curves approximating these points have been roughly sketched in by eye. Although similar in form, the curves for pelecypods and carnivores differ greatly in extent, the mean survivorship for a genus of Pelecypoda being about 78 million

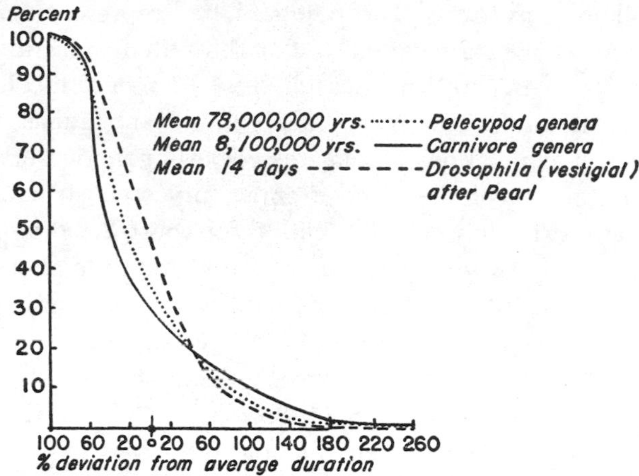

FIGURE 6. SURVIVORSHIP IN PELECYPOD GENERA, LAND CARNIVORE GENERA, AND *Drosophila* INDIVIDUALS. Reduced to comparable form with mean survivorship of the three groups coinciding on scale and time represented by percentage of deviation from this point.

years and for a genus of Carnivora only about 8 million years. The data probably exaggerate the difference, for various reasons, but it is safe to say that in terms of durations of genera carnivores have evolved faster, on an average, than pelecypods. That they may have evolved ten times as fast represents a probable upper limit for the difference.

The similarity of the curves is more clearly shown and differences between them are revealed by replacing the absolute time scale by deviations expressed in percentage of mean survivorship, Figure 6. (This method is also explained in Pearl, 1940.) An analogous curve for individual survivorship in a population of mutant *Drosophila* is also given in the figure. It is quite similar to the generic survivorship curves, especially to that for pelecypods. The *Drosophila* curve is based on life spans of individual flies and so reflects a wholly different phenomenon from the spans of genera, but in both cases what might be called metaphorically a sort of metabolism in populations is involved.[4]

Truly analogous curves can be constructed on the basis of living

[4] This does not support any vital analogy between individual and racial life cycles, a point discussed in Chapters VII and IX. Similarity in shape of curves does not reflect homology or even usefully interpretive analogical similarity in the processes involved. Pearl (1940) amusingly demonstrates this point by showing that survivorship curves for automobiles and for cockroaches have the same pattern. A similar fallacy underlies the interpretation of his "hollow curves" by Willis (1940).

genera also known as fossils. The points plotted represent percentages of such genera known to have existed at the stated times in the past. The sketched curves (broken lines in Fig. 5)[5] reflect the lengths of time living genera have now survived. If the recent faunas were random samples of populations similar, as regards generic survivorship, to the extinct genera of the same groups, curves constructed in this way should approximately coincide with those constructed in the previous way. Obviously they do not coincide, and the differences are significant for the study of evolution. These differences are perhaps shown still more clearly in Table 12, in which the expectation of survival is based on the generic survivorship curves for extinct genera. The figures for gastropods are taken from Schindewolf (1950b) and were calculated by him in the same way as the others.

Among the carnivores, survival to Recent agrees sufficiently with expectation for genera that appeared before late Pliocene, but it is much lower than expectation for late Pliocene and Pleistocene genera. The discrepancy was largely, perhaps wholly, caused by the unusually high mortality of the Pleistocene. Among recent pelecypods and gastropods, on the other hand, survival from the Tertiary agrees well enough with expectation, but survival from previous periods, back to the Ordovician for pelecypods and to the Triassic for gastropods, is greater than expectation. This means that the living pelecypod and gastropod faunas, far from having experienced increased mortality, as have the carnivores, include a large number of very slowly evolving genera and that these slowly evolving lines are less likely to become extinct than are other pelecypods and gastropods—a striking point of unusual importance, to be discussed in Chapter X.

Data on either average age or average duration of the genera simultaneously existing at different times also give a relative measure of average phylogenetic rate of evolution at those times, analogous with the phyletic rates previously illustrated. Rates obtained in the two ways, phyletic and group, are not, however, strictly comparable. As a rule, group rates will probably tend to seem somewhat more rapid than phyletic rates for the same organisms. Relative values of group rates from the same distribution are comparable.

Westoll (1949) has given data from which approximate average

[5] The fit is not as smooth as in curves for extinct genera. Smaller numbers of genera are involved and the poorer fit may be due only to sampling effects.

TABLE 12

EXPECTED AND ACTUAL GENERIC SURVIVORSHIP

Time	*Genera Appearing*	*Percentage of Approximate Expectation of Survival to Recent*	*Expected Survivals*	*Actual Survivals*
		CARNIVORA		
Early Miocene	25	0	0	0
Middle Miocene	22	2	0	0
Late Miocene	19	15	3	2
Early Pliocene	40	23	9	7
Middle Pliocene	8	37	3	2
Late Pliocene	9	90	8	3
Pleistocene	34	98	33	19
		PELECYPODA		
Ordovician	33	0	0	1
Silurian	38	0	0	4
Devonian	54	0	0	4
Carboniferous	23	0	0	3
Permian	8	2	0	3
Triassic	68	3	2	20
Jurassic	56	6	3	16
Cretaceous	61	24	15	24
Tertiary	82	88	68	64
		GASTROPODA (Schindewolf)		
Cambrian	25	0	0	0
Ordovician	100	0	0	0
Silurian	70	0	0	0
Devonian	70	0	0	0
Carboniferous	53	0	0	0
Permian	40	0	0	0
Triassic	129	0	0	7
Jurassic	97	2	2	10
Cretaceous	271	5	14	109
Tertiary	716	60	429	444

durations of some lungfish genera can be calculated, as in Table 13. The data are incomplete and very rough, but they clearly reflect the same rate trends less crudely shown by the character complex analyses of Table 6 and Figure 4.

In order to test the hypothesis of Bubnoff that there has been progressive quickening of the tempo of earth history down to the present, Schindewolf (1950b) compared average duration of (among others)

Ordovician and Cretaceous genera in various groups. The results, copied in part in Table 14, certainly do not agree with the hypothesis. (This result supports rather than excludes the probability that real changes in average rates of evolution did occur within these groups.)

TABLE 13

AVERAGE DURATION OF LUNGFISH GENERA IN EXISTENCE AT VARIOUS TIMES
(Calculated from Westoll's data)

Time	*Ave. duration of genera in millions of years*
Devonian	7
Permo-Carboniferous	34
Mesozoic	115

TABLE 14

AVERAGE DURATION OF SOME ORDOVICIAN AND CRETACEOUS GENERA
(From Schindewolf)

	AVERAGE DURATION OF GENERA IN MILLIONS OF YEARS	
Group	*Ordovician*	*Cretaceous*
Gastropoda	91	83
Bryozoa	70	60
Brachiopoda	52	80

A method similar to that of survivorship, as discussed above, but less complete and less fully enlightening, has been in use ever since Lyell defined the Cenozoic epochs on this basis. The method is essentially group study of ages of living taxonomic units. Being based on groups rather than on individuals, it gives more general and more reliable information. Familiar forms of such data are exemplified in Table 15, after Zeuner (1946a) from Boswell's data, and Figure 7, after Umbgrove (1946).

In Figure 7 sharp increases in percentage of recent species occur at different times in the Miocene for both molluscs and reef-corals. These increases obviously represent evolutionary events of some sort but, without further information, the interpretation is not evident beyond the given fact that surviving species then appeared in this region in sharply increasing proportionate numbers. Among numerous possibilities, which might apply singly or in any combination, are: immigration of recent species already existing elsewhere, with local changes of facies to one

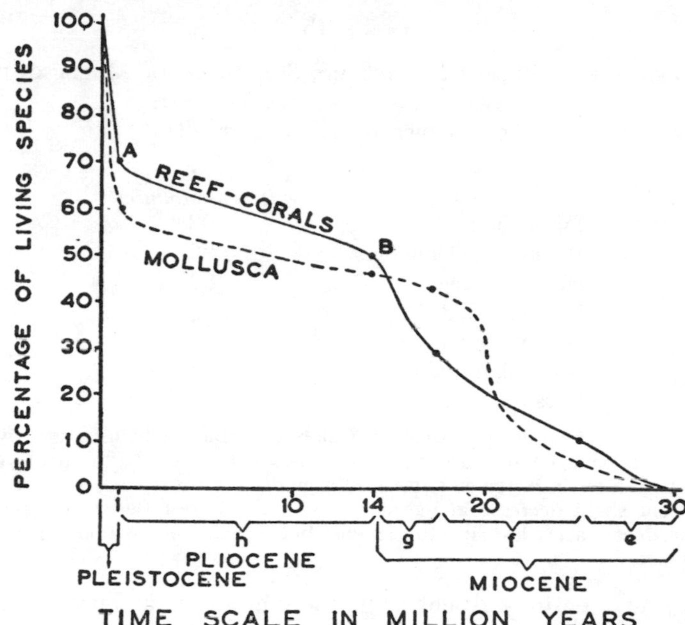

FIGURE 7. PERCENTAGES OF RECENT REEF-CORALS AND MOLLUSCS IN FOSSIL FAUNAS OF THE EAST INDIES. After Umbgrove (1946).

more like the recent; accelerated phyletic evolution followed by a marked slowing in phyletic rates; accelerated speciation followed by decrease in phyletic rates, rates of origination, rates of extinction, or any or all of these; sharp decrease in rates of phyletic evolution, origination, extinction, or any or all of these, without acceleration in other rates.

Such data as those of Table 15 and Figure 7 go far toward answering the objection, legitimate in its own terms, that one cannot really compare rates of evolution by durations of genera in molluscs and mammals because no one can say that genera in such different groups represent equivalent amounts of evolution. Both sets of data show latest Pliocene or earliest Pleistocene faunas with about 60 percent of Recent species of molluscs. This means, in effect, that the molluscs in question have changed so little in that time, or longer, that qualified taxonomists cannot see the change. But in terrestrial deposits of approximately the same age (Villafranchian in Europe, Sanmenian in China, Blancan in North America, etc.) surely identified Recent species of mammals are ex-

TABLE 15

PERCENTAGE OF RECENT SPECIES AND SUBSPECIES OF MOLLUSCS IN SUCCESSIVE EAST ANGLIAN DEPOSITS
(From Zeuner, data from Boswell)

Bed [a]	*Percentage of Recent Molluscs*
Forest Bed	90 [b]
Weybourne Crag	93
Norwich Crag	84
Butleyan Red Crag Newbournian Red Crag	73
Waltonian Red Crag	67
Coralline Crag	60

[a] The Coralline Crag is approximately Calabrian-Villafranchian in age, by some authorities classed as latest Pliocene and by others as earliest Pleistocene. Overlying beds are distributed in the given sequence through the Pleistocene.

[b] Apparent slight decrease in recent forms in the Forest Bed may represent a temporary, local facies less like the recent, but is probably within the limits of sampling error.

tremely few or entirely absent (e.g., there are none in Europe, Zeuner, 1944). Again, in the same East Anglian deposits, the relatively late Forest Bed has 90 percent of Recent molluscs and 14 percent of Recent mammals. Such comparisons could be greatly multiplied.

Now, this is not just a matter of what students call a species or a genus in two incomparable groups. A large percentage of molluscs did not visibly change at all while virtually all mammals changed quite appreciably. Nor is this just a matter of different sorts of preservation: the usual fossil mollusc specimen includes evidence of a much larger proportion of the animal's morphology, both soft and hard parts, than does the usual fossil mammal specimen involved in the comparisons. It is fair to conclude that mammals did, in a really significant sense, evolve faster during the last million years or so, and this tends to support more equivocal evidence that they did so also in earlier times.

The example of Figure 7 also bears indirectly on another set of evolutionary problems of extreme interest: development of multispecific local biotas. In a given area, evolution *in situ,* without migration, tends to produce mainly or solely phyletic evolution, with little or no increase in frequencies of taxonomic groups (see, e.g., Mayr, 1942). In highly isolated regions, such as oceanic islands, after an initial burst in speciation of immigrants, evolution *in situ* may predominate, but most faunas and floras cannot long have evolved by this process, alone. Exuberantly multispecific biotas such as the fauna of a tropical reef or the flora of a rain forest pose special and fascinating problems in this respect. In such cases, it may be necessary to postulate an existing or former condition in which there were numerous separate areas closely similar ecologically and nearly but not quite isolated from each other (that is, with strong but not quite complete or constant barriers to the spread of the organisms concerned). In tropical seas, as in the Great Barrier Reef and adjacent South Pacific or in the East Indies of Umbgrove's example, such conditions may be almost ideally realized by reeferies with numerous, separate local reefs.

It is also well known that paleontological local faunas and floras differ greatly in the numbers of species and genera present (see, e.g., Stenzel, 1949). The problems of ecological evolution involved here are of extraordinary interest and importance. For the most part they lie somewhat outside the topics chosen for this study, but they will again be touched on in Chapters VI and VII. They are mentioned at this point because they also lead into the broad subject of the last major sort of evolutionary rates to be defined and exemplified: taxonomic frequency rates.

TAXONOMIC FREQUENCY RATES

Frequency rates have been rather fully discussed and exemplified elsewhere (Simpson, 1949a, 1949b, 1952, also Newell, 1952, and other contributors to that symposium), and they are not altogether essential to many of the themes of this work, but they are involved in some of the topics of Chapter VII and should in any case be mentioned in a general consideration of evolutionary rates.

Data for taxonomic frequency rates consist of counts of taxonomic groups, at any one level in the hierarchy, at defined times or for defined spans of time. Data usually available are in terms of the sequential geological scale, but may be plotted against estimates of absolute dates for that scale. In the usual graph of absolute frequencies, underlying rate phenomena may still be obscured by marked differences in lengths of the geological time units. If distortion is really appreciable and if actual rates are not to be calculated, a means of compensating for this distortion is to make areas under the curve (or areas of rectangles in a histogram) proportional to frequency, rather than heights.

Straight time-frequency measures and curves do not directly show rates. Rate of change in total frequency is represented by the slope of a curve for total frequency and is positive when the total frequency is increasing, negative when it is decreasing. The rate of change is, in fact, a resultant of two other rates, the rate of origination, i.e., of first appearances per million years (or other unit), and of extinction, i.e., of last appearances per million years, being positive when the first rate exceeds the second and negative in the contrary case. It is thus more enlightening to tabulate frequencies of first and last appearances as well as total frequencies. (Given the initial frequency, all subsequent total frequencies can of course be calculated in a simple way from those of first and last appearances.) All three essential rates of taxonomic frequency can then be calculated in various ways, the simplest of which is to divide the relevant frequency for a given span by the length (in absolute time units) of that span. This is not mathematically the soundest measure of rate, but it gives a real value for some (not necessarily exactly determined) point in time and is generally a sufficient approximation.

Representative data and rates based on them are exemplified in Table 16 and Figure 8. Some experience in such compilation suggests that unless sampling is known to be adequate and scaling is carefully done, fluctuations of little or no real significance will appear. Then it is advisable to smooth these out by obtaining large frequencies, either by using larger time units or by using relatively small taxonomic units in a large group. Either procedure surely loses some real details of rate change, but this is preferable to having very doubtful changes or those due only to sampling. In the example of mammalian ungulate families, fossil and recent, sampling is excellent. A few more will doubtless be discovered but the roster must be nearly complete, aside from splitting by erection of families for genera now known.[6] In such a case,

[6] Although discussion of this point is outside the present theme, this prediction may be of sufficient interest to warrant insertion of the following brief tabulation of the date of definition of the oldest genus in each ungulate family here recognized:

1750–1769	11	1850–1869	5
1770–1789	1	1870–1889	19
1790–1809	4	1890–1909	14
1810–1829	8	1910–1929	2
1830–1849	9	1930–1949	3 (all before 1940)

In view of the fact that exploration has been increasingly intense, it seems fair to say that we can expect the discovery of few genera representing ungulate families of which representatives are now unknown.

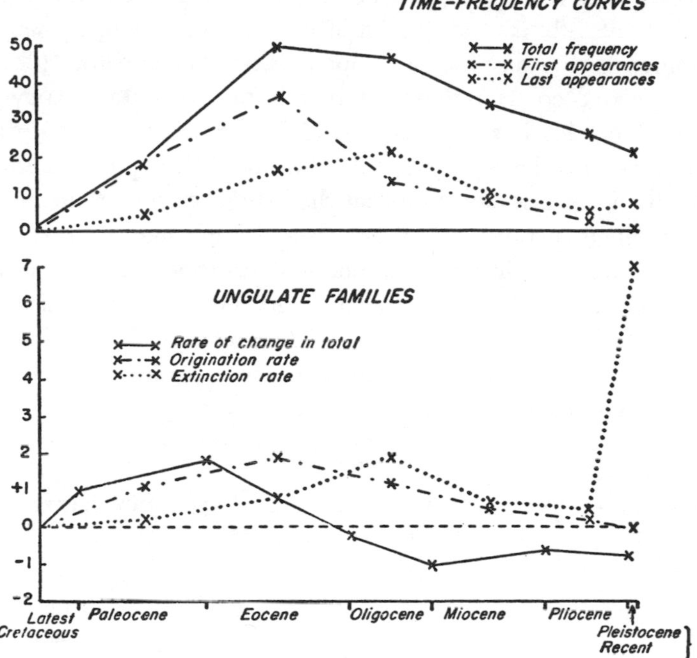

FIGURE 8. TIME-FREQUENCY AND RATE CURVES FOR MAMMALIAN UNGULATE FAMILIES. Data from Table 16.

maximum reliable information may usually be obtained by using time units such that the average duration of the taxonomic units noted is somewhat but not greatly superior to that of the time unit. This is true of the Cenozoic epochs and ungulate families. With such choice of units and good samples, even low frequencies, as in this example, give consistent and apparently trustworthy results.

It has been suggested (Zeuner, 1946b) that the time-frequency curve is a resultant of a rate of splitting and a rate of extinction. In fact, it is the resultant of initial total frequency and rate of change in total frequency. The latter rate is, in turn, the resultant not strictly of a rate of splitting and a rate of extinction but of a rate of origination and one of extinction. The rate of origination is not solely or strictly a rate of splitting, because it includes also a usually inextricable amount of phyletic change without splitting. For instance, three genera of horses first appear in the Eocene, therefore they had a rate of origination of about .15 genera per million years, but there was no splitting on the generic level. In such small, particular parts of very well-known groups, the two very different rates of splitting and of phyletic change can thus sometimes be disentangled, but it is seldom possible to evaluate their respective contributions in larger bodies of data. I know of no case in which this has been done and I cannot provide an example, although reasonable estimates, at least, should be possible for some groups. It should also be noted that the rate of extinction includes two different things, extinction without issue, which may be taken as the strict opposite of splitting, and nominal disappearance of a species or genus because it has been transformed into something else, which is an aspect of phylogenetic evolution. The rate of extinction for Eocene horses was also .15 genera per million years, but no genus became extinct in the strict sense.

TABLE 16

TIME-FREQUENCY DATA AND TAXONOMIC FREQUENCY RATES FOR UNGULATE FAMILIES

	First Appearances	Last Appearances	Totals	RATES PER MILLION YEARS [a] Origination	Extinction	Change in Total
ARTIODACTYL FAMILIES:						
Pleist.—Recent	0	1	10	0	1.0	−.2
Pliocene	1	1	11	.9	.1	−.3
Miocene	5	5	15	.3	.3	−.3
Oligocene	5	9	19	.5	.8	+.3
Eocene	14	0	14	.7	0	+.8 [b]
PERISSODACTYL FAMILIES:						
Pleist.—Recent	0	1	4	0	1.0	0
Pliocene	0	0	4	0	0	−.1
Miocene	0	1	5	0	.1	−.3
Oligocene	0	4	9	0	.4	−.2
Eocene	12	3	12	.6	.2	+.7 [b]
OTHER UNGULATE FAMILIES EXCEPT SOUTH AMERICAN:						
Pleist.—Recent	0	3	5	0	3.0	0
Pliocene	1	0	5	.9	0	0
Miocene	3	1	5	.2	.1	−.1
Oligocene	4	4	6	.4	.4	−.2
Eocene	3	7	9	.2	.4	0
Paleocene	9	3	9	.5	.2	+.5 [b]

	First Appearances	*Last Appearances*	*Totals*	RATES PER MILLION YEARS [a] *Origination*	*Extinction*	*Change in Total*
SOUTH AMERICAN UNGULATE FAMILIES:						
Pleist.—Recent	0	2	2	0	2.0	−.7
Pliocene	0	4	6	0	.4	−.2
Miocene	0	3	9	0	.2	−.3
Oligocene	4	4	13	.4	.4	−.1
Eocene	7	6	15	.4	.3	+.3 [b]
Paleocene	9	1	9	.5	.1	+.5 [b]
ALL UNGULATES:						
Pleist.—Recent	0	7	21	0	7.0	−.8
Pliocene	2	5	26	.2	.5	−.6
Miocene	8	10	34	.5	.6	−1.0
Oligocene	13	21	47	1.2	1.9	−.2
Eocene	36	16	50	1.9	.8	+1.8 [b]
Paleocene	18	4	18	1.1	.2	+1.0 [b]

[a] Origination and extinction are conventionally considered as occurring at the middle of epochs and averaged for the epoch, change in total as occurring between epochs and averaged for half each of preceding and following epochs. This does not mathematically give precise average rates or slope of curve at the indicated point, but it represents figures that were taken by the curve slopes at some intermediate time and the simple procedure suffices for figures recognized as rough approximations in any case.

[b] It is postulated that the Eocene rise can be averaged from mid-Paleocene and the Paleocene rise from about 10 million years before the beginning of the epoch. This, too, is a convention, but a reasonably realistic one.

The importance of choice of proper time units, or of allowance for their effects, must also be noted. If horse genera be tabulated by epochs, the Eocene frequency is 3, but in fact during the Eocene the frequency was never more than 1 at any given point in time. The rates, however, as opposed to the total frequency, are the same whether calculated for the whole epoch or for parts of the epoch. (This is the rationale of graphing total frequency by area rather than by a linear vertical scale.)

In spite of these complications, which are confusing but which must be faced if time-frequency curves and corresponding rates are to be used intelligently, it remains true that a rise in a properly scaled time-frequency curve indicates an episode of increased splitting and diversification and that a marked rise in origination rate *usually* does too. With due precautions, this seems therefore to be the generally best technique for determining the existence of episodes of taxonomic

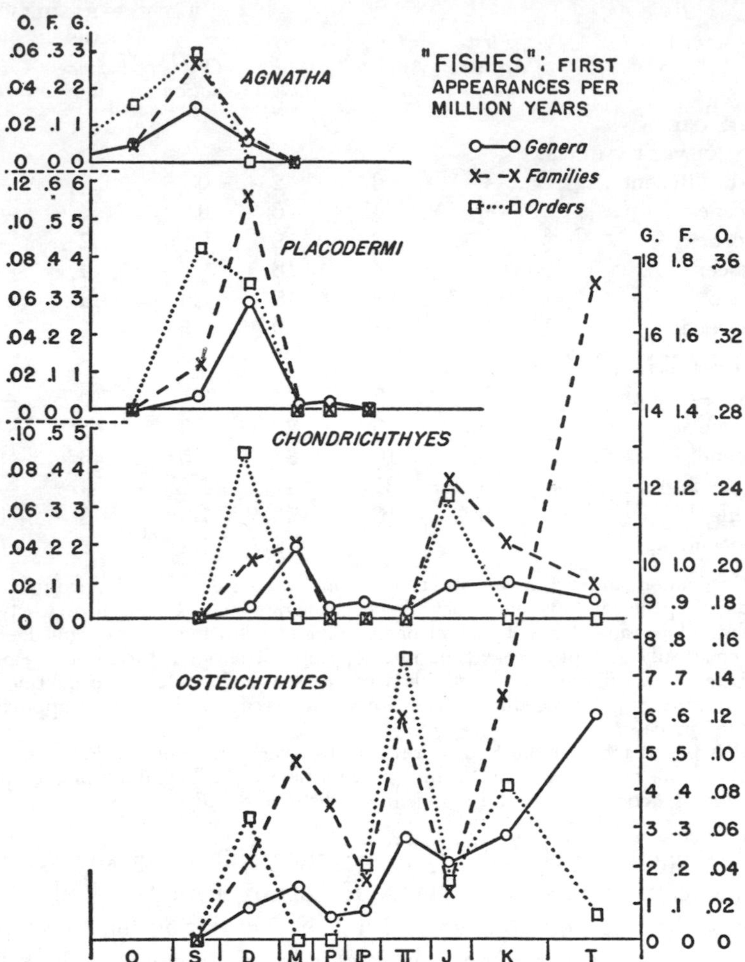

FIGURE 9. RATES OF ORIGINATION FOR ORDERS, FAMILIES, AND GENERA OF THE FOUR CLASSES OF "FISHES." O., F., and G. stand for orders, families, and genera per million years. The time scale is divided into periods, Ordovician to Tertiary.

diversification in evolution and for measuring their intensities and rates. Figure 9, for example, summarizes data in which episodes of proliferation, most of them at the different taxonomic levels of orders and genera, can clearly be seen. There is one such episode each for the Agnatha and Placodermi, two for the Chondrichthyes, and no less than three for the Osteichthyes. When strongly marked and relatively short in duration (but even "short" episodes of this sort commonly cover time on the order of millions of years), such episodes have long been discussed as examples of "explosive evolution." They will be mentioned again in Chapter VII, our concern in this chapter being only with the descriptive evidence for them and the measurement of their rates.

In the nature of evolutionary processes, and of taxonomic arrangement of their results, any group starts with a low taxonomic frequency which almost always rises somewhat or markedly as the group becomes established. In most cases, followed through any considerable length of time, there is a subsequent fall, which may drop to zero (extinction) or may flatten out and fluctuate around some value below the peak. A simple, unimodal pattern of rise and fall would be expected to be common and is so, indeed, among the examples that have been compiled. Yet even such simple time-frequency curves vary greatly and in all possible ways: in total duration, in relative height of peak, in position of peak (early, middle, or late in the total span), in steepness of rise and fall, in steepness of rise relative to that of fall, and in relative durations of low, high, and changing phases. Among, apparently, all other possibilities one may observe rather rapid rise to a sustained high and then about equally rapid fall (e.g., species of tetracorals, Sloss, 1950), steady rise to a high, brief peak followed by catastrophic or, at least, relatively extremely rapid fall (e.g., species of the sea-urchin, *Scalenia,* Zeuner, 1946a, b), more or less rapid rise to a brief (e.g., genera of Spiriferidae, Zeuner, 1946a) or more sustained (e.g., species of *Lingula,* Zeuner, 1946b) high, followed by a long decline slower than the rise, or very rapid rise followed by long fluctuation just below the peak and finally by a sharp drop (e.g., genera of Pleurotomariidae, Rensch, 1947).[7]

[7] These examples and the many others available preclude generalization from Sloss's conclusions (Sloss, 1950). He states that the patterns of time-frequency curves for species and genera of some marine invertebrates are symmetrical normal probability curves. Approximation to such curves does occur but this is not a rule and may be mere coincidence when it does occur. Even Sloss's examples do not bear out his conclusion. In one of his examples when a normal curve obviously does not fit (genera of inarticulate brachiopods), he assumes that the curve "is affected by loss of data in the early portions," although I see no necessity for this assumption. His curve for species of trilobites is also made symmetrical only by extrapolation beyond the data. Even in examples for which extrapolation is not used, his normal curves sometimes are clearly not close to the data. For instance his normal curve for species of Orthacea, obtained by differentiation from three cumulative frequencies, is (as required by the method of construction) symmetrical and drops off to or near zero in the Silurian, whereas the data clearly show that the real time-

Some special interest attaches to time-frequency curves that expand rather rapidly to a maximum and then decline over a longer span, that is, positively skewed curves with well-defined modes to the left of mid-span as usually plotted. In more biological terms, such a curve would correspond with relatively rapid (more or less "explosive") differentiation of a group following its origin and then slower decline in diversification such as might, for instance, accompany the spread of some other, competing group of later origin. Such curves and the corresponding evolutionary process are certainly rather common. Nevertheless, they are not common enough in available examples to constitute the rule, and the examples do not indicate that they will necessarily prove to be more common than other, quite different patterns.

Many time-frequency curves have two and some have three or more peaks. Among many other examples, this is true of species of the family Clypeastridae (Zeuner, 1946a), genera of the family Euomphalidae (Rensch, 1947), genera of the superfamily Spiriferacea (Zeuner, 1946a), and orders (also families and genera) of the class Osteichthyes (Fig. 9, above). A second or later peak usually represents diversification following acquisition of new structural grades, spread into new (for the given group) adaptive zones, or invasion of new (also for the given group) geographic regions. Such changes are often properly reflected in taxonomic subdivision of the group, and then by constructing curves for lower taxonomic categories the multimodal curve may be analyzed into two or more unimodal distributions. For instance the first peaks (ordinal, familial, generic) of the Osteichthyes occur mainly in the superorder Chondrostei (also the Choanichthyes), the second mainly in the superorder Holostei, and the third mainly in the superorder Teleostei.

Similarly, the second (Triassic) peak of the Spiriferacea is due mainly to the family Athyridae, which has a unimodal time-frequency curve (skewed to the *left*) if plotted alone, and Zeuner has shown that the curve for the Spiriferidae is unimodal (skewed to the *right*) if plotted separately. In these terms of families, the situation is really still more complicated, for the great Devonian peak for the Spiriferacea, coincident with the peak for the Spiriferidae, also involves several

---

frequency curve is not symmetrical but definitely skewed to the right and that the group still had considerable frequency in the Permian, at least 100 million years after the Silurian.

other families with peaks at different times, especially the Atrypidae with a Silurian peak and the Meristellidae with a peak probably between those of Atrypidae and Spiriferidae. The bimodal curve for genera of Spiriferacea is of course correct and is also meaningful at its own level, but further information is obtained by analysis at other levels.

Such examples may suggest that there is some underlying rule that time-frequency curves are "really" or "basically" unimodal and that two or more modes appear because the group involved is "composite" (Zeuner), somehow includes "too much," or has been tabulated by "wrong" categories. Although it is probably true that most multimodal curves can be analyzed taxonomically into two or more unimodal curves, it does not seem to be correct to think of multimodality solely as an artifact of the grouping and units used. It is probable that multimodality may appear with units of any size tabulated in groups of any inclusiveness. In the few examples cited above, it occurs in species of a family, genera of a family, superfamily, and class, families of a class, and orders of a class. Neither by scaling upward nor downward does it seem possible to reach a point where the curves are inherently or invariably unimodal, where this is a "law" of evolution. Consideration of recent forms strongly suggests that even demes and subspecies would *sometimes* give multimodal curves within groups of any larger size, if frequencies of these smallest units could be usefully tabulated through appreciable stretches of geological time (which is impossible at present). (It may also be remarked that in some cases two modes are taxonomically separable simply because a taxonomic separation was based on existence of the modes.)

In any case, the presence of more than one mode in a time-frequency curve, if based on adequate data, is a fact and reflects evolutionary events whether or not lesser, included groups are unimodal.

CHAPTER III

Variation

FACTORS OF EVOLUTION

MANY ATTEMPTS have been made to determine, name, and classify various factors, determinants, or causes which can be supposed, in sum, to account for the whole course of evolution. Darwin's formulation (1859, but clearer in later editions) was: (1) natural selection, (2) inherited effects of use and disuse, (3) inherited effects of action of the environment on the organism, and (4) "variations which seem to us in our ignorance to arise spontaneously." The classical neo-Darwinian school tended to study evolution under three headings: variation, heredity, and selection. Having abandoned Darwin's views as to the sources of variation and the nature of heredity, neo-Darwinians were then constrained to seek other explanations, or second-order factors, underlying these objectively present factors, and hence has arisen the modern, "post-neo-Darwinian," or I would prefer to say, synthetic school.

Cuénot (1951), whose final views might be characterized as neo-Darwinian plus an agnostic sort of vitalism, summed up the neo-Darwinian factors as: (1) variation, (2) action of the environment (not in a demonstrably neo-Lamarckian sense), (3) competition, and (4) selection, a resultant of (1) to (3). He added as "uncertainties": (5) the inheritance of acquired characters, and (6) finalism or antichance. Rensch (1947), a student fundamentally belonging to what I call the synthetic school, lists (1) mutation, (2) variation in population size, (3) isolation, and (4) selection as the factors of lower level (micro-) evolution and concludes that with differences in their interaction they also account for higher, transspecific levels of evolution. Günther (1949) modifies Rensch's list to read (1) mutability, (2) chance, (3) isolation, (4) selection, and adds (5) "ecological license."

It seems obvious enough that none of these lists really and clearly designates *all* the factors involved in evolution or represents a fully logical classification of such factors. As Rensch of course knows, size is not the only characteristic of populations affecting evolutionary outcome. As Günther would probably agree, "chance" as a factor is quite meaningless without answer to the question, "Chance of what?" "Ecological license" is not logically excluded from overlap with the factor of selection, nor do any two of the stated factors represent categories on the same level in a consistent hierarchy.

Moreover, it becomes clear as one studies these or other proposed factors that no one factor can be considered quite separately in the interplay that really determines the course of evolution. One must agree with Zimmermann (1948) that the cause of an evolutionary event is the *total* situation preceding it (and this in turn the result of total situations on back to the beginning), so that it is not entirely realistic to attempt designation of separate causal elements within that situation. At most, one may speak of "factor complexes" or "constellations."

Such a point of view obviates argument as to whether, say, mutation or selection is dominant in evolution as a whole, and it should (but will not) end the long-continued dispute on internal or autogenetic as against external or ectogenetic control of evolutionary trends. The apparent alternatives are not real and choice is not forced, indeed is meaningless in these terms. Yet after making the point that the causes, or the determining factor-constellations, of evolution are indivisible, and while holding fast to this as a basic point of view, it is necessary to divide the indivisible, to examine the factor-constellation in terms of constituent factors. The dissection for purposes of study (and Zimmermann makes it, also) must be arbitrary in part, at least, since we are only taking different aspects of a single thing. Whether the chosen aspects are mutually exclusive, without overlap, and even whether they cover the whole, may be immaterial in a given study, nor is there inherent contradiction if different aspects are designated at different times or by different students. The aim is only to select such aspects as may in a given study best clarify the phenomena in question.

In this book about only certain aspects of evolution, it is not intended or necessary to discuss all factors pertinent to all aspects. A few modern summaries from a still broader point of view are available (e.g., Huxley, 1942; Carter, 1951) as well as many studies of other, more restricted parts of the general subject. Here the aim is to specify evolutionary factors most likely to cast light on major aspects of evolution, to discuss them from this particular point of view, and to relate them to the primary evidence of long-range evolution. The factors chosen are as follows:

1. Variation
2. Mutation
3. Population
4. Time and the length of generations
5. Selection
6. Environmental factors, and interaction of all factors in adaptation
7. Isolation and splitting

Factor 1 of this list is discussed in the present chapter; 2 in Chapter IV; 3, 4, and 5 in Chapter V; and 6 in Chapters VI–VII. Factor 7 is incidentally included at several points in the ensuing discussion and explicitly in Chapter XII from the special point of view of major evolutionary patterns. In detail, isolating mechanisms are obviously of crucial importance in evolution, and they have been considered at length by students of recent organisms (e.g., Mayr, 1942; Stebbins, 1950; Dobzhansky, 1951). Their result, when it eventuates in phyletic splitting and the rise of essentially distinct, long-continued lineages, is an essential part of the present subject. The detailed mechanisms in themselves are not, and for the most part the existence and nature of those mechanisms can be taken for granted here.

THE NATURE AND SOURCES OF VARIATION

It is a truism that all populations of organisms vary. No two individuals are ever exactly alike, not even monozygotic twins. It is equally commonplace nowadays to recognize that some of this variation involves the genotype and some does not, although the still usual custom of speaking of "hereditary" and "acquired" differences does not clarify the distinction. This is another set of false alternatives setting up an unreal problem. What is inherited, mostly but not exclusively through the genotype, is known to be an organized physico-chemical system with a definite growth tendency or pattern. The organism eventuates from the interaction of this system with all environmental influences (in the broadest sense) throughout development and the whole of life.

The naive conception that genes correspond with somatic characters or that the latter are inherited as such is now well known to be untrue in most cases. It is approximately true only when the "character" in question is directly a chemical concomitant of the gene, as, for example, the ability of *Neurospora* to synthetize indole from anthranilic acid (Beadle, 1945), and even in such a case independence from the environment is not absolute.[1] For such morphological characters as here concern us, it is fair to say that they are never *strictly* inherited nor *strictly* acquired, but are both or neither, depending on the point of view.

An "acquired character," or one should say, any character is thus simply a particular variant within the reaction range determined by the given genotype. The important point is that different variants in the range of the same genotype are not materials for evolution. Similarly, differences in genotype are not materials for evolution when the same phenotypic characters develop in their overlapping reaction ranges. With, again, certain exceptions of no particular importance here,[2] materials immediately available for evolution occur only when *both* phenotype and genotype vary, and vary in a correlated way.

Although both phenotype and genotype are thus concerned, the source of variation available for evolution must be in the genotype. This is simply because differences arising in phenotypes with the same genotype produce no change in the latter whereas differences arising in genotypes with (before the change) the same phenotypes usually do produce differences in the phenotypes. (I do not here propose to discuss neo-Lamarckism or Michurinism, the schools that continue to deny these extremely probable, if not absolutely proven statements.)

Correlated variations of genotype and phenotype within a population, or lineage, are the indispensable materials without which no evolution can occur. Their sources are therefore the basic and most fundamental sources of evolutionary change. As now known to geneticists,[3] these sources are as follows:

[1] For instance, the operation depends on presence and amount in the environment of materials for prior synthesis of anthranilic acid, and on a different gene for that synthesis, and presumably the action would be modified if indole were already present.

[2] Notably cases concerning compatability of genotypes in conjugation, with selection acting directly on the genotype and not indirectly through the phenotype.

[3] In summarizing such genetic data as are here pertinent, it is assumed that authority need not be specified for facts that may be found in any competent, modern, general text on genetics, for example, Sinnott, Dunn, and Dobzhansky (1950).

A. Mutation
   1. Of genes
   2. Of chromosomes
      a. Structural (deficiencies, duplications, translocations, inversions)
      b. Numerical (polyploidy, polysomy, etc.)
B. Recombination
   1. Of genes (by crossing over in meiosis)
   2. Of chromosomes (by conjugation of gametes with unlike chromosomes)

Although recombination produces new and variant characters and is, in fact, responsible for most of the variation observed in interbreeding populations of sexual organisms, it produces no genetically new materials. Strictly new materials arise by mutation, which is much less common than recombination but which must ultimately supply materials for long-continued evolution. The genetic changes here listed do not (ordinarily) provide immediate materials for evolution unless they are accompanied by changes in phenotype, but as discussed below they may provide cryptic variation which may later become available.

Another point about variation greatly affecting its availability for evolution and its relationship to different evolutionary processes concerns the units in or between which variation is occurring. Obviously variation occurs within individuals, which are never precisely the same throughout life, or for any successive instants. Such variation affects the fate of the individual and may therefore affect evolution in his lineage, but the variation, itself, is not material for evolutionary change. Variation also occurs, so obviously that this is usually meant when "variation" is mentioned without qualification, between contemporaneous individuals within a population of interbreeding individuals or of asexually produced descendants from one (not too distant) ancestor. Variation also occurs between or among two or more such groups, with no (or little) interbreeding. Both intra- and intergroup variation provides material for evolution, but the bearing is quite different in the two cases. With passage of time, intragroup variation is a unit, transferable through the group, material for phyletic change and for initiation of speciation. Intergroup variation cannot be (or can only to limited extent be) transferred from one group to another. It may condition extinction or survival of any one group or, in combination with intragroup variation, it may be involved in divergence or other relationships among various groups. Finally there is successive variation within groups, lineages or phyla, with the passage of time. This is not material for evolution; it *is* evolution.

CORRELATED PHENOTYPIC VARIATION

Another aspect of variation is extremely important because it helps to explain some otherwise baffling evolutionary phenomena for which a great variety of speculative and frequently metaphysical hypotheses have been advanced. This is correlated variation in two or more different (or, at least, separately designated) phenotypic characters. By this is not meant that two characters tend to change by equal amounts or in the same direction, but that a variation in one is more likely than not to be accompanied by some particular sort of variation in the other within the same individual. Analogously with variation in general, the correlation is not material for evolution unless it has a genetical basis —that is, the correlation, itself, must be related to something in the mechanism of heredity, and not merely the variations that happen to be correlated. Two characters, such as teeth adapted to a plant diet, and hoofs, may tend to develop together and each may be closely correlated with the genetic mechanism, and yet their rise *together* may not be fixed by that mechanism. If selection is the effective factor, it then acts separately on the two,[4] each of which must have selective value. In fact, although the hoof-herbivore correlation is usual, as noted by the ancients and made into a law by Cuvier, there have been clawed herbivores. "Environmental correlation," or correlation by selection without genetic correlation, may be difficult to distinguish from genetic correlation, although methods have been developed for doing so in animal breeding (Lerner and Dempster, 1948).

Some truly genetic character correlations may be rather obvious and indicate plainly only that the characters, as analyzed, are parts or aspects of a larger character. Thus if an organ becomes larger as a whole, so will its height, width, and breadth in close correlation with each other. Other cases may be analogous and yet less obvious. For instance, a shift in a growth gradient over a field may produce rather

[4] Although a trend in one may affect separate selection for the other, see Chapter VIII.

complicated, genetically correlated changes in all the structures in that field, as in a series of teeth, for instance (Butler, 1939, 1946). Allometry, likewise an aspect of growth pattern, may also produce changes actually but not obviously correlated as parts of a single, genetically controlled growth pattern. Other correlations are still more complicated or inexplicable by phenotypic study but have genetical explanations.

Among the mechanisms possibly producing genetic correlation of phenotypic characters are:

1. Genetic control of growth gradients and fields, including allometry, as above
2. Pleiotropy
3. Linkage
4. Incidental inclusion of genes in a genetic system (or among polygenes) integrated on some other basis

There is no doubt that all these mechanisms occur, although there can be dispute as to their relative effectiveness or their applicability in a given example.

Pleiotropy, multiple effect of single genetic factors, is certainly common. Indeed, it may prove to be the rule rather than the exception. Among many examples are the *se* gene in mice, which produces short ears and also lowers the number of ribs (Green and Green, 1946), "vestigial" in *Drosophila,* producing small wings, lowered fecundity and life span, and changes in the halteres, spermathecae, ovaries, and elsewhere (Mohr, 1932), or a mutation in mice producing a remarkable, and remarkably variable, series of malformations in various individuals and various parts of the body (Grüneberg, 1947).[5] There is no point in multiplying examples here; they can be found in abundance in any work on genetics.

There seems to be no really clear distinction between correlation by linkage and by inclusion in a genetic system or among polygenes,

[5] The variability of expression is unusual in this case and is due to the fact that the primary effect is production of blisters under the ectoderm which migrate variously and may interfere with development elsewhere. This may not be considered "real" pleiotropy, but it is deliberately mentioned in order to point out that any or all cases of pleiotropy might well prove, if pushed back to chemical effects of the gene, to be diverse consequences of a single primary action. The point is to emphasize that genes do not control characters but are elements in a developmental system, a single change in which may produce changes on any number of different characters. See also Grüneberg, 1948.

and the latter may be a special case of the former in the sense that the correlation or retention in the system would still require some persistent linkage. The long-range effectiveness of linkage to maintain the correlation may be questioned in some cases or as a general rule, although it evidently could be effective, especially as there are mechanisms which impede or prevent crossing over. These are also more difficult cases to establish experimentally, but there are now examples for which no other explanation seems likely. Among these are correlation of physiological (salinity preference) and morphological (plate number) characters in *Gasterosteus*, interpreted as due to linkage groups with impeded recombination (Heuts, 1947), and selection for high and low numbers of abdominal chaetae in *Drosophila* which also produced changes in the spermothecae and other parts, interpreted as incidental to selection on a polygenic system (Mather and Harrison, 1949). Whatever the precise mechanism involved, long experimental selection for a given character is commonly found to have produced correlated changes in other characters. As one more example, MacArthur (1949) selected for large and small size in mice and found that changes in coat color, proportions, litter size, and temperament had also occurred. He interpreted these changes, as far as they were consistent trends, as involving allometry, pleiotropy, and linkage, all three, with drift or sampling effects also producing some nontrend differences between the stocks.

Several students, among them Lerner and Dempster (1948) and Mather and Harrison (1949), have noted that such correlations provide a mechanism by which selection for one character (or character complex) might move another character *against* the pressure of selection. We will return to this point later (Chapter IX).

THE VARIABILITY POOL AND BALANCE

With the unimportant exception of clones still quite near their point of origin, every natural population has a store of correlated genotypical and phenotypical variation. That phenotypical variation is always present is a matter of observation, to which there is no established exception. That a certain proportion of this reflects genotypical variation is also beyond any doubt. Experimental association of genetic variants with phenotypic variants is the basis of the whole science of genetics. It is also often evident that high morphological variation is accompanied by high cytological variation (in the chromosomes, e.g., in Salmonidae, Svärdson, 1945). Such studies as have been made of genetics in wild populations suggest that these have in particular groups, at least, large stores of genetical variability (e.g., in *Drosophila*, as reviewed by Spencer, 1947[6]; see also Dobzhansky, 1951).

It is this variability that is used, so to speak, in any evolutionary change. This fact has led to the suggestion that variability may eventually be used up in the course of long-range evolution, and a supposed "law of reduction of variability" has been promulgated from time to time, perhaps first by Rosa (1899). In fact, Rosa's classic formulation, whether true or not, is somewhat beside the point or, at least, confusing in terminology. Rosa's "effective variability" is for the most part successive intergroup variation, which is not what modern students would consider the, or the most, effective variability in a majority of evolutionary processes. Thus by "reduction of variability" Rosa meant primarily restriction in the number of possible directions of structural modification. This is an aspect of reduction of adaptability with increase in specialization, an important empirical principle known before Rosa and still accepted, as a generalization subject to exceptions, but not particularly or necessarily related to loss of variability.

Evolutionary change can occur by utilization of existing variation and without the introduction of any new hereditary factors. Aside from the influence of extrinsic factors on phenotypic variability, which is immediately reversible and hence of less long-range significance, the store of genetic variability in a large, widespread population can be unequally distributed among descendant units, which thus come to differ from their ancestors and from other contemporaneous descendants. In secular shifts within one population the proportions of the different variants in the population may change, and some variation may be eliminated. The latter is true reduction of variability, as the words are used here. It will be shown that apparently this is not a normal long-range phenomenon of phyletic evolution and that there is, under these definitions, no "law of progressive reduction of varia-

[6] This study, based on a limited repertory of easily visible mutations in wild flies, seems to make its general point adequately, i.e., that a large supply of mutant forms occurs in nature. It has, however, one peculiarity perhaps inadequately accounted for: most (although by no means all) of the variants failed to reproduce in breeding tests over two generations and were therefore presumed to be phenocopies. Dobzhansky (1951) has compiled evidence somewhat more convincing in this respect.

bility." In the differentiation of one population into several, the same processes occur, but the typical result is the unequal distribution of ancestral variants among the descendant groups. Thus, part of the intragroup variation becomes intergroup variation, and its status in and effect on further evolution become radically different.

The reality and importance of these phenomena have been abundantly illustrated both by experiment and by observation in recent animals and plants and hardly need further emphasis in this field. Paleontological evidence is rare and less clear, largely because of the limitations of the record and because of the deficient techniques of "standard" paleontological practice, but it can be found. Thus, I have elsewhere (Simpson, 1937a) discussed an example of fossils, collected at exactly the same horizon and locality and almost certainly belonging to one population of the extinct notoungulate mammal *Henricosbornia lophiodonta,* which showed extraordinary structural variability. Some of the differences between individuals within a single group of these primitive animals are analogous to and some homologous with differences segregated in other and in allied more advanced types and then characterizing [7] distinct species, genera, or even families.

An example of a different sort—less clear-cut and open to alternative explanation, but highly suggestive—can be extracted from data of Brinkmann (1929) on the evolution of the Jurassic ammonite *Kosmoceras.* Within the stratigraphic span of his rich and essentially continuous sequence of materials, one species, *Kosmoceras castor,* is slowly transformed into another, *K. aculeatum,* and also gives rise to a second species and lateral phylum, *K. pollux,* which in turn is transformed without further branching into another species, *K. ornatum.* The *castor-aculeatum* line is more abundant and longer lived and may be considered the main stem from which the *pollux-ornatum* line branched. I have analyzed Brinkmann's data on the following important characters: (1) terminal diameter; (2) greatest diameter of outer whorl; (3) diameter of umbilicus; (4) number of inner ribs on outer whorl; (5) number of outer ribs on outer whorl.

In all these characters except the last, which distinguishes *pollux* by its sharp progressive reduction, the earliest individuals of *pollux* are within the probable range of variation of immediately antecedent and ancestral populations of *castor.* But the variation, as measured by

[7] But not alone and in themselves defining.

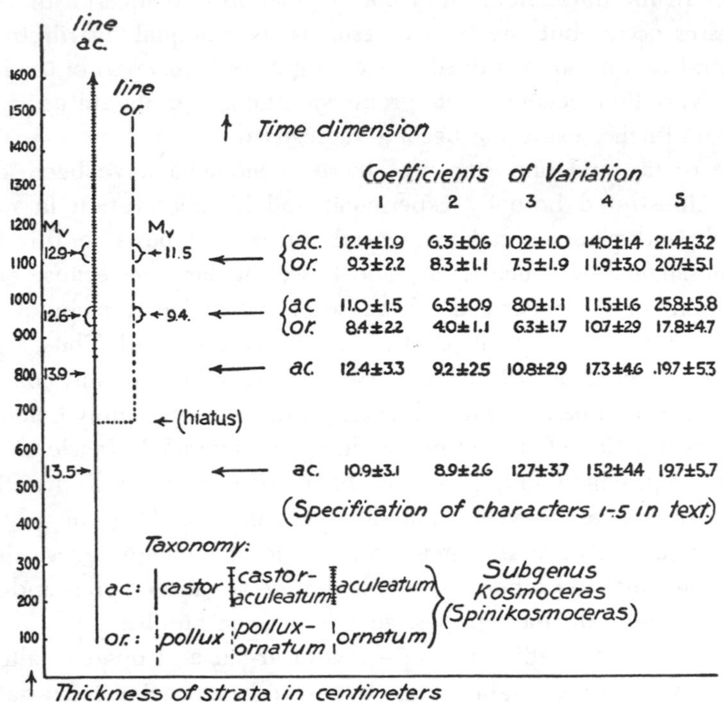

FIGURE 10. VARIATION IN A BRANCHING PHYLOGENY. Data from Brinkmann, rearranged. M_v, mean of the five coefficients of variation for each phylum and level. There is no abscissal scale. For further explanation see text.

the Pearsonian coefficient of variation ($V = 100 \times$ standard deviation / mean) was less for all five characters of *pollux* soon after its origin than in the *castor-aculeatum* line either then, earlier, or later.[8] The amount of variation within the more abundant *castor-aculeatum* line shows only random fluctuations. Only a little later, about 140 cm. in terms of thickness of strata, the coefficients have all increased in the *pollux-ornatum* line and have become comparable with those of the continuing ancestral stock. The pertinent data are shown in Figure 10.

In terms of evolutionary processes the most probable (although not the only possible) interpretation of these facts appears to me to be as follows: From a small section of the great, far-flung *castor* popula-

[8] Individually the differences between the V's for the two lines at the same time are not statistically significant, but the consistent relationship for all five characters may safely be called significant.

tion a less abundant group was cut off and became morphologically differentiated, in part by minor qualitative genetic change, not pertinent at this point, but mainly because it received only part of the store of variability in the main population. It therefore showed at first considerably less variation than the latter. Subsequently it reestablished the amount of variation usual in these groups, but did so about different means. The remaining, greater portion of the population was sufficiently large so that (statistically) random withdrawal of variants into the branch unit did not wholly eliminate representation of these in the main fraction or permanently change its equilibrium, so that its variation was not noticeably affected.[9]

The store of variation available for evolution either by shift of mean within a population or by parceling out between descendant populations may be considerably greater than is evident in the observable phenotypic variation of the ancestral population. Insofar as it reflects genetic variation, this existing phenotypic variation represents, at any one instant, the free or immediately available material for such changes. There is generally, however, a considerable amount of potential or cryptic genetic variation which can become available by later appearance in, or in a sense, by release to the phenotypes. There is also variation that is in part visible phenotypically, hence not entirely cryptic but that may be called static because it is potentially but not immediately available, e.g., variation between incompletely isolated populations.

Among the more important sorts of cryptic or static genetic variation, and means of their release, are the following:

Wholly or partially cryptic or static genetic variation	*Mechanism of release to phenotype*
1. Unrealized genetic permutations and combinations in population	1. Recombination
2. Separate occurrence of complementary genes	2. Recombination

[9] The principal objection to this interpretation is based on the improbability of sympatric speciation in these animals and the fact that the second lineage appears after a brief hiatus in the record. It may therefore be probable that *pollux* arose in isolation elsewhere, and its variation after spreading to the sampling locality might have little bearing on its origin. Nevertheless it would remain a reasonable hypothesis that this spread occurred very soon after origin of this lineage (early samples are still very like *castor*) and that lowered variation incidental to its origin still persisted, to be built up only gradually in the larger populations of the expanding lineage.

3. Genes with action suppressed or depressed by epistatic genes or modifiers	3. Inactivation of epistatic genes or modifiers, evolution of increased penetrance, recombination without epistatic genes or modifiers
4. Recessives partially or wholly masked in heterozygotes	4. Recombination with increased homozygosis of recessives, evolution of increased penetrance or of dominance
5. Limitation of combinations by linkage, polygene systems	5. Recombination
6. Different genes and gene systems with similar phenotypes under existing developmental conditions	6. Change in developmental conditions, producing or increasing difference of phenotypes on different genetic backgrounds
7. Genotypic differences between separate but not permanently or wholly genetically isolated populations	7. Migration, outbreeding, hybridization, and introgression (producing recombination *between* populations)

The different categories are not absolutely clear-cut. For instance, 1 and 2 as well as 3 and 4 may be considered different aspects or special cases of essentially the same genetic phenomenon in each case. Most of these genetic factors simply represent the store of potential variability from which recombination continually replenishes the actually available variation, genetic and phenotypic. Some of the factors, however, notably the last two, represent a somewhat different sort of phenomenon.

The great importance of these factors and processes, taken collectively, is that they supply materials and means by which selection can build genotypes and phenotypes that are actually new. They demonstrate that selection, in the sense discussed in Chapter V, is not limited to a negative role, that it does not merely eliminate variants actually present in a population but also produces variants not previously present. They also demonstrate that a shift not only of the mean but also of the range of variation may occur without mutation. Paleontological data usually reveal only the actual or free phenotypic variation available and none of these sorts of potential or cryptic variation, which must also be assumed to exist in interpreting paleontological sequences. Nevertheless, most of these sorts of potential variation may tend, roughly, to have some sort of proportional relationship to the visible variation in a given population, and the one sort of which this is plainly not true, 7, is in part visible as between populations, including paleontological populations under favorable circumstances.

Asexual reproduction severely limits or even eliminates cryptic genetic variation and certain cytogenetic mechanisms also considerably modify its nature and extent (White, 1951). There is, however, no serious doubt that each of these genetic factors does exist in many or most natural groups. The geneticists do not agree as to their relative importance over-all or their pertinence in particular cases. Having established their collective significance, we are here interested in but not greatly concerned with specific identification of one or the other as *the* factor in a given example. As throughout this book, it is not the genetic processes in themselves that are the primary concern, but only their adequacy to explain long-range phenotypic evolution and to assist in the interpretation of this. We may mention but need not take sides in some of the differences of opinion among geneticists. Thus Chetverikov in 1926 (see Sinnott, Dunn, and Dobzhansky, 1950) and, I would judge, most students since him have emphasized the role of recessives in storing variability, but Muller (1949a) considers this "fallacious." Mather (e.g., 1949, also Darlington and Mather, 1950) strongly emphasizes potential variation tied up in polygene systems, while Wright (1945) thinks this has little significance and emphasizes the potentialities of differences between incompletely isolated populations. But all these and other geneticists agree that large amounts of potential variability *are* stored, in one way or another.

There is an interesting and important relationship between the pattern of expressed and potential variability of a population, its current adaptation, and its adaptability to change, diagrammatically shown in Figure 11. A population may have low variability and narrow range, *A*. Then a maximum proportion of individuals will be at or near O_1, the optimum under existing conditions, and therefore this represents maximum adaptation to those conditions. Such a condition implies a high degree of homozygosis, low recombination or mutation, and little "contamination" from other populations. It would be most fully realized in inbreeding (especially in self-fertilizing) and in asexually reproducing groups, but (especially after long and highly effective selection) may be approximated also in others. *B* represents a population with high variability and high range, both real and potential. A rela-

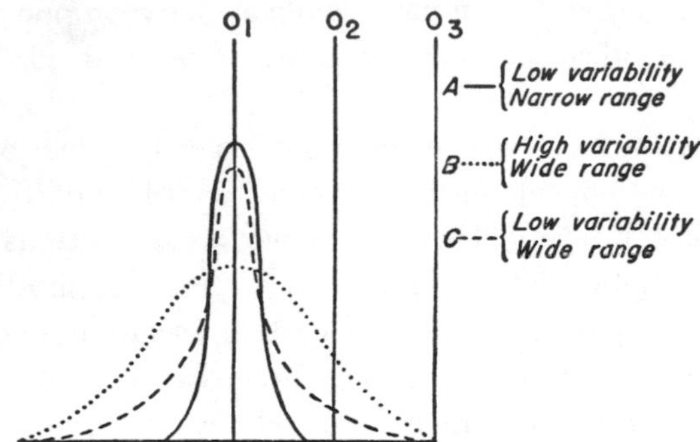

FIGURE 11. VARIABILITY, RANGE, AND ADAPTABILITY. Modified from Mather. For explanation and discussion see text.

tively small part of the population is near O_1 and adaptation to existing conditions is poor. Such a situation corresponds with great genetic heterogeneity of any or all sorts such as, among other possibilities, would be expected with free hybridization, extensive genetic migration, or panmixia in a large population spread over a varied area.

C corresponds with a population with low realized variation but wide range, corresponding with equally wide or wider potential variability. This group also has a large proportion of its population near O_1, although the proportion will usually be somewhat less than for A. Adaptation is good, but not, as a rule, maximal. Such a condition implies high potential variability but low release in existing circumstances. A population of the sort could arise in a number of different ways. Mather (1941), to whom the essentials of the diagram are due, postulated correspondence between C and the occurrence of linked multiple genes, or polygenes. This explanation was not generally accepted at the time and although Mather has since adduced strong evidence of the reality of this factor (as already variously cited) it still may not be universally accepted. But the condition represented in C could also arise in other ways considered effective by those who minimize Mather's explanation, notably in an inbreeding local population with possible but at the given time slight genetic interchange with other, different local populations, a pattern emphasized by Wright. Variation more or less as in C also arises in certain cases of balanced polymorphism (recently emphasized by Dobzhansky).

If, in the diagram, the optimum shifts to O_2, population A can adapt only by rapid mutation, will be at a great disadvantage, and will probably become extinct. Population B, with a relatively large proportion of its variability near O_2, will adapt readily. Population C, with some but less variability near O_2, can also adapt to the change but probably not as quickly as B. If, finally, the optimum should shift to O_3 or beyond, the existing variability (at O_1) of all three populations becomes inadequate for full response and good or even perhaps viable adaptation will require mutation. Thus A has excellent adaptation but very poor adaptability, B has poor adaptation but excellent adaptability, and C has good adaptation and adequate adaptability. All three sorts of populations exist and each is favored under certain conditions. In the long run, C would be expected to be most often successful.

So far, discussion has been in terms of variability actually present in a population (or group of adjacent and related populations). Even when this variability, including both realized variation and potential variability, is high, it is strictly limited. Without replenishment, it could not allow truly long-continued progression or diversification.

In passing, it should be mentioned that in spite of this, as it seems to me, obvious limitation of evolution by expenditure of existing variability, such a process has been hailed as the dominant or only essential factor of evolution. For instance, Clark (1930) maintained that an essentially similar process is the only mode of evolution within the major phyla, and he went so far as to state that "all mutants arise through the subtraction of something from the usual form." Lotsy's theory (1916) of evolution by hybridization is another allied generalization, with the *petitio principii* of explaining the origin of hereditary differences by saying that their determinants have always existed and that the succession of life as we see it results merely from their shuffling and dealing.

Evolution by segregation of previously existing variability would eventually become a degenerating kinetic system, limited by and tending toward a condition either of cyclic repetition or of dead level, all variability having been segregated into isolated invariable groups (transferred to the intergroup level from which no new group could be differentiated). The paleontological record as a whole shows that no such cycles have as yet appeared in the history of life on the earth. At least one paleontologist, Broom (e.g., 1933), has maintained, on other and mainly metaphysical grounds, that evolution has now almost reached a dead level and cannot progress further; but (as he notes) few students agree. Huxley (e.g., 1942) does agree that evolution has essentially stopped (except in man) but, again, on other grounds than depletion of the stock of variability in all populations. These are, rather, generalizations (which I consider dubious) partly on the basis of the previously mentioned loss of adaptability through specialization. Special discussion of the probabilities or possibilities of future evolution is not pertinent here in spite of its great interest.

Even though their characters evolve far beyond earlier limits of range, evolutionary sequences in the fossil record do not, as a rule, show depletion of variability or progress by expenditure of unreplenished variation.

Brinkmann's data on *Kosmoceras*, already discussed, also bear on this question. In the main *castor-aculeatum* line the coefficients of variation themselves vary, but they show no significant trend and no evidence of secular depletion of variability. In the *pollux-ornatum* line the evidence suggests unusually low variability at the time of origin and a secular *increase* in variability, until it is built up to about the level normal in the other branch.

In the horse sequence illustrated from a different point of view in Chapter I, the coefficients of variation shown in Table 17 were found.

TABLE 17

COEFFICIENTS OF VARIATION FOR PARACONE HEIGHT AND ECTOLOPH LENGTH IN FOUR SPECIES OF HORSES

	V of Paracone Height, Absolute Value of which Evolves from $M = 4.7$ to $M = 52.4$	*V of Ectoloph Length, Absolute Value of which Evolves from $M = 8.2$ to $M = 20.8$*
Hyracotherium borealis	6.2 ± 1.3	5.7 ± 1.2
Mesohippus bairdi	4.8 ± 1.1	4.6 ± 0.9
Merychippus paniensis	5.9 ± 1.2	5.3 ± 1.0
Neohipparion occidentale	4.6 ± 1.5	5.3 ± 1.7

Despite the great amount of evolutionary change in both these characters, their variation remains about the same throughout, with only random fluctuations as far as the data show.

These figures are typical for linear dimensions of functional hard parts in mammals, for which the great majority of such coefficients lie between about 3 and 10, as further exemplified by the following examples taken at random from the many available.

Analogous variates in widely disparate mammals:

Didymictis protenus (creodont), length M_2	5.2 ± 0.7
Notostylops murinus (notoungulate), length M^3	7.1 ± 1.6
Ptilodus montanus (multituberculate), length M_1	5.7 ± 1.4

Homologous variates in ordinally related (condylarths) but otherwise very different mammals:

Phenacodus primaevus, length M_2	4.5 ± 1.0
Haplomylus speirianus, length M_2	6.7 ± 1.0

Variation does, of course, fluctuate, but for analogous characters of animals even distantly related it appears as a rule to be remarkably circumscribed in its fluctuations in time or differences in contemporaneous groups. Exceptionally low values usually indicate a sample more uniform than is the whole interbreeding population, and exceptionally high values usually indicate either characters for which the mechanics and function have been less rigidly integrated or characters that are degenerating and have lost all function. This last point has probably become apparent to every paleontologist who has handled large collections. For instance, a functionless tooth, such as P^2 in *Hoplophoneus,* may vary from a well-developed state to complete absence, not only within one race but also within one individual (i.e., the left side may differ from the right side). In *Ptilodus montanus,* in which functional teeth have much the same variability as in other mammals (as exemplified above), the length of P^3, which is degenerating and losing function, has the high value $V = 18.5 \pm 2.8$. It is so commonly true that degenerating structures are highly variable that this may be advanced as an empirical evolutionary generalization.

That generalization, published in the same words in the predecessor of this book (Simpson, 1944a), was made the subject of special, critical study by Goldschmidt (1946) in line with his objections to the "neo-Darwinian" interpretations of evolution here advocated. He pointed out that a number of known mutations, notably homoeotic mutants in *Drosophila,* also are extremely variable in their phenotypic expression, i.e., in penetrance. He speculates that "the decisive determining substance, whatever this means, is produced by the alleles in question in an amount near the threshold value. . . ." Variations in genetic and external milieu during development may then bring about action of the gene before, during, or after a critical time zone so that its phenotypic expression is complete, partial, or absent. This seems to me to be at least a possible physiological mechanism for production of the observed variations, but it seems to have no particular bearing on the evolutionary significance of such events. The evolutionary process could well involve relaxation of selection against or positive selection for a mutation tending by some such mechanism of variable penetrance to bring about loss of a tooth. (Incidentally, this seems to be a somewhat shaky parallel with a homoeotic mutation which tends to make one member of a series resemble another, different member). On Goldschmidt's hypothesis there is, indeed, a gradual change in average penetrance, and the structures in question eventually disappear entirely. As a matter of fact no proposed genetic or evolutionary mechanism was, or is, offered at this point, but some discussion on that aspect will be in order later in connection with the variable rise of new structures, to which Goldschmidt has also alluded.

The general tendency for amount of variation to remain roughly constant through long evolutionary sequences and over broad groups of contemporaneous animals indicates that the pool of variability is not steadily and usually drained either by phyletic evolution or by splitting. It may be drained, or nearly so, in such groups as approximate A in Figure 11, but this represents a type of adaptation and narrow specialization that is particularly vulnerable to extinction. Groups that are successful in long phyletic progression or in abundant splitting clearly must replenish their variation and must do so, in many cases or on an average, at rates approximating its expenditure.

With exceptions hardly significant in the over-all picture, the sources of new variation to replenish the pool must be gene and chromosome mutations. It is usual to add that chromosome mutations only rearrange, multiply, or divide existing genes and that therefore the ultimate source of new variation is gene mutation alone. The distinction may not prove to be quite as absolute as it sounds, and in any case it does not matter particularly for present purposes. Mutation in the broadest sense is the source of new variation. It seems extraordinary that mutation may neatly balance reduction of variation over long periods of time, since it can hardly be supposed, on any present evidence, that reduced variation directly stimulates mutation, or that increased variation inhibits mutation. There must be some other factor capable of reacting to both of these and tending to balance them. The probability that this factor is selection will be developed as discussion proceeds.

VARIABILITY AND RATE OF EVOLUTION

Since variation is the raw material for evolution, a certain minimal amount, at least, must be present if evolution is to proceed. In short-range processes depending mainly on the existing pool of variation, the amount and sort of variation evidently may partly determine, or at least limit, the rate of evolution. In any usual circumstance a more variable population should be capable of more rapid adjustment to rapid environmental changes and often also to more rapid splitting and divergent speciation.

The extent to which this sort of process may go, just how much change may actually occur on the basis of existing variability, has been only lightly and perhaps ambiguously touched on above. Indeed, I do not find any data on which to base a reasonably specific estimate. Wright (1945) thinks that "under exceptionally favorable conditions great evolutionary advance is possible at an explosively rapid rate" without new mutation but making use of a large store of potential genetic variability, which Wright thinks would be in the form of intergroup differences in a large group of small, nearly but not quite isolated local populations. He also thinks that I formerly (Simpson, 1944a) somewhat underestimated the amount of evolution that could occur in this way. I frankly see no way to make an estimate that could be validly tested at present. Perhaps no one would hesitate to say that successive species can arise in this way or even that contemporaneous species frequently do so. Probably, too, few would claim that new mutations were not necessary between *Hyracotherium* and *Equus* or, somewhere along the lines, between *Bos* and *Ovis*, the vertical and horizontal differences, respectively, being conservatively evaluated as at about the subfamily level in both cases.

Moreover, it seems probable that any new evolutionary trend might have been *initiated* on the basis of existing variability, or again that it might not in a given case. (That important changes are *always* initiated by new mutations is an extreme view, to which I do not subscribe, to be discussed later.) Yet wherever the limit may have been in any particular instance, a limit must have existed if evolution continued long enough.

and thereafter new mutation must finally have been required for further change.

There is some evidence that the rate of long-range evolution is not generally determined by or correlated with amount of variation. In the example of the horses in Table 17, we know that tooth evolution, particularly paracone height, was proceeding at an exceptionally high rate in *Merychippus* (see Fig. 3), but variation is not significantly different from that in the other groups sampled. The variation in *Merychippus* happens (without statistical significance) to be somewhat lower than for *Hyracotherium* in these samples, although the rate of change of these characters in *Hyracotherium* was relatively very low.

Another important line of evidence on the relationship between variation and the control of evolutionary rates would be a study of variation in living groups known to have had exceptionally fast or slow rates of evolution; but as far as I know little attention has as yet been given to this point. I have examined samples of several low-rate vertebrate groups, including crocodiles (little changed since early or middle Cretaceous), opossums (since late Cretaceous), armadillos (since late Paleocene), and tapirs (since about Oligocene). These slowly evolving forms all show intragroup variation at least as great as in allied, more rapidly evolving lines (e.g., lizards, sloths, kangaroos, and horses, respectively) or in reptiles and mammals in general. There is even some not wholly conclusive evidence that conservative groups are sometimes exceptionally variable. For instance, in a very homogeneous sample of opossums, V for tail length is 15.4, while in a similar sample of the more rapidly evolved group of white-footed mice it is 5.0, a figure usual for mammals with undegenerated tails.

Man must be considered zoologically a mammal that has evolved at more than average rate and, moreover, he has breeding habits and other characters that must tend to increase variability. It is therefore pertinent that the variation in his homologous structural characters is not noticeably greater than in allied groups or in mammals in general. Pearl (1940, pp. 356–59), in his compilation of coefficients of variation in man, gives 70 values of linear dimensions of nonpathological groups. Sixty-three of these V's lie between 3 and 10, and 45 of them between 3 and 6. The greatest value given, 18.99 (for neck length in the Swiss) and the least, 2.35 (internal maximum length of skull in male Australians), are still within the range of V for analogous characters of more slowly evolving mammals, and the whole distribution demonstrates quite usual variability for a mammal, similar to that in single races of horses, cats, and many other groups of about average evolutionary rates within the Mammalia.

Huxley (personal communication) has suggested that man is unusually variable if the whole species is considered and not only such local populations as are included in Pearl's tabulations. In terms here used, Huxley's suggestion is that man possesses an unusually large pool of potential or static variability in the form of differences between groups that can interbreed but that usually have not done so freely. Another possibility is that this variability has been parceled out from a more unified ancestral pool, which would then have been unusually large. Yet it seems to me more probable that the high, present intergroup variation, which clearly exists, was developed by divergent evolution of local populations each of which was (and is, unless highly hybridized, *vide* Pearl's data) of quite usual variability. The group differences then do not reflect a correlation of high variability with rapid evolution, but rather the fact that man has adapted to an altogether uniquely large number of different regions and habitats. Of course the variability, however acquired, is available today, but I know of no conclusive evidence that it is now involved in rapid evolutionary change.

On theoretical grounds maximum rates of evolution would probably be accompanied, not by high variability, but by exceptionally low variability at any one time. This may be demonstrated by a hypothetical case, which could not, indeed, occur in nature, but represents the limiting extreme for rate of evolution depending on selection of favorable mutations. Suppose that a series of progressive mutations occurs in a gene: A_1, A_2, A_3, A_4, and so forth. Suppose that one such mutation appears in each generation and that the selection value for each is 100 percent, i.e., that only those animals inheriting the most progressive mutations of one generation survive to breeding age in the next. Then each successive mutation in the chain will survive for three and only three generations. The F_3 generation will be entirely different from the parental generation with regard to this gene, which (for bisexual reproduction) represents the most rapid possible evolution of the gene. Moreover, in each generation the population will be almost invariable in genotype: there will be only two genotypes, one possessed by the population as a whole, the other by a single individual.

Continued progression of a population over long periods of time must, in more complex and necessarily slower form, approach the model of this limiting case. Earlier mutations must be eliminated from the population on one side as new progressive mutations are added on the other. Mutations are, so to speak, being run through the population, and the number present at any one time is a determinant of the variability of the population. In a large population the number of alleles actually present at any one time may be relatively independent of the rate of their loss and addition as long as these tend toward a balance and the variability pattern itself is subject to selection or some other control. This is the situation suggested by most paleontological series from large populations with moderate evolutionary rates. Decreased loss, e.g., by relaxation of selection on nonfunctional characters, would tend to increase variability, and this, too, agrees with observation. In small populations, in which the most rapid sustained phyletic evolution is theoretically possible, rate of evolution would tend to be inversely proportional to variability, because more rapid "running through" of mutations would lower the number of alleles present in the population at any one time and slower transformation would increase the number.

It seems to follow from all these examples and considerations that above a possibly very low minimum the amount of variability present at any one time does not determine or even limit the rate of long-range evolution, although it certainly limits and may help to determine rates over relatively shorter periods of time. On the other hand, if evolution is long sustained its rate must eventually be limited (although, again, not necessarily determined as regards values above the lower limit) by rate of supply of *new* variation, that is, by rate of mutation. The next step in this inquiry is to consider mutation from the present, special point of view of evolutionary rates and patterns.

CHAPTER IV

Mutation

THE TERM "mutation" was first applied by a paleontologist, Waagen (1868), to a recognizable stage in a continuously evolving lineage. In modern systematics, this would imply that a mutation is a population and a taxonomic unit, of more or less subspecific rank in terms of morphological distinction, but differing from a neontological subspecies by being delimited from others in time rather than in space.

Then De Vries (1901), referring to the already long-known fact that individuals may suddenly appear with hereditary characters sharply different from those of their ancestry, applied the same word, "mutation," to this altogether different phenomenon. Mendelian genetics took over De Vries' term in a similar but more precise sense, applying it to a distinctly variant phenotypic character associated with a change at some one locus in a chromosome and segregating according to determinable rules. Then it was found that more or less similar phenotypic changes may be related, not to such a locus, but to various rearrangements within a chromosome or to changes in the number of chromosomes. (De Vries' mutations later proved to be chromosomal, not genic.) These changes were, and still often are, also called "mutations," or "chromosome mutations," structural or numerical, to distinguish them from the point, locus, or gene mutations. It is sometimes difficult to make a sharp distinction between the three established sorts of genetic mutations or to determine which is involved in a given case. "Mutation" may then be used as a general term for the appearance of any character not inherited from the ancestors of an organism but heritable by its descendants.

Some paleontologists (e.g., Swinnerton, 1947) still use "mutation" in the sense of Waagen, and others (e.g., Osborn, 1927) continued to use it in the sense of De Vries after this had been importantly altered by the geneticists, giving rise, in both cases, to considerable misunderstanding between paleontologists and geneticists. Most of us now recognize that

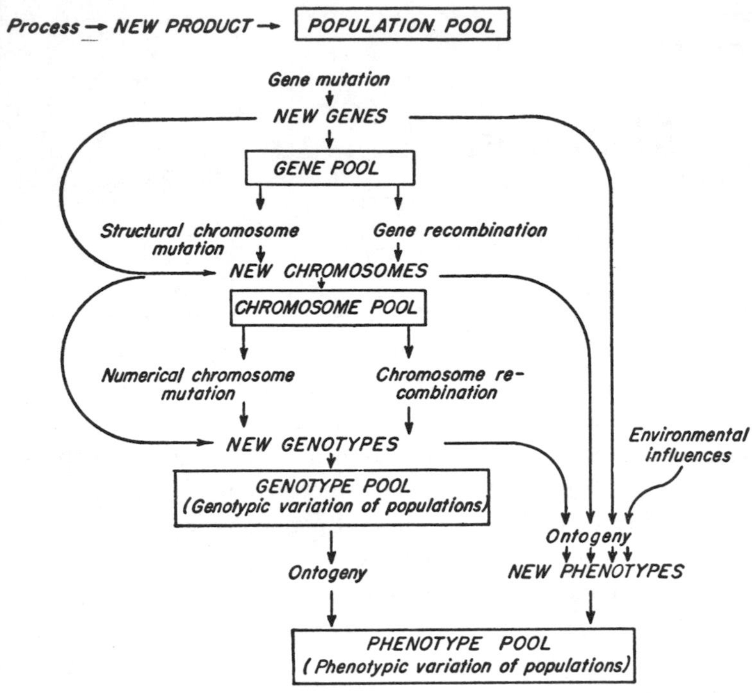

FIGURE 12. DIAGRAM OF THE RISE, FLOW, AND STORAGE OF MATERIALS FOR EVOLUTION.

it is futile to stand on prior definitions and that overwhelming usage compels us to define "mutation" according to current consensus among geneticists.[1]

The general relationship of mutation, among genetical processes, to the production of phenotypes is diagrammatically suggested in Figure 12. (It is, of course, the origin and destiny of phenotypes that are the subjects of this study and, as elsewhere, the genetical factors are here important only to the extent that they help us to understand phenotypic

[1] This, incidentally, deprives evolutionary studies of a term for a mutation, *sensu* Waagen. "Waagenon" has been suggested but is not widely used. There is here a real problem not only of terminology but also of concept, which has not been usefully resolved: the problem of distinguishing vertical from horizontal systematic units. The usual, but sometimes confusing, practice is to use Linnaean terms for both, even though they are quite different things. Solution is made difficult by the facts (1) that paleontological material often makes it impossible to be sure which sort of unit is actually in hand (see Simpson, 1951b), and (2) that the problem is not confined to the level of "waagenon" = (essentially) "vertical subspecies" but involves all levels of the hierarchy.

evolution.) The diagram does not represent a cycle that must start at one end and flow to completion at the other. In given cases it may and does start with any process and end with any product. For instance a structural chromosome mutation may (and usually does) occur without involving a newly mutant gene, and the new chromosome then produced may not become involved in either numerical chromosome mutation (it is unlikely to be so involved in most groups) or chromosome recombination (although this is more likely than not in most groups). That any new genetical product gives a new genotype and may, by the process of ontogeny, lead to a new phenotype is indicated by arrows bypassing other processes and products. Regardless of how much of the chain is involved in any given case, what does occur is in the direction and sequence shown.

Beyond this basic diagram of evolutionary materials, their origin and flow, are other processes still more important in some ways and inseparable in consideration of the whole problem of evolution. On one side are the factors initiating and controlling the genetic processes. On the other side are the factors influencing the abundance and availability of genetic products in the various pools. Study of these factors, as far as they are essential to the themes of this book, was begun in the last chapter and continues in this and following chapters.

One other point to be made about this diagram and the whole discussion of genetic factors in this book is that cytoplasmic heredity is generally ignored. This is not because cytoplasmic inheritance does not occur—of course it does (see, e.g., Caspari, 1948)—but because it seldom significantly changes the particular picture we are looking at and generally has little effect sharply separable from that of nuclear inheritance. There are some exceptional cases, especially among the protistans, where the whole organization may be directly inherited and not only a mechanism for the development of this. (For this and other reasons protistan evolution has aspects peculiar to that group, but there is no reason to doubt that the general principles involved are the same as in the, or the higher parts of the, plant and animal kingdoms.)

ORIENTATION AND LIMITATIONS OF MUTATION

It does not much matter from the present point of view what goes on inside the nucleus when a mutation occurs, beyond the fact that certain nuclear events, notably the origin of polyploidy, tend to have characteristically different consequences in phenotypic populations. It is fortunate that our need goes no further than this, for not a great deal more is definitely known about the intimate and immediate aspects of mutation. Changes in chromosome number can usually and changes in chromosome structure can often be seen cytologically. Changes within genes cannot, and the results may intergrade so that listed gene mutations seem to be only a residuum of mutations not (or, perhaps, not yet) known to be chromosomal.

These considerations and the further, more important fact that specific gene action is often only arbitrarily to be isolated from the interaction of the whole genome have even led to questions whether the gene is not a place rather than a thing, the chromosome being the smallest "real" unit involved. That view goes against strong evidence and the consensus of geneticists (see, e.g., Muller, 1947, or, indeed, any recent textbook of genetics) that genes are structural, chemical entities tied together in some way longitudinally into chromosomes. Fortunately, again, for our purposes this does not matter. The observed effects of mutation, recombination, linkage, genetic interaction, and so on, are solid facts whatever may be the nature of the gene. In long-range studies the effects of mutations naturally have great importance, but their cytological basis does not—and of course the cytological basis is forever indeterminable in any particular case with most of our materials.

Whatever its cytological or chemical nature may be, any mutation occurs within a genetic system that is variable in the population but that is in every case integrated to a remarkable degree. This places quite rigid limitations on what a mutation can accomplish, so to speak. Homoeotic mutants have special interest because (at full penetrance) they have spectacular phenotypic results, but their limitation by the system in which they occur is obvious. They do not produce new structures but only change the places where structures occur. "Tetraptera" in *Drosophila,* for instance, does not produce new structures, wings, which of course are a normal result of the *Drosophila* reaction system, but only produces the wings characteristic of the system in an additional place. That the result is four wings instead of two and that four wings occur in nonmutant or, better said, average members of other orders of insects clearly does not mean that the tetraptera mutation, so plainly operating in the *Drosophila* reaction system, has changed the system to that of another order. Nor does it show that an order may arise as the result of a single mutation. The conclusion that homoeotic mutants demonstrate how new taxonomic groups can arise by single mutations, expressed in a review of homoeosis by Villee (1942), involves, as Davis (1949) [2] has said, a "rather naive interpretation."

No one has more strongly stressed the integral nature of genetic reaction systems or produced stronger evidence for it than Goldschmidt (especially 1938, 1940). The obvious conclusion would seem to be that since single mutations occur within an established reaction system and do not immediately convert this into another, distinctly different, integrated system, the development of such a new system must involve more than one mutation and a period of reintegration. Goldschmidt nevertheless draws from his evidence a conclusion exactly opposite to what seems to me its most probable bearing. He argues that since a new reaction system cannot arise from one mutation of the sort known to occur therefore it must arise from one mutation of a sort not known to occur, a mutation of the whole system or "systemic mutation." The rejection of this view by the consensus of geneticists [3] in the twelve years since it was most fully expressed makes it unnecessary to belabor the point now as much as when I first discussed it (a critique written in 1941–42 and published in Simpson, 1944a). It must, however, be mentioned again below, especially in connection with "size" of mutations and Schindewolf's belief in the origin of all systematic categories by single mutations.

Even within the limitations of a given reaction system, it is evident that not all conceivable mutations occur and that those that do occur do so with very unequal frequencies. In organisms extensively studied in this respect, notably *Drosophila,* of course, mutations are known affecting practically every part of the body and numerous different physiological characters, but effects do not necessarily take all directions. Met-

[2] Davis also remarked that they had "contributed to misunderstanding" exemplified by Simpson (1944a). I do not know what the misunderstanding is, since Davis evidently agrees in rejection of Villee's interpretation. It is also true, as Davis stresses, that homoeosis bears on the whole question of growth fields and their genetic control, but that is not pertinent at this point.

[3] As a nongeneticist, I of course do not pretend to be competent to judge Goldschmidt's thesis except by its consistency with evidence on phenotypic evolution. An early reaction by a competent geneticist was that of Wright (1941), which seems fairly typical of the consensus of Goldschmidt's peers: profound respect and admiration for his genetical work but disagreement with his hypothesis on the origin of species and higher categories. Wright (1949) later expressed agreement with Goldschmidt on one aspect and in a limited sense; and, indeed, with abandonment of the purely hypothetical "systemic mutation," the disagreement becomes less profound than at first appeared.

rical characters (length of wings, numbers of bristles) can only be affected in two ways, increase or decrease. Direct observation of single mutations and artificial selection in one or the other direction show that mutants in both directions and consequent alleles and gene combinations producing both sorts of change in metrical characters do occur with some regularity. But in cases where multiple directions of physically possible change exist, all are not necessarily followed by known or even, as far as one can determine, by genetically possible mutants. The well known multiple alleles affecting eye color in *Drosophila* are a case in point. Many colors are produced by known alleles, from white through yellowish to red and related hues, but blue eyes, for instance, never occur. The reason is that the mutations are acting on, or in, an established chain of biochemical reactions leading, normally, to red eye pigment (see Beadle and Ephrussi, 1936) and that no interruption or branching in the chain can (presumably, or in any case does) produce a blue pigment. This does not mean that other chains built up by accretion and integration of mutations might not lead to blue pigment or might not be subject to further mutations to blue, as does, indeed, occur in other organisms. Again, the effects of any one mutation are limited by the existing gene (or reaction) system in which it occurs. A more profound reorganization is required to make possible other directions of mutational change.

This sort of limitation and the fact that different mutations may have widely and characteristically different rates of incidence show that mutations are not random in the full and usual sense of the word or in the way that some early Darwinists unrealistically considered as fully random the variation available for natural selection. I believe that the, in this sense, nonrandom nature of mutation has had a profound influence on the diversity of life and on the extent and character of adaptations. This influence is sometimes overlooked, probably because almost everyone speaks of mutations as random, which they are in other senses of the word.

There is, on one hand, a randomness as to where and when a mutation will occur. Mutations induced by hits by X-rays, hits that must be statistically random, tend to have about the same distribution as naturally occurring mutations although the frequencies are higher (Muller, 1947, and numerous references there). This indicates a "molar indeterminacy" (Muller), a randomness of energy supply or stimulus (whatever the source may be in spontaneous mutation), although still not a wholly random reaction, since different genes still mutate at greatly different rates.

On the other hand, the term "randomness" as applied to mutation often refers to the lack of correspondence of phenotypic effect with the stimulus and with the actual or the adaptive direction of evolution. Heat-induced mutations do not produce phenotypic change related to heat tolerance.[4] It is a well known fact, emphasized over and over again in discussions of genetics and evolution, that the vast majority of known mutations are inadaptive. They are almost always disadvantageous to the individual and (although this is not always so clear) to the population under the actually existing conditions. Just as often, it has been pointed out that this is precisely what would be expected. If adaptation is perfect, any change is disadvantageous. Even if adaptation is imperfect, mutations tending to improve adaptation and occurring frequently enough to be detected would in most cases already have been incorporated into the genotype of the population, i.e., would have become wild type genes and not mutants with respect to the existing situation. This consideration applies to any population that has achieved a considerable degree of adaptive stability, which is true of most laboratory organisms and specifically of *Drosophila,* otherwise such enlightening experimental material.

A population in process of adapting to change in its environment or to an environment new to it may be expected to have some adaptive instability. It may be adapting by utilization of expressed and potential variability but it may also be adapting in part by adaptive mutations. Sooner or later and in some changes of adaptation, if it is true that mutation is the ultimate source of material for evolution, adaptive mutation must be involved. In spite of the general "randomness" of mutation in the special senses noted, there is adequate evidence that adaptive mutations are often available under such circumstances. In the laboratory, mutations may decrease viability under some conditions and increase it under others. The now classic example is the mutation "eversae" in *Drosophila funebris* (Timofeeff-Ressovsky, 1934). Mutant flies are inferior to wild type in viability at 15°–16° and 28°–30° C., but superior at 24°–25°. Numerous similar examples are now known. In nature, popu-

[4] There remains a possibility that mutation induced chemically by reaction with the gene might produce physiological changes related to the inducing chemical. This, however, would only be the sort of exception that proves the rule. Its occurrence is still doubtful in the laboratory and unknown in nature.

lations adapting to new conditions, either imposed on them or involved in their geographic expansion, often show adaptive increase in frequency (within the population, not evident or known increase of mutation rate) of what are, with respect to the ancestral population, mutant forms. Classic examples here are increased melanism of certain butterflies in industrialized areas (Ford, 1945 and earlier) and expansion of melanic mutants of *Cricetus* in the U.S.S.R. (Kirikov, 1934; Timofeeff-Ressovsky, 1940). Many other examples are known of this sort, too (see, e.g., fuller discussions of the subject in Dobzhansky, 1941, Huxley, 1942, Carter, 1951). Dobzhansky (1941) has further pointed out that what appear to be monstrosities when they arise in mutants away from wild type may nevertheless be wild type in other organisms, of different adaptive type, and that "the value of anthropomorphic judgments on what constitutes a malformation is spurious."

The conclusions are: (1) that mutations are not strictly random in range of phenotypic effect or in relative frequency, (2) that given the limitations of range of effects and of characteristic rate, mutations are random in incidence, (3) that the nonrandom tendencies of effect and rate do not correlate with present adaptation or with past or present changes in adaptive type, so that mutations may also be said to be random from this special point of view, but (4) that mutations adaptive for changing or for different conditions do occur and are involved in adaptation.

THE "SIZE" OF MUTATIONS

The concept that there are two distinct sizes, so to speak, of mutations has a strong bearing on what is here the main reason for discussing mutations at all, i.e., their relationship to the origin of new taxonomic groups and to rates of evolution.

Darwin and most of his contemporaries believed that most variation blended in inheritance, for instance that the offspring of one short and one tall parent would be intermediate in height and would pass on a hereditary factor for medium height. Mendel's experiments and the rediscovery of Mendelism showed that this is not true in some cases. As every student now knows, in Mendel's low and high peas the f_1 hybrids were not intermediate but all high. The next generation (f_2) was either low or high in the ratio of 1:3, and certain of the f_2 high plants, bred among themselves, produced a similar proportion of low offspring. In such cases inheritance obviously does not blend but must be particulate. The striking nature of such results and the development of elegant experimental methods for their study focused the attention of geneticists on inheritance that does plainly "mendelize." This left untouched a large number of characters that seem to vary continuously and for which the f_1 generation, at least, is more or less intermediate between the parents. Naturalists noted that it is frequently precisely these characters that distinguish taxonomic groups and that have been most important in evolution. Thus arose a conflict, which has continued to the present time and is still pertinent to our theme in somewhat different guise, between those who thought that new groups usually, or always, arise by abruptly discontinuous change and those who thought that such change is usually, or always, more gradual or continuous.

In work strongly instrumental in the rise of the modern synthetic theory of evolution, Fisher (1930) showed that blending inheritance cannot be effective material for natural selection. Since not only discontinuously varying, demonstrably mendelizing characters but also, and even more particularly, seemingly continuously varying and apparently blending characters have been involved in evolution, one must either discount the effectiveness of natural selection or postulate that the hereditary factors underlying apparently blending characters are also particulate. There was already some evidence, due especially to W. Johannsen, H. Nilsson-Ehle, and E. M. East (all working separately on different organisms) that the heredity of continuous variation is particulate. This conclusion has since been completely established, and special experimental methods have been devised for the study of inheritance not subject to Mendelian methods (see especially Mather, 1949; Darlington and Mather, 1950).

In one of the two sorts of inheritance thus distinguished, a single mutation produces, and consequently a single gene, chromosome configuration, or chromosome set controls, at least one phenotypic expression sufficiently distinct to be readily recognized and to permit counting the individuals that have this single genetic factor. In the other case, there is no one outstanding phenotypic effect from one genetic factor, but several or many genes have roughly comparable degrees of control over the characters in question. On this basis, two students recently much concerned with the latter sort of inheritance, Darlington and Mather (1950), have set up a rather elaborate distinction between two sorts of genes, and therefore also of gene mutations:

"Major genes"	*"Polygenes"*
1. Producing discontinuous or qualitative variation, analyzable by Mendelian methods	1. Producing continuous and quantitative variation, analyzable by biometric methods
2. Acting separately on one or, if pleiotropic, more characters, "monogenes"	2. Acting together, "polygenes"
3. With relatively radical or independent effects	3. With less radical effects, often as modifiers of "major genes"
4. Usually destructive in mutant form	4. Often constructive in mutant form
5. Less linked, or linked at random	5. More and systematically linked
6. Localized entirely in euchromatin	6. Localized mainly in heterochromatin

Darlington and Mather have not listed the distinctions in just this sharply contrasted way and they recognize some intergradation between the two, but they do state and emphasize these differences. I observe that other geneticists and cytogeneticists do not wholly agree with these partly speculative specifications, but even if we took Darlington and Mather's conclusions as entirely factual, their "major genes" and "polygenes" would still represent only the terminal members of a continuous series and not a real dichotomy into two distinct sorts of genes. Note the following authoritatively established points, numbered as above:

1. The genes of Mendelian experimentation vary greatly in phenotypic expression, from strong and obvious down to (and it is only reasonable to conclude, below) barely visible. They produce both qualitative and quantitative effects. Even in Mendel's experiments, one of the genes controlled an obviously quantitative character and other characters could be expressed as either qualitative or quantitative. In fact the distinction is often spurious. We may call a character "qualitative" merely because its variation has convenient discontinuities or because we do not know how or do not want to bother to measure it.

2. It is only a convention to state that any gene acts separately. All certainly act as parts of the whole genetic system. Within this system, presence or absence of one gene may accompany presence or absence of one character. But one character may also be affected by two, three, or

more genes. In their status as separable units, "major genes" and "polygenes" intergrade completely.

3. As to more or less radical effects, see 1. As to independence of action, see 2.

4. The larger the phenotypic effect of a mutant gene, the more likely it is to disturb the integrated system but, again, magnitude of effect is a completely graded sequence. Moreover, as previously noted, even maximal effects may be "constructive" in a different genetic and adaptive context; they are merely less likely to be so in any given context.

5. All genes have some linkage and all linkages may be broken. This is another fully graded sequence.

6. Darlington and Mather conclude that both "major genes" and "polygenes" occur in euchromatin. That polygenes occur in heterochromatin and that no "major genes" do seem at present two propositions neither of which is established.

The distinction is useful in some respects, but it seems fair to conclude that "major genes" and "polygenes" are not two different sorts of genes but merely extreme aspects or instances of the graded series of gene effects and gene integration in general. Of course this conclusion does not deny, in fact it even emphasizes, the point well taken by Darlington and Mather and long generally recognized that genes do differ greatly in the magnitude of their effects and in the extent to which they interact as regards particular characters. Such differences certainly make a difference in phenotypic evolution, but they are not to be summed up as involving two, and only two, distinct sorts of genes.

Similar recognition of real differences in the bearing of different mutations on evolutionary processes with similar confusion in dividing a continuous series into two and only two categories is widely involved in the recent, remarkably fine book by Carter (1951). Citing Goldschmidt (1940), Carter defines "macro-mutations" as "the large, easily recognisable mutations of genetic experiments" (i.e., of the classical Mendelian methods; of course many genetic experiments involve Carter's "micro-mutations") and "micro-mutations" as "small mutations caused by modifying genes, position effects, etc." Carter believes that small scale evolution (microevolution, on the specific level or lower) is largely based on micromutations but may also involve macromutations. He ascribes some of the evolution of higher categories to accumulation of microevolutionary change, but reserves opinion about the origin of distinctly new patterns. He does not explicitly ascribe the evolution of new patterns to macromutation.

Certainly there is no objection to the use of such contrasts as "major genes" vs. "polygenes" or "macromutations" vs. "micromutations" if the distinction is understood to be one of degree and not of kind. A distinction in degree is useful and significant. Goldschmidt, however, meant the distinction between "micro-mutation" and "macro-mutation" to be one of kind, between the gene mutations recognized in the laboratory and a postulated radical reorganization of the chromosomal system as a whole. Such a postulate is open to discussion in any defined terms, but different usage of those terms—as for ordinary gene mutations with different degrees of effect—confuses the issue.

The difference between gene mutations and chromosome mutations and that between structural and numerical chromosome mutations are differences of kind, as far as concerns their cytological basis. The distinction may also affect evolutionary events, for instance in the fact that polyploidy commonly sets up an immediate, isolated, new system and gene mutation does not. In magnitude or character of phenotypic effect, however, there is no clear difference (see, e.g., Stadler, 1932, and the extended review of effects of chromosome mutations in plants by Stebbins, 1950, with many references). Nor, with certain reservations connected with polyploidy in plants, are the differences consistently related to taxonomic distinctions or levels. In *Drosophila* the two similar but distinct species *D. melanogaster* and *D. simulans* have almost the same chromosome arrangement (Sturtevant, 1929; Kerkis, 1936), while the equally similar *D. pseudoobscura* and *D. miranda* have radically different arrangements (Dobzhansky and Tan, 1936) and nearly indistinguishable forms of *D. pseudoobscura* also differ markedly in arrangement (Sturtevant and Dobzhansky, 1936). The possibility of intergeneric hybridization (e.g., *Bos* × *Bison*) also shows that organisms certainly different in many genes may have no marked chromosomal differences. There is therefore no warrant for thinking that the known sorts of chromosome mutations are necessarily involved in the origin of new taxonomic groups, to the level of genera and possibly well beyond, or that they provide a mechanism for the instantaneous or rapid origin of such groups with, again, the exception of some polyploid species of plants.

One is led again to the view that as far as phenotypic characters are concerned—and it is these characters that interest us here and that are, in the main, the *direct* materials of evolution—there is usually no significant *qualitative* distinction between types of mutations, even when such a distinction does exist genetically. Quantitative distinction, along the continuous scale from imperceptible to quite radical effects, nevertheless exists and is important. On this point, certain conclusions may be stated at once even though the evidence and argument for them are cumulatively given in later pages of this and subsequent chapters:

1. Both small and large mutations (in the sense of evident degree of phenotypic effect) and all grades between smallest and largest are involved in evolution.
2. All sizes of mutations may be concerned at all levels and in all sorts of evolutionary processes, e.g., the origin of taxonomic categories from subspecies to phyla or the splitting of groups into diverse species, genera, etc.
3. All these processes may occur on the basis of relatively small mutations only.
4. It is exceptional in nature for any considerable evolutionary change to occur, or for a taxonomic group to originate, on the basis of relatively large mutations alone. When relatively large mutations are concerned, they usually do not persist or do not come to characterize whole populations unless they are accompanied by more numerous smaller mutations tending to modify and to integrate the effects of the larger mutations.
5. Smaller mutations tend, although also with exceptions and irregularities, to be more frequent than larger mutations in rate of origination and, especially, in abundance in the genetic pools of populations.
6. As an over-all tendency, subject to exceptions and irregularities in some particular cases, the importance of mutations in all sorts of evolutionary processes tends to be inversely proportional to their size. (There may, however, be a critical lower limit, among nearly or quite imperceptible mutations, below which their importance decreases.)

Evidence bearing on the sizes of mutations involved in particular cases of evolutionary change is largely given in the next section of this chapter and in Chapter XI, on higher taxonomic levels of evolution. One example may be briefly discussed here as one of the few cases where a unit mutation, with clear-cut phenotypic effect, can be identified in fossils with some probability and as a typical case of continuing change not involving such mutations. Osborn (1915, p. 217) has noted the symmetrical appearance of a new tooth cusp in an individual of *Phenacodus.*

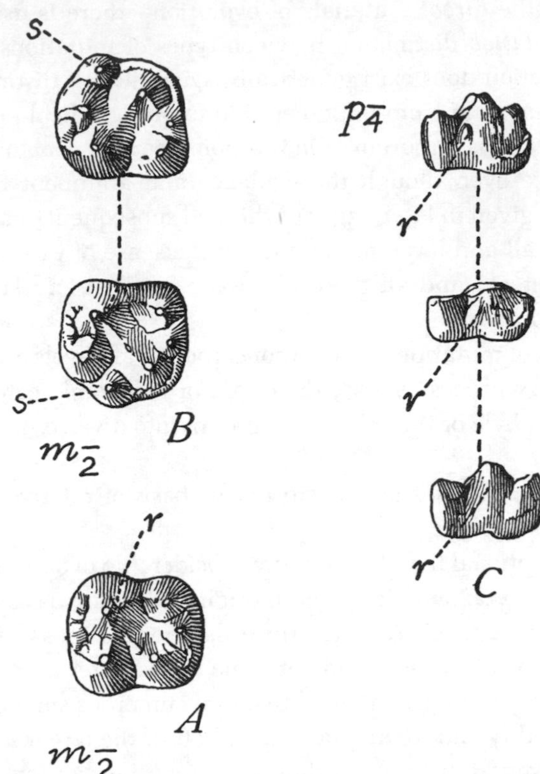

FIGURE 13. CONTINUOUS AND DISCONTINUOUS PHENOTYPIC VARIATION IN A FOSSIL MAMMAL. A, normal second left lower molar of *Phenacodus*, an early Eocene condylarth, crown view; B, right and left lower molars of an individual of *Phenacodus* (Amer. Mus. No. 15274), with symmetrically developed supernumerary cusps, *s*, presumably due to a single mutation with relatively large, definite, discontinuous phenotypic expression—"mutation" of Osborn; C, external views of right fourth lower premolars of different individuals of *Phenacodus* showing a region, *r*, in which there may be no cusp (lower figure), or a well-developed cuspule (upper figure), or any intermediate stage (e.g., middle figure)—"rectigradation" of Osborn. After Osborn, 1915.

which almost surely represents a new unit mutation. Osborn's original figure, despite its great interest for theoretical paleontology, was not labeled or discussed. It is here reproduced with the pertinent data (see Fig. 13). This mutation led nowhere: it does not appear as a normal character in any known descendants of this group or even in its distant relatives. On the other hand, in some cases it can be shown that cusps that did become integral parts of a normal tooth pattern (i.e., that characterize a whole population) have arisen by minute, practically imperceptible, steps, and as far as I know there are no cases in which it can certainly be shown that such cusps have arisen by a single mutation (although this remains theoretically possible).

Observations like these led Osborn (e.g., 1927) to the conclusion that "mutation is an abnormal and irregular mode of origin, which . . . is . . . a disturbance of the regular course of speciation." His statement has been ridiculed by geneticists who perhaps were not familiar with the checkered history of the word "mutation." Having abandoned the original, paleontological usage, Osborn was using a genetical definition of the word, but one then already out of fashion with geneticists: he meant a large, discontinuous phenotypic change resulting from some genetic event, a mutation in the sense of De Vries. Osborn applied other terms, notably "rectigradation" to changes that geneticists would now variously call "small mutations," "polygene mutations," "micromutations," etc. In discarding "mutation" as part of the normal or regular course of speciation he was simply expressing the conclusion that species more often arise by an accumulation of small changes than by one large change. Most geneticists now seem to agree, perhaps with inadequate recognition of the great accumulation of paleontological data by Osborn in support of this view (e.g., Osborn, 1915, 1925a and b, 1926a and b, 1927, 1929, 1934).[5]

The frequent discussion of size of mutations has generally been related to the question as to whether they are involved in or are the cause of saltation, the origin of new taxonomic groups by a single step.[6] This question is discussed in the next section and in Chapter XI. Here it remains to be added that a large mutation, even in the general sense including all kinds of chromosome mutation as well as gene mutation, is

[5] In fairness to those who have misunderstood or neglected his work, it must be admitted that Osborn's great contribution was obscured not only by a peculiarly personal terminology but also by a vitalistic and finalistic theoretical interpretation which, as he noted, found few supporters.

[6] The old term "saltation" has fallen into some disrepute. It does not appear in recent works that discuss the problem in other words, or it is somewhat condescendingly mentioned. (Carter, 1951, uses the word twice and in quotes only.) Nevertheless it is a useful, simple, and unambiguous way to designate a phenomenon the reality and frequency of which are still widely discussed. It is not synonymous with "large mutation."

not the only conceivable way in which a large phenotypic change, such as might be involved in saltation, could arise in one step. Among other possibilities are the simultaneous occurrence of multiple smaller mutations or the sudden segregation or recombination of genes with strong, additive, total effects. In fact, among these processes only mutation seems at all likely to produce an instantaneous change of phenotypic scope comparable to mean differences in taxonomic populations, if, indeed, mutation can do this aside from exceptional cases and low categories.

Segregation and recombination are normal processes constantly going on in all sexually reproducing populations. They are the main genetic processes involved in population variation, as has been seen. They can, without mutation, carry the population beyond its former range of variation by utilization of the pool of potential genetic variation. These processes are, however, gradual in a population as well as fluctuating and, in the main, reversible. There is no evidence known to me of segregation or recombination in nature occurring in a single step and carrying an individual well outside the former normal range of variation, even to the extent that a large mutation can do this. If such an event did occur, its results would almost always be quickly damped or lost by backbreeding of the individual or its descendants. The fluctuating variation of a population does not seem to include any self-contained mechanism for permanent, one-step changes in the mean and range of that variation.

The chances of multiple, simultaneous mutation seem to be even smaller, indeed negligible. Postulation of a mutation rate of .00001 and of each mutation's doubling the chances of another in the same nucleus would correspond with the most favorable circumstances warranted by laboratory evidence. Under these postulates, the probability of five mutations in the same nucleus would be approximately 10^{-22}. With an average effective breeding population of 100 million individuals and an average length of generation of one day, again extremely favorable postulates, such an event would be expected only once in about 274 billion years, or about a hundred times the probable age of the earth. Obviously, unless there is an unknown factor tremendously increasing the chance of simultaneous mutations, such a process has played no part whatever in evolution. The chances of five different mutations occurring separately in a population and eventually being found together in one zygote may be appreciable under favorable circumstances, but this is a gradual, not a one-step process; it is part of the normal mechanism of nonsaltatory evolution.

The reasons why some mutations have small and others large effects are to be found in the process by which mutations acquire phenotypic expression, i.e., in development of the individual. It is known that some mutations affect rates of development and times of termination of development, changes in which may have distinct or even radical effects on adult form. It is also known that some genetic effects, presumably subject to mutation as all genetic effects seem to be, may appear only or, at least, most strongly in one stage of development. As De Beer (1951) has emphasized, these facts might make possible quite sudden and radical changes in adult structure. Caenogenesis (appearance of youthful characters not affecting the adult) followed by paedogenesis or neoteny (retardation of development, cessation of development at an earlier stage than in the ancestry, or both) would result in paedomorphosis, appearance of the formerly juvenile characters in the adult. This might seem to be abrupt if the earlier juveniles were unknown and if the neoteny resulted from a single large mutation (a point on which De Beer does not commit himself).[7]

Another possible and on the whole more probable way in which a mutation may produce phenotypic results of varying magnitude, including large, is by "deviation," the turning of development into an alternative channel, also discussed at some length by De Beer (1951). As would be expected, phenotypic changes are relatively slight if the turning point comes late in development, greater if it comes early in development. De Beer suggests that this process has been particularly important at high taxonomic levels, and he gives examples of differences at levels from family to subclass in which he thinks it has played a part.

The classification and rather cumbersome designation of these embryological processes is clearly valid and useful as applied to the description of some features of comparative embryology. They help to describe the way in which developmental systems and therefore their results, the

[7] Schindewolf's "proterogenesis" (e.g., 1950a) is the same phenomenon, long known under various other names (e.g., recognized as a synonym of "paedomorphosis" by De Beer, 1951, and of "bradygenesis" by Ivanow, 1945). Its significance seems to be the same as that of neoteny in general, as discussed briefly above. The fact that the direction of change has been mistaken in some instances (see Wood and Barnard, 1946) does not alter the fact that the phenomenon does occasionally occur. Nor does the fact that it does occur give it a role in evolution different from or more important than other ontogenetic processes involving genetic change.

adults, differ, but there is considerable doubt whether any light is thereby shed on how those developmental systems came to be different. They do not help to determine, for instance, whether neoteny of echinoderms, giving rise to chordates, was by one mutation or a million, instantaneous, fast, or slow. Nor, supposing that neoteny of echinoderms really was involved in the rise of chordates (which is far from certain), does such a classification take into account how, when, or why the manifold chordate characters that obviously are not neotenic echinoderm characters arose. From the present point of view (entirely different from De Beer's, so that no criticism of his fine work is involved) a descriptive classification of embryological phenomena has little importance except as it may indicate mechanisms by which large and small mutations lead to altered phenotypes. The real evolutionary process, the change in populations, goes on from there and could be studied about as well if nothing were known about the individual developmental steps intervening between mutation and observed phenotype.

Schindewolf (beginning in his pioneer study of 1936 and culminating in his recent book of 1950a) has proposed a theory of saltatory origin of taxonomic groups at all levels by single mutations of varying size, from what a geneticist would probably already call large, giving rise to new species, to mutations so enormous in phenotypic result that the genetical vocabulary would have no word for them, giving rise to new phyla. As mechanism he suggests, essentially, mutations acting by deviation: mutations affecting latest stages of development for origin of species and affecting the very earliest stages for phyla. The suggestion of a possible mechanism is of course supportive for the hypothesis of such tremendous mutations, but it does not demonstrate either that they occur or that if they do, they could or did really produce a taxonomic group at any level—and that is the real problem from Schindewolf's as from the present point of view. Known sorts of mutations, whether acting early or late, produce no taxonomic differences above the level of species and seldom that high. This is evident from (among much other evidence) Goldschmidt's work on physiological genetics, and it caused him to exclude known sorts of mutations as agents in the rise of taxonomic groups from species upward, which he also believes arise by saltation.

The issue is not, or certainly not only, whether large or small mutations are normally involved in evolutionary processes, as some discussants seem to think (e.g., Davis, 1949). It is rather whether a real or possible mutation, of any size, can give rise to a higher taxonomic category in one step and whether single mutations can occur and have occurred producing differences, genotypic and phenotypic, corresponding to those between families, orders, classes, etc. These questions are among those next to be discussed.

MUTATION AND THE ORIGIN OF TAXONOMIC GROUPS

Another, perhaps still more enlightening approach to the sort of problem just mentioned is to ask whether differences between taxonomic groups are usually of the same sort as those involved in variation within a population. If they do seem to be of the same sort, although perhaps more numerous or quantitatively increased by additive effects, then it is likely, at least, that the taxonomic categories in question have built up their differences from intragroup variation and therefore have not arisen by saltation. In such a case the roles of larger and smaller mutations will still be of interest, but the question of the large mutation will not be crucial. The large mutations known to geneticists definitely may occur as variants within even the smallest definable breeding populations and hence are possible materials for evolution of populations without saltation. The hypothetical "systemic mutations" of Goldschmidt or "typostrophic" mutations of Schindewolf are postulated as, in themselves, removing the mutants from the population and hence, if they occur, they have a different bearing from merely large mutations.

The answer to the question whether differences between groups are of the same sort as those within groups is almost certainly "yes." One line of evidence is based on the fact that phenotypic variation within populations often involves the same character differences that distinguish other populations and higher categories from each other. The homoeotic mutants mentioned above as producing in *Drosophila* some of the characters of quite different insects do not show that those other groups of insects arose at one step by such mutations, but they do show that single characters distinguishing higher categories, up to orders, at least, may be morphologically identical with variations within one population. (Of course the orders have differences additional to these single characters, and it does not follow that the single characters as they occur in other orders arose by single mutations although such may have been the case.)

This relationship is not, of course, confined to homoeotic mutants but is general for characters varying in populations, any or all of which may also distinguish taxonomic categories. A good paleontological example is the variation of *Henricosbornia lophodonta,* also previously cited (Simpson, 1937a). One needs only to take the diagnosis of virtually any taxonomic group to find that the diagnostic characters are variable in other and related groups of the same or lower hierarchic rank and that eventually they can be traced down to variation in populations. (The process of tracing variation in this way is, incidentally most enlightening and would make a good class exercise as corrective for the common textbook impression that taxonomic categories are fixed and given by some sort of fiat.) As one simple example, an authoritative diagnosis of the Equidae may be given (translated from Weber, 1928):

"$I\frac{3}{3}C\frac{1}{1}P\frac{4}{4\text{–}3}M\frac{3}{3}$, brachydont, premolars simpler than molars, gradually becoming hypsodont with molariform premolars. A wide diastema is gradually developed, with the canine in the middle of it. The orbit gradually achieves an orbital ring. The cusps of the lower dentition become loop-shaped. Manus 4, 3, or 1-toed, pes 3 or 1-toed."

In the first place, the diagnosis shows how hard or impossible it is to designate even one *constant* character in a group of fairly well-known history. In the second place, every character given can be shown to vary at different levels down to populations. Extreme deviation in dental formula in related (perissodactyl) families occurs in the Rhinocerotidae, $I\frac{0}{0}C\frac{0}{0}P\frac{3}{3}M\frac{3}{3}$, and all variants up to the primitive equid $I\frac{3}{3}C\frac{1}{1}P\frac{4}{4}M\frac{3}{3}$ occur in that family. Variation from $I\frac{3}{3}C\frac{1}{1}P\frac{4}{4}M\frac{3}{3}$ to $I\frac{3}{3}C\frac{0}{0}P\frac{3}{3}M\frac{3}{3}$ is common, indeed usual, in populations of recent horses. Reduction of incisor numbers is not common but can occur. Relative height of teeth (brachydont-hypsodont) is variable in populations at all stages, as was exemplified in Chapter I, and especially in those involved in change-over from low to high. The degree of molarization of the premolars was quite variable in populations of eohippus (*Hyracotherium*) and continued so through most or all of the Eocene. (Thereafter the last three premolars were essentially molariform and varied little in this respect.) The length of the diastema and the position of the canine in it was also quite variable in eohippus and still is, around a different mean, in *Equus* (in which the presence of the canine is also variable). Post-orbital and jugal processes varied markedly, with a trend toward greater length, until they met and completed a closed orbital ring. Incipient looping (or transition from bunodonty to lophiodonty) is markedly variable in populations of eohippus. By *Epihippus* lophiodonty was well established and did not vary much thereafter (although the shapes of the lophs did). Transition from 4 to 3 toes on the manus does not happen to be demonstrated by known fossils (it occurs in the most poorly sampled part of the whole sequence) but the presence of 3 or 1 functional toe is known to have varied within populations at the time when the change was occurring (in early *Pliohippus*).

It may fairly be concluded that all the characters diagnostic of the Equidae and differential from related families were variable within the populations in which they were arising. The recent phenotype of the family could have been, it is even fair to say that it definitely *was,* built up from what arose as intrapopulation variation.

Another approach is the comparison of inter- and intragroup genetic variation in recent organisms. Analysis is limited by the fact that reliable and detailed comparison requires hybridization and therefore cannot, at most, go above the generic level.[8] Around the level of species there is good evidence that intergroup differences are of the same sort as intragroup and represent segregation and combination of the latter. Here also another important point appears: even at the relatively low level of species, differences between groups do not, as a rule, represent a single mutation but almost always involve numerous genes. (The relatively unimportant exceptions will be mentioned later.) Two species of cotton, analyzed by Harland, Hutchinson, and Silow (cited and discussed by Dobzhansky, 1941), were alike in eight of sixteen genes studied, different in the other eight. Alleles prevailing in one species are more or less rare variations in the other. Although the specific differences again appear as derivatives of variations within populations, the species are found on further analysis to have differently integrated genetic systems with different modifiers such that a gene transferred from one to the other does not necessarily have identical effects in both.

In two species of goldenrod, closely similar and hybridizing freely (although with reduced fertility in f_2), Charles and Goodwin (1943) found that at least 21 genes are involved in the differences in leaf characters and that as "at best a very crude minimum estimate" about 40 gene differences are required to explain all the obvious morphological differ-

[8] Evidence that homologous genes and mutations may occur in different families or even higher categories is not quite conclusive and is insufficient for useful comparison of the genomes.

ences between the two species, the actual number being probably much larger. Between *Drosophila pseudo-obscura* and *D. miranda* there are 49 structural differences in chromosomes in the parts that could be compared (Dobzhansky and Tan, 1936). Some other species show relatively few differences in chromosome structure, but certainly have multiple differences in genes.

Clearly, as Timofeeff-Ressovsky (1940) has further exemplified and emphasized, the differences between natural taxonomic groups normally depend on many gene and other mutations and cannot be ascribed to the effects of single mutations. The direct evidence from lower levels warrants concluding from the indirect evidence of phenotypic differences at higher taxonomic levels that even more genetic differences, ultimately deriving from many separate mutations, characterize the latter. At upper levels, the number must run into thousands in many or most cases.

Viewing the same problem from a different angle, paleontologists have long been concerned with the question of continuity and discontinuity in phylogeny. Thus Spath (1933) wrote: "We know now that new types appear as saltations and that it would be as idle to insist on seeing a complete series of passage forms from *Orthoceras* to the belemnites, as a progressive modification of the shell in a baculite or other hamitid."

And thus Schindewolf (1950a), now the leading paleontological exponent of this opinion: "The organizational structure of a family or an order has thus not come into being by progressive specific change in a long chain of species but *has arisen discontinuously, by saltation, through the remodeling of the type-complexes from family to family, order to order, class to class.*"[9]

But many paleontologists, I believe a large majority at present, agree with Matthew (1926): "The more complete the record, the more abundant the material, and the nearer we are, judging from the available evidence, to the probable centre of evolution and dispersal of any race of animals, the more continuous does the succession become, the more it appears to evolve through a succession of minute changes which lie within the limits of ordinary individual variation."

Matthew of course knew that sharply discontinuous changes can occur in single characters. Like Osborn and following De Vries, Matthew

[9] Original in German. The italics are in the original.

called these "mutations" and distinguished them from "minute heritable variations." [10] Matthew agreed with Osborn in thinking that large, Devriesian mutations are not the usual stuff of evolution (although he disagreed with Osborn as to what constitutes the real stuff) and he added the important point that evolution normally occurs by "ordinary individual"—i.e., intrapopulation—"variation."

Although the point is often obscure in arguments about continuity, it is only in the context of populations and their changes that an issue between continuity and discontinuity makes any sense. The developed organism is absolutely discontinuous from its parents (with exceptions not essentially modifying the conclusion) and any real difference between the two, however small it may be, is discontinuous. A morphological continuum is impossible. Most of us who speak for usual essential continuity in evolution believe that mutations are the ultimate source of evolutionary change, and mutations are, by definition and in observed character, discontinuous. That many have phenotypic effects so small and so affected by developmental influences as to require the methods of continuous variation for their analysis does not alter the fact that the separate mutation is a discontinuity in heredity.

The continuity in question is necessarily the only one that does make sense: continuity of populations. Such continuity means that a species or higher group does not arise by sudden appearance of an individual or more than one from which develops the new group, outside the range of the parental population from the start. When origin is continuous, the new group arises on the basis of variation (wide or narrow, involving large mutation or small) in the parental population. Variation of parental and new groups then overlaps at first and develops discontinuity only in the course of generations.

The paleontological evidence for discontinuity consists of the frequent sudden appearance of new groups in the fossil record, a suddenness common at all taxonomic levels and nearly universal at high levels. Since the record is, and must always remain, incomplete, such evidence can never *prove* the discontinuity to be original. It might always be due to nondiscovery of the ancestors. Schindewolf (1950a and earlier) has claimed that the sudden appearance of the Equidae, *Hyracotherium,* is

[10] Matthew also used the word "mutation" in the sense of Waagen, thus confusing some less erudite geneticists who have thought that Matthew did not know what a mutation is—because they did not know the original and historically correct meanings of the word.

absolute proof of discontinuity in the family's origin, but this is fallacious (see Simpson, 1949c). On the other hand, when there happen to be no gaps in the record paleontological evidence can prove continuity of origin, and it does so in numerous cases. Aspects of this evidence, bearing especially on the rise of high categories, will be discussed in Chapter XI. It suffices here to state that paleontologists have in hand many examples of the continuous origin of species and genera and progressively fewer of still higher categories.[11] Reference may here merely be made to the many such transitions known in the Equidae, not because these are the most numerous or even the best transitions on record, but because they are the most widely familiar (see Simpson, 1951a).[12]

In the nature of the record, there are no paleontological examples of a taxonomic group that *must* have arisen by saltation. There are likewise no examples known to me among recent animals. Wheeler (1942) claimed that this must have been true of *Gorgonorhynchus*, a nemertean, in Bermuda, but Zimmerman (1943) and I (Simpson, 1944a, written before Zimmerman's note appeared, although published later without opportunity for revision) have shown that Wheeler's conclusion does not follow from his evidence.

On the other hand, it is well known that numerical chromosome mutation, especially polyploidy, may in one discontinuous step give rise to new groups of plants. Examples are given in all discussions of general or plant genetics and the whole subject has recently and thoroughly been reviewed by Stebbins (1950). No one can doubt that numerous species of plants have arisen in this way, and it remains possible that some species of animals have, also, although no example seems to be surely known. In plants the novelty is at a low level, the polyploid descendants

[11] Schindewolf, an experienced paleontologist, is of course aware of this fact although he still insists that when we do not have the transition it did not exist. He has explicitly stated that many such transitional sequences are known (Schindewolf, 1950a, p. 292).

[12] Correction may here be made of a serious error before it is more widely copied. Dietrich (1949) has recently reviewed the question of continuity and discontinuity in the Equidae. Among other misstatements, he says of *Parahippus* and *Merychippus* that "no intermediate form bridges the gap between the two, no gradual transition can be established." The statement has already, and quite excusably in view of Dietrich's authority, been seized on and elaborated by Schindewolf (1950b), but it is false. There are unified samples, surely representing local populations, perfectly intermediate between *Parahippus* and *Merychippus* and so varying in the "diagnostic" characters that assignment of individuals in the single population could be made to both genera and assignment of the population to one or the other is completely arbitrary.

being rather closely similar to their diploid ancestors and generally referred to the same genus. Whether genera, also, arise in this way is a matter of judgment as to the degree of change meriting generic rank. At significantly high levels, polyploidy may have been one of a number of cumulative changes involved but it may be doubted whether polyploidy alone has given rise to high categories. The evolutionary importance of the phenomenon is not great and is concerned largely or entirely with low-level splitting. Stebbins concludes that "polyploidy, although it multiplies greatly the number of species and sometimes of genera present on the earth, retards rather than promotes progressive evolution."

It is a different question whether evolution on the basis of variation in populations, and continuous on that basis, occurs mainly or solely by large or by small mutations. As previously quoted, Osborn and Matthew, both of whom piled up much supporting evidence, believed that the usual process was change by variations, and hence inferentially by mutations (in the modern sense), with steps so small that they seem to be essentially continuous in distribution. This is, I believe, the usual experience and opinion of paleontologists as regards sequences with transitions actually represented. Change by very small steps is the rule not only for metrical characters, as would be expected, but also frequently for qualitative characters and the appearance of new structures and tissues, which might be expected to originate by large mutations.

As typical of many cases, an example will be briefly given of characters which have essential evolutionary importance and which subsequently become associated with others in such a way as to be definitive of generic units, but which arise singly and by small mutations within a population in which they are not modal, definitive, or "normal." One of the essential differences (for purposes of practical identification) between the Oligocene horse genera, *Mesohippus* and *Miohippus,* and their immediate successor, *Parahippus,* is that the latter has and the former typically lack a tiny spur or crest on the upper cheek teeth, which in later members of the line to *Equus* gradually became a large, essential element in the pattern, the crochet. But even in the earlier *Mesohippus* there are occasional forms with a small crochet, found in direct association with and otherwise indistinguishable from more numerous individuals without the crochet. Evidently these are single populations in which the mutation "crochet" had occurred, but had not yet become abundant or segregated (see Schlaikjer, 1935;[13] Stirton, 1940; also American Museum of Natural History collections, which confirm and reinforce the conclusion). In some earlier species of *Miohippus* the same situation obtains, and among latest Oligocene forms there is some evidence, still inadequately analyzed, that segregation was occurring and that closely related species were in part distinguished by relative constancy of presence or absence of the crochet. In the early Miocene the segregation is almost complete and constant. Some well-defined species (evidently the type of *Miohippus, M. annectans,* which is not an average or modal species in the genus) then apparently never had a crochet, and they gradually gave rise to later genera, e.g., *Anchitherium,* in which this structure did not develop. Other early Miocene species, e.g., *Parahippus pristinus,* then constantly had a small crochet and are, partly on that account, placed in a different genus, having intergraded almost imperceptibly with earlier *Miohippus,* but later becoming more distinct in this and other characters, so that they unquestionably warrant generic separation. From these species came the later true equine horses, in which the crochet became large and important. It can be seen that other tooth characters, helping to distinguish between the typical forms of the genera, arise in the same way, but separately from the crochet. For instance, the first mutations for united metaloph begin to appear as fluctuating rarities in populations normally lacking them, probably slightly later than the first crochet mutations occur. This character, unlike the crochet, did not become segregated as a distinction between contemporaneous groups, but spread and became universal in later Equidae.

Another example in the horses is the appearance of cement on the teeth of *Parahippus* and its immediate successor *Merychippus.* Cement first appears as a mere film on the teeth of some members but not others in single populations. Gradually it comes to characterize whole populations and, still varying, it increases in average thickness until it reaches

[13] Aside from specimens referred to *Mesohippus barbouri* some of which do have, but most of which do not have a crochet, Schlaikjer set aside a few closely similar specimens from the same quarry and erected for them a different genus, *Pediohippus* (rejected by Stirton), defined principally by the constant presence of a crochet. He explained the intergradation of the supposedly distinct genera by the hypothesis that they hybridized. The explanation seems to me improbable, but I have not reexamined the specimens and Schlaikjer does not give sufficiently full data to permit critical revaluation. Even if his alternative explanation be granted, it does not alter the conclusion here reached except to claim that some of the forms with a crochet did become segregated earlier than other students believe.

an evident optimum about which it fluctuates without further secular change down to recent *Equus*.

When so-called qualitative characters first appear they often cannot be classed, even in an individual, as "wholly present" or "wholly absent," but vary both in size and extent. This point is exemplified by the data in Table 18 (from Simpson, 1937d) on occurence of a cingulum on lower cheek teeth of a late Paleocene condylarth. Some allied groups lack this structure; in others it is relatively constant.

TABLE 18

OCCURRENCE OF A CINGULUM ON LOWER CHEEK TEETH OF LITOLESTES NOTISSIMUS

SEPARATE TEETH

Teeth	*No. of Observations*	*No. with Distinct Cingulum*	*Percentage*
P_4	24	2	8½
M_1	35	11	31½
M_2	36	8	22
M_3	24	3	12½
Totals	119	24	20

ASSOCIATED M_{1-3}

Cingulum on	*Number*	*Percentage*
M_{1-3}	1	5½
M_{1-2} only	2	11
M_{2-3} only	0	0
M_1 and M_3 only	0	0
M_1 only	3	16½
M_2 only	0	0
M_3 only	1	5½
None	11	61
Total	18	

Goldschmidt (1946) has criticized the statement, repeated above (from Simpson, 1944a), that the mutation "crochet" (or "small crochet," as of then) had not yet become abundant or segregated in *Mesohippus,* and also the suggestion that in examples like that of Table 18 the mutation produces a new tendency in a complex developmental field, fluctuating under the influence of other genes, which may be "buffers" in genetical terms. He suggests that in such cases "the evolutionary process begins with mutants of low penetrance and specificity" and prefers this explanation to "the complicated constructions of the Neo-Darwinian scheme." But the mutation "crochet" did become segregated in descendants of *Mesohippus;* some had it and some did not. To say that the mutants had (at first) "low penetrance and specificity" is only to repeat in genetical terms the fact that the character fluctuates in expression.

Since in some lines "penetrance and specificity" did increase, some genetic change took place. That the change was further progressive mutation at the same locus is possible, but this does not accord with Goldschmidt's views. If the change did not occur there, it must seemingly have involved other genes, perhaps buffers in some sense or degree. Incidentally, Stern (1949a) cites the cingulum of *Litolestes* and its relatives among probable examples of mutations with variable expression later, or in other lines, buffered by other genes to yield constant results. If, on the other hand, Goldschmidt's point is that the mutations of these examples are after all "large" ones starting with small effects, this is incomprehensible because the "size" of a mutation can only be rated by its effects. That a small mutation may become large as a result of *additional* genetic changes is, however, both possible and important.

Certain mutations must by their very nature be large, that is, the only alternatives as regards a given character are strikingly different. Then the relative and sometimes almost meaningless nature of "size" in mutations is especially evident. Even such characters may be involved in transitional, not saltatory evolution in populations. A character of this sort is dextral or sinistral coiling in gastropods. As shown by Crampton (1932), races of *Partula* can be defined by the occurrence and frequencies of dextral and sinistral shells, a marked morphological difference controlled by a single gene. Variation in the character itself has to be discontinuous: there are only two possible classes. But variation and change in the frequency of the character in populations is perfectly continuous. There are races purely dextral or purely sinistral, but there are also races with all proportions of dextral to sinistral in single populations. Note also that intergroup variation is again found to be identical in kind with intragroup and may be inferred to arise from the latter. Further, one species of gastropod may be sinistral and another species dextral, but the appearance in a sinistral species of a dextral mutant does not *ipso facto* produce a new species, nor is the direction of coiling in any known case the *only* difference between taxonomically valid races or species.

That mutations of clear-cut effect, such as are studied by Mendelian methods and reasonably called "large," do play a part in evolution is highly probable and not denied by statements either that evolution is normally continuous rather than saltatory or that small mutations are over-all more common in evolution. While recognizing that large and small mutations differ only in degree of effect and not in kind, most recent students think that both are involved in many or all evolutionary processes, see, e.g., Muller (1949a), Stubbe and von Wettstein (1941), and Stebbins (1950). Fisher (1930) has shown that the average [immediate] selective value of mutations is inversely related to their size and diminishes from a limit of 0.5 as the phenotypic effect increases. The students just cited agree that a large mutation is more likely to upset an integrated system or otherwise to be disadvantageous, but they reasonably hold that when large mutations are involved they are subsequently integrated by modifying and buffering genes.

It is, moreover, known for some experimental (e.g., Timofeeff-Ressovsky, 1935) and natural (e.g., Dubinin and others, 1934) populations that small mutations are much more frequent than large, and this is probably true of most or all populations. There is thus a wide concurrence of paleontological, neontological, and experimental data that small mutations are the common elements in the origin of new taxonomic groups, at lower levels, at least, and that they are more frequent and less likely to be seriously disadvantageous, but that large mutations may also be involved. No warrant is here found for the view that large mutations produce, at one step, new taxonomic groups except in such relatively unimportant cases as some plant polyploids, nor is there evidence of a level at which small mutations are ineffective or excluded from the process.

That small mutations are sufficient to account for large-scale evolution can be demonstrated by consideration of the time available for some known sequences. From *Hyracotherium* to *Equus,* for instance, there must have been at least 15 million generations. The average effective total breeding population can hardly have been less than 100,000 individuals in the successful phyla, giving at least 1,500 billion individuals in the real or potential ancestry of the living horse. As minimal postulates, suppose each tooth character were controlled by one gene mutating at the rate of .000001, a moderate rate for large mutations and probably a very low rate for small mutations. At least 1,500,000 mutations would then have occurred in the postulated locus. For dimensions, only two directions of change are possible and with fully random mutation half would be in the favored direction. Certainly to suppose that one-fifth were favorable would be a minimal postulate, giving at least 300,000 favorable mutations. If only a thousandth of these represented new steps, there were at least 300 steps.

Divided into 300 steps, change in any one character from *Hyracotherium* to *Equus* becomes almost imperceptibly small and almost incomparably smaller than the intragroup variation at any one time, even though this line evolved more rapidly than some others in the same family. In ectoloph length, for instance, the maximum change was from about 8 to about 40 mm. and each of 300 steps would be about 0.1 mm. In the smallest species the intragroup range for this character was about 3 mm.

Zimmermann (1948 and earlier work there cited) has made somewhat similar estimates for the evolution of plants. He finds that the average time available in each case for the origin of all categories from races to orders, at least, is fully consistent with evolution by small mutations, only.[14]

A final question about mutation and the origin of taxonomic groups concerns initiating or key mutations. An example is provided by Patterson's work on the taeniodonts, extinct early Cenozoic mammals. Patterson (1949) believes that most of the differentiation of the Stylinodontinae from the less progressive Conoryctinae followed from two modifications, acquisition of claws and enlargement of canines, other divergent trends being adaptively related to these, and that the acquisition of claws was earlier. Hence he concludes that "a single orthodox mutation resulting in the development of claws is believed to have been the starting point of the new adaptive shift represented by the Stylinodontinae." This could be interpreted as the origin of a subfamily by a single mutation, in Schindewolfian fashion, but in fact Patterson's data do not support such an interpretation and he, himself, did not make it. Granting his conclusion full value (although it may be slightly oversimple or arbitrary), the mutation arose in a population of Conoryctinae which did

[14] Schindewolf (1950a) criticized some of Zimmermann's earlier figures because they contained an arithmetical error and because they allowed for only 60,000 progressive mutations from *Rhynia*, Devonian, to *Pulsatilla*, Recent, which may be too few. Zimmermann had earlier allowed only one progressive mutation per 1,000 generations, and Schindewolf's objection takes this figure as established. It seems certainly to be far too low, and this and Zimmermann's later treatment rob Schindewolf's criticism of its force.

not forthwith become Stylinodontinae but only after isolation and splitting, not due to the same mutation, and a transitional shift, occupying an appreciable number of generations and surely coming to involve other mutations.

In a sense, every well-marked change of adaptive type might be traced back to some one mutation that initiated or made possible the change, but it still would not be true that the new taxonomic group characterized by the adaptive type arose when the mutation occurred and as the effect of that sole cause. In fact, as before stressed, evolutionary cause is not and cannot be so simple. The key mutation, if there was one, so resulted only because it occurred in a given genetic environment, which in turn resulted from previous mutations and other events. Logically traced back, the real key mutation will probably be found at the origin of life.

Schindewolf (1950a) has suggested that the concept of key mutations fits in with his idea of single-mutation origins for all categories, but I think him mistaken for the reasons already stated. He further suggests that the "Grossmutation" at the origin of, say, a class or order may be a key mutation, not necessarily bigger than an ordinary genetic mutation and possibly of the same sort, its effect in initiating an order resulting from later building on the new basis. This could be true in the sense, again, that a single mutation might start the transition to a new group, but not that it gave rise, at once and all alone, to the new group.[15]

In essence, Schindewolf is supporting the Devriesian theory of evolution by mutation, which has long been abandoned for cause by most geneticists and which never found much favor with other paleontologists or with systematists. This is the opposite viewpoint from Osborn's and Matthew's; they rejected mutation (*sensu* De Vries) as important in evolution and Schindewolf makes it all-powerful. The possibility has deserved a hearing but now—more than sixty-five years since De Vries began work on *Oenothera* and fifty since he elaborated the theory (1901), and after due consideration of the later arguments in the light of later fact and theory—it seems that the old discussion might come to an end (although it would be unrealistic to suppose that it really will). As would be expected, Schindewolf, an experienced and able student,

[15] This suggestion by Schindewolf also seems to me quite inconsistent with his repeated statements elsewhere about the catastrophic effects of the postulated typostrophic mutations. In the same work he says, for instance, that what is involved is "Umbau von Grund auf," rebuilding from the ground up, and he accepts as literal Garstang's dictum that the first bird hatched from a reptile's egg.

bases his conclusions on real and important phenomena, but he seems surely to be neglecting still more important phenomena and to be overgeneralizing on the basis of old genetical theory now requiring much modification. The concept of key mutations may be also the key to what is sound in his theory.

RATES OF MUTATION

If, as is reasonably certain, all basic materials for evolution ultimately arise by mutation, mutation rate is clearly a factor in evolution and might set limits, at least, to rates of evolution. Unfortunately the study of mutation rate is difficult and its results to date are suggestive rather than conclusive. Laboratory data refer to few genes in few organisms, which may be atypical in this respect.[16] Estimates of mutation rates in wild populations are even fewer and involve may uncertainties and difficulties (but see Wright, Dobzhansky, and Hovanitz, 1942, and for rates in nonexperimental populations of man, Stern, 1949b, and his references). Such estimates for fossil populations seem to be impossible. It is highly improbable that phenotypic variation is reliably correlated with mutation rate, and to postulate that rapid evolution reflects high mutation rates is to assume what should be proven and what seems again, on balance of all evidence, to be improbable. Moreover, even in the laboratory small, polygene, or modifier mutations, which are evidently highly important in evolutionary processes, can seldom be definitely counted or have their rates directly determined. They are probably much more frequent than the major or Mendelian mutations for which rates have been determined.

Data on major mutations (reviewed, e.g., in Dobzhansky, 1941, 1951) show that mutation rates differ greatly in different genes, in different populations, and even in the same population at different times. Exceptionally unstable genes may have spontaneous rates on the order of 10^{-2}, but the usual magnitudes for the mutations of Mendelian experimentation are around 10^{-5} or 10^{-6}. There is reason to suspect that these rates may be lower than the average for all genes ("large" and "small") concerned in rapid progressive evolution, but it was suggested in the previous section that they are quite high enough to account for known evolutionary sequences and would, indeed, be consistent with consider-

[16] Fenton (1935) has suggested that *Drosophila* may have an unusually low rate of evolution, and it may also have minimal mutation rates.

ably faster evolution than really occurred in the Equidae, for instance. There is, at least, no reason to postulate exceptionally high mutation rates as concomitants of any high rates of evolution demonstrated or likely in nature. On the other hand, it has often been suggested in various terms (e.g., by Ruedemann, 1922) that very low rates of evolution, or instances of arrested evolution, imply low mutation rates. These low (bradytelic) rates of evolution will be discussed later. In fact they do not seem to require low mutation rates and there is no evidence that they are indeed accompanied by low rates.

Another speculative suggestion commonly made after Muller had found that radiation increases mutation rates was that periods of exceptionally rapid evolution corresponded with periods of high natural radiation at the earth's surface. There is, however, no good evidence that such episodes of rapid evolution, as they really occurred over spans in the millions of years, would require increased mutation rates, and Muller (1930 and elsewhere) has shown that natural radiation can account for only an insignificant fraction (on the order of 1 percent) of observed spontaneous mutations.

Recent discovery of mutagenic chemicals enhances the probability that the greater part of the stimulus and energy for mutation is chemical and intracellular. There is also increasing evidence that mutation rate is itself, to some degree, a genetically controlled character (e.g., Rhoades, 1938; Mampell, 1945; Ives, 1950). It thus seems probable that the suggestion made long since by Sturtevant (1937) is correct and that mutation rate is a character that evolves and that is subject to natural selection. In that case, rates might be expected to be minimal in currently well-adapted populations (perhaps including populations of *Drosophila*). In such populations the rates may be reduced below the average for actively changing populations and to a point so low that mutation pressure is insignificant in comparison with selection against mutants. Yet we have seen that mutation rates even in such populations are sufficient to allow for rapid, sustained evolution. Strong selection for increased mutation rate seems probable only in a crisis so severe that almost any change would have selective value, a situation that must be extremely unusual and that would almost certainly be more effectively met by utilization of existing variability. If the latter were insufficient, extinction would probably occur before significant increase in mutation rate.

Mutation rate is theoretically a limiting factor in evolution, but it seems probable that it is generally well above any really limiting value. In that case, mutation rate can rarely be an effectively determining factor in rate or direction of evolutionary change. This is also the conclusion of Muller (1949b), leading student of mutation rates.

There remains, nevertheless, the possibility of rare exceptions, cases in which mutation rate is so large and other factors, especially selection, so weak that mutation might control the direction of evolution in spite of the evidence that this is not usually true. Wright (1940) has shown that the situation could arise in small populations. The systematic effect of mutation in such cases would be degenerative but there is some, although extremely small, chance that it might lead to progressive change.

CHAPTER V

Population and Selection

SOME OF THE GENETIC PROCESSES involved in evolution have now been reviewed and the basis laid for including them in broader synthesis and applying them to interpretation of long-range and large-scale evolution. As has been emphasized, these processes operate in populations. Their interaction and outcome are largely conditioned by the nature of the population involved in each case. The type of reproduction has a strong influence, both individual method of reproduction and such population factors as the sex ratio. Most of the discussion here can be confined to sexually reproducing populations, because they are much the commonest among recent organisms and almost universal among the groups adequately known as fossils (including most of the Foraminifera, the protistans best known as fossils). When change to asexual reproduction has occurred, which is common only among plants, this has usually led into an evolutionary blind alley and, in taxonomic terms, has inhibited further progression even to the level of genera or subgenera (Stebbins, 1950).

Profound influence is further exerted by the size of the population and by its structure: whether, for instance, it is panmictic or subdivided into local breeding groups and in the latter case how large the groups are and what degree of genetic interchange occurs among them. The bearing of these factors has been extensively studied theoretically and, lately, in the laboratory and in nature. It has, indeed, become a new subscience variously called statistical, mathematical, or population genetics. Among these studies are the now classical books of Fisher (1930) and Haldane (1932) and a whole series of shorter contributions by Wright (1931, 1932, 1935, 1937, 1939, 1940, 1942, 1948, 1949a, 1949b, and other papers mostly cited in these). There are also theoretical works devoted to more specialized approaches (e.g., Hogben, 1946) and many studies of experimental and natural populations, for instance by Dobzhansky and his associates (Dobzhansky, 1941, 1947, 1951, earlier work there cited and extensive work still in progress).

It is not necessary here to review the whole field of population genetics. Its pertinence to evolution in general has been adequately summarized by others (e.g., Dobzhansky, 1951; Stebbins, 1950; Carter, 1951). Certain aspects are, however, of vital importance to the themes of this book and these must be briefly discussed from the present special point of view. Such necessary discussion follows, mainly on the basis of the works already cited and without separate reference for each point.

POPULATION SIZE AND STRUCTURE

For most studies of evolution the significant population is the breeding population, which is usually somewhat smaller than the total census figure and may be very much smaller. This is not to say that nonbreeding members are without effect on evolution. In societal groups, especially, they may have a strong or decisive influence on evolutionary outcome, but in studies bearing directly on genetic change the effective size of the breeding population, symbolized as N, is involved. In panmictic populations with approximately equal numbers of breeding males and females, N is the number of individuals actually engaged in reproduction at any one time. When, as in some ungulate groups and others, breeding males are much less numerous than females, N approaches four times the number of actually reproducing males (Wright). In populations with large fluctuations in numbers between different breeding seasons, the value of N effective over a sequence of breeding seasons is near the lowest figure for any one of them. Over short periods with rather regular fluctuations N for the period as a whole is given by Wright as

$$\frac{n}{\Sigma\left(\frac{1}{N}\right)}$$

in which N is the effective breeding population in any one generation and n is the number of generations. If a population tends to maintain an average size but has rare great reductions in numbers, the evolutionary effect is best studied by considering the reduction as a nonrecurrent episode rather than by attempting to arrive at an estimated average effective N.

The situation is further complicated by the fact that large populations are seldom fully panmictic and that their essentially panmictic local units, local populations or demes, are seldom highly isolated. *N* for the population as a whole, as estimated in the ordinary ways, is then larger than a value really effective in evolutionary change, and *N* for a single deme, in which genes are really being drawn from some individuals outside as well as from those inside the deme, is lower than the effective value. Wright analyzes this situation by considering *N* for local populations and making additional analysis of a factor *m*, for cross-breeding pressure or (genetic) migration (which does not necessarily involve the bodily migration of the individuals concerned). One of the effects of *m* is to increase the really effective breeding population of a deme although, if adjacent populations are genetically similar, the effective increase may be considerably less than the actual interchange would suggest.

Having noted these complications in application to real populations, we may simplify such discussion as is necessary for present purposes by assuming that populations do have an average effective value of *N*, which is influenced by fluctuations in population size, by sex ratio and breeding pattern, by cross-breeding, and by other factors important in particular instances but not requiring separate analysis in generalizations involving what might be called a global or over-all value for *N*.

Among the important factors making for genetic change in a population are mutation rate and selection. Mutation rate, *u*, is the average number of mutations (to a given allele) per individual, a subject already discussed. The selection coefficient, *s*, is in the simplest case the relative increase from one generation to the next of one allele over another. If, for instance, for every 1,000 gametes with allele *A* there are only 999 with allele *a*, the value of *s* for A is .001, and that for *a* is —.001. (By this convention, *s* represents selection *for* a given genetic element; some students prefer to evaluate it as selection *against*, in which case it would be positive for *a* and negative for *A* in the example.)

The reason for again mentioning mutation rate and for anticipating fuller discussion of selection at this point is that the effectiveness of both in real evolutionary processes depends not only on their own values but also and very markedly on the value of *N* in the given case. Disregarding, for the moment, drift or sampling effects which are to be considered later, the chances that a mutation will become established in a population depend in considerable part on the absolute number of occurrences of the mutation, which in turn depends on mutation rate and population size. As a general tendency, mutations of equal rate will have better chances of establishment in larger than in smaller populations. In either case a nonrecurrent mutation or one with very low rate has little, often negligible chance of establishment, but mutations at what seem to be usual rates, on the order of 10^{-5} or 10^{-6}, may readily become established in populations of any size (assuming absence of effective selection against them).

The effectiveness of selection tends to vary inversely with the size of the population. Fisher holds that for a constant value of s selection is effective if $s > 1/N$. Wright gives $1/(4N)$ as the critical value. The discrepancy is not vital, since both figures have the same order of magnitude and it cannot be supposed that this is an absolutely fixed threshold. The remarkable thing about these figures is the exceedingly small values of s that may be effective in natural populations. In populations with $N = 10{,}000$, for instance, populations such as are surely abundant in nature, selection pressure on the order of $s = .0001$ would almost surely be and much weaker selection could well be effective. By present techniques, it would be quite impossible to observe such weak selection either in the laboratory or in nature. Since selection may be highly effective although quite beyond our powers of observation in contemporaneous populations, no credit can now be given to the formerly usual criticism of selection as a factor in evolution on the basis that some observed variation is "too small to be of any selective advantage." On the contrary, it must be concluded that in populations of usual size, at least, any variation large enough to be observed is more likely than not to be large enough for effective selection. Indeed, Fisher has concluded that no gene can remain permanently neutral to selection; the threshold is so low that fluctuations in the most stable real environment are sure sooner or later to bring s above the critical value. (This does not mean, however, that an advantageous mutation is *certain* to become established or a disadvantageous one to be eliminated in all cases and under all circumstances.)

Since theoretically effective very low values of s cannot be directly observed and determined, it might be argued that the theory is quite unsubstantiated and has status only as a speculation. In fact there is excellent indirect evidence on this point, from both neontology and, particularly, paleontology, the bearing of which has often been completely misunderstood. These facts will be considered in later separate discussion of selection.

Pooled genetic variability, v, tends to be proportional to N, although there are critical points above which little further increase in variability is to be expected (Wright). If selection rate is low relative to mutation rate, this value is about $1/(4u)$. If, for instance, $u = .0001$ and $s < .0001$, increase in N above about 2,500 will tend to produce little increase in variability, but the combination of so high a mutation rate and so low a selection coefficient must be extremely rare in nature. In more usual circumstances, with $s > u$, the critical value is about $N = 1/(4s)$. With $s = .001$, probably a frequent value, little increase in variability would be expected above $N = 250$, a figure so small that it might be questioned whether the increase of variability in large panmictic populations is really of much importance in nature.

Wright suggests that in populations with very small values of N evolution will tend to be slow and more or less inadaptive. Most of their genes will be fixed, variability will be low, and transfer from one fixation to another, depending on mutation and random elimination (to be discussed below), will be slow. In panmictic populations with large N, on the other hand, variability is greater and selection more effective, but Wright points out that although existing variation may be relatively large the building up of new variability will be very slow because in such populations strong selection limits variation and weak selection tends to produce equilibrium with fixed gene ratios. Although evolution on the basis of existing variability, particularly in the form of splitting or speciation, might be favored by larger populations, as Fisher has maintained, on Wright's premises populations of intermediate size would be most favorable for rapid and flexible evolutionary change. The most favorable size would be such that u is not much greater than $1/(4N)$ if s is not much greater than u. With different values of these and other variables, such a favored size might be from about $N = 250$ to about $N = 25{,}000$. This conclusion is, however, still based on the frequently invalid assumption that the populations are panmictic. Wright has emphasized that there are more frequently realized and more complex population patterns that may be much more favorable for rapid and effective evolution than any panmictic population. This point will also be discussed later.

In addition to systematic effects and interactions of N, u, s, v, and other factors such as dominance or expressivity of genes, breeding types, and so on, there are in sexually reproducing populations random or stochastic processes which lead to nonsystematic changes, individually unpredictable although their average effects on variation may be predicted (their effects on gene frequencies are "indeterminate in direction but determinate in variance," Wright). Wright has devoted particular attention to these effects, and his results have given rise to considerable misunderstanding and controversy. The misunderstanding has been belief that Wright assigned preponderant roles to small, fully isolated populations and to gene drift in evolution, which he explicitly has not done. The controversy has been over whether random genetic changes, regardless of selection, do occur and have played a part in evolution. There now seems to be no real doubt that they occur and have *some* effect on evolution. Their effect in particular cases and the extent of their involvement in evolutionary processes are of course still subjects for discussion.

In any population, random loss or fixation of an allele is possible. The probability, however, varies enormously and is profoundly affected by characteristics of the population. In general, and rather obviously, the probability of loss varies inversely and the probability of fixation directly with u and s. The probability of random change, either loss or fixation, varies inversely with N. Even in cases where u and s are in effective control of genetic change, there is some influence of random change or of accidents of sampling, although in given cases this may be wholly negligible in comparison with the systematic factors. It is Wright's position that the results of accidents of sampling (his phrase) or of genetic drift, as this factor is also commonly called,[1] may become appreciable under some (not necessarily common) circumstances, particularly when N is small.

Other things being equal, random loss of alleles tends to be from about $1/(2N)$ to $1/(4N)$ or less per generation, depending on sex ratio and other factors in the breeding pattern. In large populations this figure is very small and is almost sure to be offset by repeated mutation,

[1] It has also been called the Sewall Wright effect.

but the effect increases markedly as populations become smaller. This result is enhanced by the fact that absolute incidence of mutations for given values of u is also reduced in small populations and that effectiveness of selection for given values of s follows the same course. With respect to selection there is a critical value of N at about $1/(2s)$ above which selection effectively influences gene ratios and below which accidents of sampling begin to outweigh selection. With $s = \pm.0001$, which may be less than average (an uncomputable average because the separate selective values of small mutations cannot be calculated), $N = 5{,}000$. Many natural populations have N this small or smaller, indeed most students would consider this a moderate or even rather large population—it will be remembered that N is usually well below the census figure for population—and it therefore seems probable that random effects are quite common as regards small mutations with relatively slight advantages or disadvantages. Since small mutations certainly play a large part in evolution, it seems probable that accidents of sampling also play an appreciable part, although hardly a dominant one except in relatively rare combinations of circumstances.

On the other hand the selective value of large, discrete mutations may rise to $s = \pm.01$, or even $\pm.1$ and higher. Critical values of N then range from 50 to 5 or still less, values lower than in any normal real populations except in rare, brief episodes or on the verge of extinction. Accidents of sampling for such mutations, which are usually the only ones that can be readily distinguished and counted, can thus hardly be expected to be common or obvious in nature. Such change as is due to this cause will normally be hidden under the much stronger effect of selection. Thus in theory drift should be important for genetic effects that can rarely be separately observed and should usually be obscure for effects that can be easily observed, a circumstance that doubtless accounts for most of the controversy about drift in natural populations. Indeed, it becomes rather surprising and strongly supportive for the theory of drift that its effects can, with reasonable probability, be recognized in any natural populations, as is the case (e.g., Miller, 1947; Spencer, 1947; see also discussions by Carter, 1951, and Günther, 1949, both of whom strongly emphasize the role of drift in evolution).

Wright has repeatedly emphasized that the effects of drift are usually degenerative and that in wholly isolated populations below the critical size the almost inevitable result will be extinction. This is among the reasons why he considers small populations unfavorable for evolution. These conclusions may, however, be somewhat qualified not in contradiction of Wright but in some expansion of the application of his principles to the present different inquiry. The fact that the critical population size is different for different intensities of selection means that a population of given size will frequently if not usually be below the critical size for some genes (and concomitant phenotypic characters) and above it for others. Thus its general evolutionary course may be adequately adaptive while considerable nonadaptive or even inadaptive change also occurs. This is the more true since selection intensities, on which the critical value of *N* depends, will rise if a seriously disadvantageous combination of genes appears and will thus tend to reduce accidents of sampling for that particular combination without doing so for others. In populations under varied or varying conditions, especially but not exclusively if isolation is incomplete, drift may not be predominantly degenerative but may have survival value as an exploratory device leading to new adaptations, a fact that Wright does emphasize for semi-isolated populations.

It is further evident that absolutely maximal rates of phylogenetic evolution could occur only in very small populations. Maximum rate would depend on the number of generations required to change a gene frequency to O, everything, including chance, favoring its elimination, or to 100 percent, everything favoring its spread. The number of generations required varies inversely with *N*, and rates can therefore be truly maximal only in the smallest populations consistent with continuity. Evolution in such small populations doubtless would lead to extinction in most cases, but in rare instances small populations would provide for possible rapid changes over short periods of time.

Sustained evolution in very small populations would ultimately be limited by incidence of mutation, lower in small populations than in large. In, again, exceptional circumstances this could well be compensated by more efficient use of mutation, so to speak, in small populations. The chance of fixation of a favorable mutation may be considerably larger by accident of sampling in a small population than by selection in a large population—another point generally overlooked in the arguments about drift vs. selection. The time required for spread throughout the population, say for rise to frequency of 90 to 100 percent, may be only a few generations in a very small population but may be tens of thousands of generations in a large population.

A model stressed by Wright and apparently equivalent to many real populations involves a large total, specific, population split up into numerous partially isolated local populations or demes. The different possibilities for both rapid and sustained evolution inherent in large and small populations may then be combined within a single species. Each deme tends to have low variability in a balanced combination adapted to local conditions, but (as already mentioned in discussion of variation) the different genotypes of the whole series of demes involve a large pool of variability available immediately for intergroup selection and potentially for intragroup selection. Incidence of mutation in the population as a whole is large, and by transfer between demes a mutation may enter advantageous new combinations and become rapidly fixed within the small population of a deme. Exploratory new combinations can arise by drift within a deme without exposing other demes or the whole species to the degenerative effects of drift.

The main weakness of this model is probably the fact that *m*, transfer of genetic materials from one deme to another, must be very low, on the order of .01 to .001, or the separation into demes will be ineffective for evolution and the species will evolve as if panmictic, i.e., slowly, adaptively, but with little potentiality for rapid change or shift in direction of adaptation. It is a moot question whether such low values of *m* tend long to continue between adjacent demes in nature without either dropping to zero, by the intervention of complete isolation, or rising to values equivalent to panmixia, by fusion of adjacent demes or frequent shifting of the boundaries between them. For full effectiveness, not only should *m* be small, without being too small, but also total *N* for the species should be on the order of 100,000 to 1,000,000 with local *N* for the separate demes only on the order of 100. In other words, there should be from 1,000 to 10,000 very small local populations with some but very little genetic exchange among them and this condition should persist for considerable lengths of time. That such situations occur in nature is probable, but that they are very common may well be doubted.

Other situations of frequent natural occurrence may provide small populations with greater variability and more extensive evolutionary materials than could arise and be retained in isolated populations of constantly small size. In a model stressed by Elton (1924, 1942) the population fluctuates greatly in size. A good summary of such cyclic variation is given in Hamilton (1939) and the subject is reviewed, with numerous references, in Allee, Emerson, Park, Park, and Schmidt (1949). The incidence of mutations will be in proportion to the average population, while the breeding population effective as regards the subsequent fate of these mutations will be a smaller number, near the minimum population size. Haldane (1932) has criticized Elton's conclusion (congruent with, but different from, that here reached) that random gene extinction may be important in such cases by showing that in the Arctic fox of Kamchatka, varying approximately from 800,000 to 80,000 every three years, a gene would have an even chance of extinction only after 330,000 years. In reply it may, however, be pointed out (a) that this is not really a long period as species go, (b) that much greater fluctuations with lower minima occur (e.g., in varying hares, see MacLulich, 1937), and (c) that probability of random extinction less than .5 may strongly influence evolution without dominating it. In any case, the fact remains that in such a group the effective mutating population is considerably larger than the effective breeding population.

Regardless of whether accidents of sampling occur at the low point of a population cycle, the cycle, itself, probably tends to accelerate evolution. In the declining phase negative selection is particularly severe. In the expanding phase there is relaxation of selection and great release of variability for subsequent action of positive selection. An example of such increased variability with recognizable consequent change in the characters of a population of butterflies has been given by Ford and Ford (1930).

Estimates of actual population sizes are difficult and usually subject to large errors, but much attention has recently been paid to such estimates for recent animals, especially birds and mammals (e.g., Blair, 1941), and a large number of census estimates are available. Census methods and some of their results have been reviewed, with copious citations, by Allee, Emerson, Park, Park, and Schmidt (1949). Less attention has been paid directly to the problem of effective breeding populations but here, too, many estimates await synthesis, and data are available in many groups for approximate conversion of census to breeding estimates.

As regards fossils, few numerical estimates have been attempted, and these are so uncertain that the use of numbers may be more misleading than helpful. If only as a curiosity, an example may be derived from Broom's discussions (1932) of fossil reptiles in the Karroo. He estimates that there are remains of about 800,000,000,000 animals preserved in this formation over an area of about 200,000 square miles, and he considers the time involved to be at least 40,000,000 years. As an average, about 20,000 animals per year are thus supposed to have been buried and preserved as fossils. Going beyond Broom, it might be estimated, with even greater uncertainty, that about 1 in 5,000 of the animals living in this area in any one year died and was preserved as a fossil. Then there were 100,000,000 of these reptiles living over the 200,000 square miles at one time on an average—a population density of about 500 per square mile for all species (most species were large in comparison with modern reptiles, but many were comparable to living lizards in size). Another, and still more uncertain, estimate is that the number of distinct breeding groups in the area at any one time was on the order of 10,000, each of which would thus include about 10,000 individuals. In each group about ½ may have been actively breeding in any one year, giving an average $N = 5{,}000$ for the species found as fossils in the Karroo. The estimate is not unreasonable, but the true figure may have been 500 or 50,000, as an average, and individual species probably varied enormously, not improbably from as low as 50 to as high as 5,000,000 at various times.

Although such estimates are really futile, it is often possible to infer in a qualitative way whether fossil populations were more probably large or small. The enormous variations of fossil sampling under the influence of chance and many other factors may make valid inference impossible, especially the negative inference of small populations, but a reasonable conclusion may frequently be reached. As regards single species or genera at one locality or a few localities, the negative inference is never warranted. By chance alone a large population may leave very few recoverable remains in a small number of deposits in which it is, nevertheless, represented. For instance, late Pliocene horses of the subgenus *Equus* (*Plesippus*) were very rare in collections until 1928, when a deposit was discovered near Hagerman, Idaho, that contained remains of hundreds of individuals of this group. On the other hand, a species with consistently very small populations can leave no such an accumulation unless under exceptional and usually recognizable conditions that result in the concentration of the remains over a long period of time. Thus, in the above example, as in analogous cases, no direct inference as to population size was permissible until the discovery of the rich deposit, which warranted the inference that the population size was large, at least in some localities.

Surer inferences, both positive and negative, can be based on larger and more widespread series of occurrences within a phylum. If an animal group is continuously present in areas of deposition suitable for its subsequent recovery in fossil form and if the average population is large, single occurrences may be found, but among many different discoveries, one or more is very likely to include numerous individuals. On the other hand, under the same conditions a phylum with consistently small average populations will eventually leave a sequence of remains, but all occurrences will consist of one or a few individuals. Other factors must also be considered, especially the chances of fossilization, which vary greatly so that an abundant group with small chance of fossilization can leave a record similar to that of a rare group with good chance of fossilization. With reservations introduced by such effects, it is often a valid inference that a sequence of isolated occurrences represents a group with small populations and that an equal sequence of more numerous occurrences, some of which are abundant, represents a group with large populations, the inference being more probable the longer the sequence. Representative data of both sorts are exemplified by the American occurrence of Apatemyidae (an unusual family of rodentlike insectivores or primates) and early Equidae over roughly equivalent spans, Table 19.

The data for the table were compiled in 1941. Extension of sampling since then has produced a few more apatemyid individuals, not yet published, but has only confirmed the general conclusion by the fact that these animals continue to represent a very small fraction of those collected in each deposit. When a large number of formations and local faunas have been sampled, consistently low percentage figures for any one group present in many of them is probably reliable evidence of small populations for that group unless there is reason to suspect that *all* collections are from facies unfavorable to the group. Patterson (1949) has assembled interesting data of this sort for the two taeniodont

TABLE 19

OCCURRENCES IN AMERICA OF APATEMYIDAE AND EARLY EQUIDAE

	APATEMYIDAE		EQUIDAE	
Age	*Known Specimens*	*Localities*	*Specimens in American Museum*	*Localities*
Late Oligocene			39	Several
Middle Oligocene			125	Several
Early Oligocene	1	1	30	Several
Late Eocene	1	1	11	Few, 1 field
Middle Eocene	7	2	54	Several, 2 fields
Early Eocene	1	1	397	Many
Late Paleocene	5	2		
Middle Paleocene	1	1		
Maximum from one local fauna	4		About 200–300	
Average Individuals per local fauna	2		About 20–40	

TABLE 20

RELATIVE ABUNDANCE OF TAENIODONTS
(Data from Patterson)

	PERCENT OF TAENIODONTS AMONG ALL MAMMALS FOUND			
	RANGE IN LOCAL FAUNAS		FOR WHOLE AGE	
Age	*Conoryctinae*	*Stylinodontinae*	*Conoryctinae*	*Stylinodontinae*
Late Eocene		.... [a]		.1
Middle Eocene		.... [a]		.1
Early Eocene		.5–3.1 [b]		1.0
Late Paleocene	.... [c]	.5–6.7 [d]	.4	2.1
Middle Paleocene	.2–1.1 [e]	.3–1.8 [f]	.6	1.2
Early Paleocene	.... [g]	.... [g]	2.8	.5

[a] One formation, local faunas not distinguished.

[b] Three formations, local faunas not distinguished.

[c] Present in only one local fauna.

[d] Five local faunas.

[e] Four local faunas.

[f] Three local faunas.

[g] One local fauna.

subfamilies Conoryctinae and Stylinodontinae, summarized in somewhat different form in Table 20.

Patterson's figures apply only to local faunas or formations in which the subfamily in question does in fact occur. The stated percentages are therefore maximal for the various ages and the range in local faunas really starts at zero in most cases.

With data like those for the Apatemyidae there can be no serious doubt, and little as regards the taeniodonts, that the populations were small throughout. The continuity of record in both cases, the wide range of facies sampled, and the consistently low recovery of these groups for any locality make it unlikely that inadequate sampling of really abundant groups is involved.

The horse record in Table 19 is also continuous, with great fluctuation of recovery in the various local faunas and with large (relative and absolute) numbers from some faunas. Such a record almost certainly represents sampling from large populations. In this case the sampling is less intense and uniform because only one museum's collection is tabulated and its field operations have not been equally divided over the several ages. Its parties have, for instance, spent much time on the early Eocene and little on the late Eocene.[2]

A record like that of the Apatemyidae, which is exceptionally good for rare fossils, gives a fair idea of actual structural change in the phylum, but provides almost no direct evidence as to variation and other essential factors underlying this morphological evolution. Most of the principles and theories of evolution advanced by paleontologists have been based upon groups with long, continuous, and relatively abundant records; hence, on groups that probably had large breeding populations. Important theoretical studies in this field involve, for instance, the abundant fossils of trilobites, graptolites, brachiopods, gastropods, ammonites, oreodonts, titanotheres, and horses and within these the particular phyla and stages that are most abundant.

The recent work on relationship of population size to evolution shows that principles so based cannot reasonably be accepted as generalizations. As far as these principles are logically derived from adequate data, they are valid descriptions of evolutionary processes in certain sorts of populations, but as far as these facts can show, the processes may have been quite different or have had quite different results in the groups that are poorly recorded as fossils.

Recognition of this limitation does not make paleontological data less useful in the study of evolutionary theory. On the contrary, it permits better coordination of paleontological and experimentally devel-

[2] Unfortunately this makes the apparent drop in abundance in middle and late Eocene unreliable. Patterson (1949) comments on this decrease and correlates it with relatively rapid evolution of the brain at those times. The decrease may have been real, but these figures do not warrant that conclusion.

oped theories, which have sometimes seemed radically discrepant. It gives a better basis both for the checking of neontological observations against the fossil record and for the interpretation of that record in harmony with what is known of the genetics and other physiological processes of living animals.

LENGTH OF GENERATIONS

Most of the changes involved in evolution occur as between generations. The genetic processes affecting frequencies and combinations of genes and chromosomes occur as incidents in reproduction, and selection, also, in its current definition is a matter of reproduction, i.e., is change induced from one generation to the next. It has been suggested that mutation depends on passage of time rather than on number of generations, but this now seems unlikely (Muller, 1945, 1949b). It remains possible that in some cases mutation rate per year (or other unit of absolute time) is more constant than rate per individual, and therefore per generation if population size is constant, but this is uncertain and is unlikely to override the general fact that genetic processes in evolution are more naturally evaluated in rates per generation than in rates per year.

That being so, it follows that evolution can be faster in organisms with short generations than in those with long. Differences in this respect are great, generations varying in length from less than a day to more than thirty years. It might be expected that as a general rule evolution would have been faster in short-lived than in long-lived organisms. In fact there seems to be no good evidence that this is so as a rule and little evidence suggesting that length of generations has had a decided influence in any particular cases. Again we seem to have a factor that sets theoretical limits to rates of evolution but again it seems that in most recorded cases those limits have not been reached or their effect has been wholly obscured by other, more active determinants of evolutionary rates. This observation was perhaps first made by Zeuner (1931, and also, e.g., 1946a and b) and has since several times been repeated, sometimes without knowledge of Zeuner's priority.

The lack of any clear and usual correlation between length of generations and rate of evolution is quite evident in nature and in the fossil record. Among mammals, opossums have short generations and elephants long, but elephants have evolved much more rapidly. Man has extremely long generations relative to other organisms, but he has also evolved more rapidly than most. Among horses, it is unlikely that the ancestors of *Hipparion* had shorter generations than those of *Hypohippus,* but they too evolved more rapidly. In the great incursion into South America, the quick-breeding cricetines changed no more radically in the same time than did slower-breeding carnivores and ungulates. (They did split into more species, but that is an effect of size and other factors, not of shortness of generations; the carnivore and ungulate species although fewer are just as distinct.) As nearly as one can compare, the short-lived Foraminifera change no more rapidly in fossil sequences than do contemporaneous forms, vertebrates and others, surely with much longer lives. On an average, invertebrates have shorter generations than vertebrates but they seem, on an average, to have evolved more slowly. Herbaceous plants have shorter average generations than woody plants but, although there are cases in which the herbaceous plants have apparently evolved more rapidly, there are others in which this clearly is not the case and the (negative) correlation of rate of evolution with life span remains quite doubtful (Stebbins, 1950).

As Muller (1949b) has pointed out, organisms with shorter generations are often the smaller, "lower" forms, those with longer generations the larger, "higher" ones (although this generalization has numerous exceptions), and survival and reproduction tend to be more selective and less accidental in the latter. An additional speculation would be that in adapting to the same or similar cyclic and secular changes in environment it would often be advantageous for organisms to respond at comparable temporal rates regardless of lengths of generations. Many environmental changes proceed in accordance with absolute time, and groups with short generations may be eliminated if they respond too rapidly and make "short-sighted" commitments. This factor need not apply to organisms such as pathogenic bacteria, for instance, and their short generations may really be a factor in their occasionally rapid evolution.

The relationship between length of generations and changes in the environment may have other and perhaps more profound influences on evolution. Almost all organisms do have to meet cyclic, secular, or fluctuating environmental changes, the selective effect of which depends in good measure on their relationship to length of generations. Thus, land animals in the temperate zone must survive both hot summers and cold winters. The many forms with one-year generations can meet the situation by structural, as well as physiological, adaptations in different stages of the life cycle, as do many insects. Others, with longer generations, must as a rule develop wide physiological tolerance, as do many mammals, or hibernate, as do other mammals, or migrate, as do many birds.

Similar coordinations of life-span with climatic cycles are even more striking in plants, with their well-marked physiological differentiation between one-year generations and longer generations (annuals and perennials). Normally the annuals meet the temperate-zone winters as do annual insects, by a dormant or protected early period in the life cycle (dormant seeds). One evidence that this is genetically controlled is the occurrence of mutant maize, in which the seeds do not have a dormant period (Eyster, 1931). Deciduous perennials are analogous to hibernating mammals and, like the latter, often have nondeciduous (nonhibernating) relatives in regions of less severe winters.

Some animals with short generations, notably insects, can be adapted in various stages of the life cycle to one particular food, available only seasonally. Others, with longer periods of activity at stages of the individual cycle, cannot be so specialized, but must adopt either a continuously available diet or a diet varying with the seasons. In general, the individuals of successful phyla must be capable of adjustment to environmental changes that occur during their lifetime, just as a surviving phylum as a whole must be capable of adjusting to the changes that occur during its span. The two periods involved may even be of comparable length in extreme cases. It has been observed that some populations have made genetic adjustments to changes in environment within periods of five years and even less (e.g., *Carinus,* Weldon, 1899; *Aonidiella,* Quayle, 1938), a period exceeded by the length of generations in many animals. The specificity, as well as the nature, of population (genetic) and of individual (structural-physiological) adaptation is profoundly influenced by this relationship.

It is just possible that this effect was involved in the differential extinction at the close of the Pleistocene, although again it was not the only factor and probably not the principal factor. Then a greater proportion of large animals than small animals became extinct (e.g.,

Simpson, 1931a). Throughout the long Pleistocene epoch (a million years, more or less) these large forms, with long generations, would tend to make a slow genetic adjustment to the lower average temperature, or generally inclement climates, in a specific region. Readjustment to the more moderate recent climates would be slow. Smaller animals would adjust and readjust more rapidly and might tend to adapt primarily, not to a long average environment, but to a short summer-winter swing within the life cycle, more severe, but of the same nature as the shifts in their present life cycles.

In spite of the lack of evidence for any general and usual control of evolutionary rate by length of generations, it remains a hardly disputable fact that, other things being equal, shorter generations must permit higher *maximum* rates. This may have, and seems to have, no particular bearing on the great majority of evolutionary sequences in which maximum rates were not closely approached. It may, however, have a bearing on the relatively fewer, poorly recorded, and shorter but extremely important episodes that did involve rates approaching a maximum.

THE NATURE AND ROLE OF SELECTION

No theorist, however radically non-Darwinian, has denied the fact that natural selection has some effect on evolution. An organism must be viable in an available environment in order to reproduce, and selection inevitably eliminates at least the most grossly inadaptive types of structure. Aside from this obvious fact, theories as to the role and importance of selection range from belief that it has only this broadly limiting effect to belief that it is the only really essential factor in evolution.

Attacks on selection as an important evolutionary factor have been based on diametrically opposite considerations. Many real evolutionary occurrences have been, or have been believed to be, nonadaptive and hence inexplicable if selection is a controlling influence. On the other hand, various evolutionary phenomena have been considered so minutely adaptive that it was asserted that selection cannot have been efficient enough to produce them. Thus, it has been argued that there is no conceivable selective advantage of one precise ammonite suture pattern over another of equal complexity and that the evolutionary process may even result in structures that are positively disadvantageous

—the antlers of the extinct Irish stag are a classic example. Again, it is asserted that structures which are probably eventually of selective value, such as titanothere horns, arise gradually as rudiments that apparently had, at first, no use and therefore no selective advantage. Similarly, mimicry has been supposed to be of no advantage until well developed and also to be carried to extremes well beyond the point where greater perfection would give greater selective value.

The weight of such objections has led to a series of alternative theories that have in common only the minimizing of the effects of selection. Three extremes are especially outstanding among such theories. The first, antedating Darwin (and not entirely rejected by him, despite his emphasis upon selection) was the Buffonian-Lamarckian theory that adaptive structures are not selected, but are caused by environmental influences and by individual efforts to meet the exigencies of life. Although open to some of the same objections as is selection, this provides a logical alternative explanation of some but not all instances of adaptation. Particularly appealing to paleontologists, neo-Lamarckian theories reached considerable subtlety and complexity at their hands (e.g., Cope, 1886) and seemed at one time, and to some thoroughly competent students, sufficient to explain all the essential morphological observable facts. Experiments in heredity in the present century, however, not only have failed to corroborate that there is such a process but also have shown that it is highly improbable, if not impossible. I do not propose to consider the theory further in the present study. It is, perhaps, out of the question to offer a rigid proof that Lamarckian evolution could not occur and never did occur, but it is neither a necessary nor a sufficient hypothesis to explain the pertinent facts of evolution as seen by the paleontologist.

Another extreme is seen in various theories that suppose adaptive characters to have arisen, not in response to external influences and individual needs or at random with respect to them, but in anticipation of them. These theories avoid the necessity of providing a mechanism, because of which the neo-Lamarckians broke down, by begging the question. Concluding that the controlling factor of adaptation and of evolution in general is nonmechanical, they name this factor ("entelechy," "aristogenesis," and so forth), but no explanation is provided, and the definitions of their terms say little more than that they designate unknown causes of known phenomena. This appeal to the unknown, inherent in all such theories, however obscured by the naming fallacy, is metaphysical, not scientific. The categorical rejection of all physical hypotheses does not warrant the proposal of a metaphysical hypothesis as a supposed part of scientific interpretation. Recourse to such devices, impelling as is their fascination, has always resulted in stultification. The metaphysical explanation might, indeed, be true, but no hypothesis for which there can be no rigidly objective test should be accepted as a conclusion or working principle in research.[3] The greatest vogue for such theories has now passed, although they always have adherents and may well come into fashion again. Emphasis upon spiritual values, called for in times of stress like the present, sometimes has an undesirable and unnecessary concomitant of depreciation of scientific values, even among scientists.

Some of the objections to neo-Lamarckian explanations (which lack known mechanism) and vitalistic explanations (which deny mechanism) are met by a third extreme that turns out, nevertheless, to have equally serious objections of its own. This third non-Darwinian school involves theories, under various names and in various forms, that evolutionary changes arise spontaneously, from within the organisms, and that their adaptive nature is incidental or accidental. Following De Vries, this view was rather widely current among early geneticists but was greatly modified or entirely abandoned by most of them as the first, oversimple, strictly Mendelian genetics developed more realistic complications; the beginning of the swing is interestingly evident in earlier and later work by Morgan. As previously noted, the geneticist Goldschmidt still advocates this sort of evolutionary theory, although on more sophisticated grounds than did, for instance, De Vries, and another variant of it is still maintained by the paleontologist Schindewolf.

Strictly mutationist theories involve preadaptation in a rigid sense, and aspects of them have been extensively discussed under that term, particularly by Cuénot (1921, 1925, 1951). Cuénot, himself, was unable to reconcile the random nature of mutational preadaptation with the obviously nonrandom nature of adaptation, thus stumbling against the strongest (indeed, I hold, quite insuperable) objection to the

[3] Although a metaphysical theory is not open to direct physical test, it usually demands physical consequences that are subject to test. On this score, also, the various vitalistic theories must be rejected. I have elsewhere (1949a, 1950a) discussed this point and associated philosophical problems of materialism and vitalism at sufficient length, and need not go into them further here.

strictly mutationist point of view. He inclined for a time to solve the dilemma by postulating metaphysical finalistic control of mutation (1941), but he recognized the logical difficulty of this position and in the end (1951, published posthumously) was inclining toward a more selectionist interpretation even though this failed fully to satisfy him.[4]

With few exceptions (notably Schindewolf in recent years) paleontologists and systematists rejected mutationist and preadaptationist theories even while these were general among geneticists. This was one of the reasons for the former rift between observational and experimental students of evolution. Experimentalists could find no adaptive control of directions of change, but observation of nature makes it obvious that such control exists. Paleontologists, particularly, observed that many phyletic lines progress in a regular manner toward a condition that has appeared to some students so fixed as to be predestined—a condition that was not the only one possible for the phylum. If, then, the direction was determined by mutations alone, and limited only by a viable relationship between organism and environment, it seems impossible that mutations should be random, and one is driven back to the idea of purely directional mutation, which is physically unestablished and unexplained and therefore encourages metaphysical speculation.

The last word will never be said, but all these disagreements can be reconciled, and the major discrepancies can be explained. In the present synthesis adaptation, preadaptation (in a broadened sense), and nonadaptation all are involved, and all can be assigned immediate, if not ultimate, causes. That all basically depend upon mutation is partly a matter of definition. That the mutations are spontaneous and random, at least in the special sense elsewhere defined, is a conclusion warranted and, with some restrictions, demanded by the experimental data. Calling them "spontaneous" and "random" means simply that they are not orderly in origin according to the demands of any one of the discarded theories. The incidence in time and the individuals affected seem to be random or nearly so in the same sense—that they do not agree with hypotheses that assume a more specific incidence. That the direction of mutations is entirely random is certainly not true; but

[4] Probably because he never fully credited recent demonstrations of the great power of selection or quite realized the bearing and force of the theories of population genetics. "Je m'incline devant ces résultats mathématiques, mais sont-ils aussi démonstratifs qu'on veut bien le dire?"

neither is it true that mutations regularly occur in one direction only. Given a certain hereditary type of developmental pattern, the changes that can occur in it and their effects upon the structures developed are strictly limited, and alternative changes are not introduced in exactly equal numbers; but in almost every case the change can be and is in at least two, frequently more, different directions.

Many of these directions, most of them if conditions are stable, are nonadaptive. If these changes were neutral with regard to selection, they would present no particular problem. The number of changes that are really completely neutral is surely smaller than critics of natural selection have granted, but some are probably neutral in the sense that they are below the threshold of effective selection in a given population. Robson and Richards (1936) have gathered examples, some but not all of which seem to be of this nature. Some students hold that the differences between genera and higher groups may be adaptive or have selective value, but not the differences between subspecies or species (e.g., Jacot, 1932); others use similar data to attack the same theory by the diametrically opposed conclusion that local differences between subspecies are adaptive, but that generic distinctions are not (e.g., Goldschmidt, 1940). Some particular objections of this sort have been met by demonstrating that characters hitherto supposed to be nonadaptive are quite definitely adaptive (e.g., Dunn, 1935), and it has also been proved beyond much doubt that characters of unknown adaptive value are nevertheless definitely selected by differential survival (e.g., Dunn, 1942). In the nature of things it is quite impossible to establish that every single genetic difference between two populations has selective value, but neither is it possible to prove that any are really indifferent.

In a preceding section there were summarized studies by Wright that show that in small populations selection may be relatively ineffective in controlling the spread of gene frequencies. Under these conditions even disadvantageous mutations may spread rapidly and come to characterize a whole population, as long as they do not reach the rigid limit of complete inviability. Such random effects allow for the occasional rise of inadaptive characters, although populations so small as to suffer such changes in considerable number or as regards major mutations would usually become extinct within a relatively short time as disadvantageous genes accumulated. Theoretically such occurrences should be rare in large populations, and in fact they seem to be so. Thus, Osborn (e.g., 1927), who drew his conclusions from an enormous body of data relating almost entirely to large populations, was able to believe that "speciation"—hence all evolution, which was for Osborn a multiple of speciation—"is apparently always adaptive." Some recent geneticists would agree with Osborn, although for quite different reasons.

Perhaps a majority of students now believe that both adaptive and nonadaptive characters may characterize any taxonomic level, but that adaptive characters are generally much more numerous and that such nonadaptive characters as do occur are more common at and tend to be confined to low taxonomic levels. That, at least, is my opinion—I am aware that it is easy for an author to mistake himself for a majority.

Whatever may be the proportions of adaptive and nonadaptive characters in any given case, it is certain that a high degree of adaptation is common, indeed universal, in nature. This relationship between organisms and (in the broadest sense) their environment cannot now be considered as due primarily to the environment, as the neo-Lamarckians thought, nor to the organisms, as the mutationists thought, nor yet to a supernal plan, as the finalists have insisted. It has always been clear to some students, notably to Darwin, that adaptation results from material interaction of organisms and environment. The mechanism of the interaction is selection, and the role of selection in evolution is the production of adaptation. This is not to say that selection is all-powerful, so that all evolutionary change is adaptive. The opinion has already been expressed that change unguided by selection also occurs, and it is also known that selection, itself, may produce nonadaptive change as a secondary effect, although its primary effect is necessarily adaptive in some way. It is also probable that some adaptive change may arise without benefit of selection, although it is most improbable that this can persist or lead to significant evolutionary movement without involving selection.

Among such modern students as are free to inquire without deference to a political or theological ideology, there is a clear consensus that selection does have this role. Disagreement exists, but it is only sufficient to stimulate thorough examination of selection and production of added evidence as to how selection operates. Yet there is some lack of clarity, even among those who agree about the role of selection, as to just what selection is. The Darwinian concept, carried over into much of our thinking without reexamination, is that among all the individuals produced in nature some die sooner while others survive longer, and that the differential mortality is selection. Although one sometimes wonders whether all of them are clearly conscious of this fact, the population geneticists are using the term "selection" for quite a different concept, distinctly non-Darwinian although not anti-Darwinian. Their concept is that between one generation and the next there is (or, at least, there may be) a change in relative frequencies of genetic units and that selection is anything with a systematic effect on this change. It is necessary to specify that the effect be *systematic*, because even the critics of genetic drift admit that some nonsystematic change occurs and that this is not to be ascribed to selection.

Like any other word, "selection" may properly be defined in any way we choose, as long as those we address know our definition. I propose slightly to extend the definition used in population genetics and to define selection, a technical term in evolutionary studies, as *anything tending to produce systematic, heritable change in populations between one generation and the next.* This is an aspect or a sort of systematic differential reproduction and as such it contrasts with differential mortality, the Darwinian natural selection. Differential mortality may result in systematic differential reproduction. When it does, it is one sort of selection, but clearly not the only sort. That the concepts are different and that Darwinian selection is only one special case of, let us say, genetical selection is quite easy to see. Suppose that all the individuals in a population lived for precisely the same length of time, with no elimination of the unfit or survival of the fittest, hence no Darwinian selection. Suppose, further, that those among them systematically definable as, say, the taller ones, or those with an allele A, or a chromosome arrangement M, or a hereditary fondness for apples, had twice as many offspring as those without these characteristics. Then there would be very strong, clearly non-Darwinian selection and under its influence extremely rapid (although short-range) evolution of the population.

The proposed definition, which only makes explicit one of several meanings already current, is very broad and it requires that the particular sort of selection involved be specified when possible and pertinent. Under this definition, for instance, it is no longer true that selection can act only on phenotypes. Genetic factors without consistent phenotypic correlation may nevertheless influence fertility and therefore be subject to selection not acting on or through the phenotype. That the definition has appropriate evolutionary significance is evident in such cases as f_1 hybrids with vigorous phenotypes but seriously reduced fertility. Differential mortality may favor the hybrids, but genetical selection will not; the common mule is a good, even though artificial, example.

Selection, so defined, certainly exists in nature. By definition, it tends to move populations in the direction of relative reproductive efficiency for that population. (Note, in passing, that this does not mean that it invariably or without limit favors increase in individual fecundity or in total population.) This is in itself a sort of adaptation, and by definition selection favors reproductive adaptation. It does not follow automatically that it also tends to favor adaptation in a general sense or in any other particular sense. That it does so is a conclusion, not a premise or a matter of definition of "selection."[5] The conclusion rests on a vast body of evidence, some of which will be adduced as the discussion proceeds.

The criticisms mentioned at the beginning of this chapter as opposing the role of selection as the effective process in adaptation and oriented evolution will also be met in the course of this and later chapters. Only one point requires further mention in this section, and that is the old and oft-repeated objection that selection is not "creative," that it eliminates but does not produce. Even if this were true, selection would be an essential evolutionary factor, but it would have a more limited role and we would be forced to look elsewhere for some orienting factor involved in the production of new forms and adaptive types. As against Darwinian selection, which is what the critics have usually had in mind, the criticism has some force although it is not strictly true even of that.[6]

[5] It is, obviously, in part a matter of definition of "adaptation," but that is a point that need not be pursued here; see Chapter VI.

[6] Muller (1949, and earlier work there cited) has developed a way of looking at or expressing the action of Darwinian selection that assigns it a creative role and that extends the concept to near equivalence with genetical selection. He pictures the genetical mechanism as capable of producing an extremely large number of combinations, of which a "staggeringly minute" proportion corresponds with the "complexly working adaptive organizations" actually produced. The function of selection is to eliminate all but this staggeringly minute proportion, a process said to be as "creative" as the selection by a poet of "his particular words out of an almost infinite number of possible combinations." Without questioning the general and analogi-

The criticism has no real application to genetical selection, which clearly plays an active role in producing the particular genetical combinations that do come into existence. Therefore Darwinian selection does also, since it is a special case of genetical selection or, perhaps more accurately, one of the ways in which genetical selection is brought about.

EFFECTIVENESS AND INTENSITY OF SELECTION

Selection is a vector with both direction and intensity. A priori, the expected effect of selection is simple enough. In a group not already at the selective optimum, selection should, given a sufficient store of appropriate variation, move the group to that optimum. As a rule, the rate of movement should be higher the farther the group is from the optimum, should decrease as the optimum is approached, and should drop to zero when it is reached. Thereafter selection should tend to hold the group at the optimum, i.e., to inhibit evolutionary change, and to limit variation around that point. The direction of selection should be constantly toward the optimum and its intensity should vary with distance from the optimum.

In nature such does appear to be the basic tendency of selection, but the actual situation is almost always more complex. What is an optimum for one character may not be for another, the change of which is in part dependent on that of the first. Even a single character may have different optima representing, for instance, maximal adaptation to different elements of the physical environment, or to inter- as opposed to intragroup adaptation. For the group as a whole there may still be alternative optima representing different possible adaptations both or all of which are adequate compromises. These points may and in most cases will change, requiring for survival renewed evolutionary movement of the group and a compromise between selection for momentary adaptation and for successive adaptability. The pool of variation may not contain appropriate materials and probably seldom con-

cal validity of this statement, I must confess that it does not seem to me the most enlightening or even the most adequately descriptive way of looking at the process. In a sense, genetical selection as here defined is additive, and Darwinian selection, which seems to be Muller's selection, is subtractive. Genetical selection builds up particular combinations, an action more obviously "creative" than merely permitting them or preventing the others. In fact, this is what the poet does, too. It is poetic to say that he selects one out of a tremendous number of possible combinations of words, but it is not an accurate description of the real process of writing poetry.

tains those most ideally suitable. It may be depleted and effective selection may have to await its replenishment.

In an attempt to clarify and to understand such complex phenomena there is a difficulty of analysis that has already bothered us and will continue to do so throughout this book. It is a major thesis of the study that the main and basic factors of evolution are really inseparable in their action and that all are involved in practically every evolutionary process. It is therefore a falsification to discuss selection without simultaneously considering (among many other things) variation, or intensity of selection without considering its direction, just as it will later be a falsification to speak of the evolution of lower and of higher categories or of phyletic and quantum evolution as if they were qualitatively different or clearly separable in nature. Nevertheless, it is impossible to pile up words in layers, and linear sequence requires taking up one aspect of the whole after another and using different names for the aspects chosen. The only corrective is continual reminder that the analysis is artificial, while hoping that the parts can eventually be gathered mentally into the natural whole.

Let us, then, pluck out from the complex process of evolutionary movement the one aspect of selective intensity for summary. It has already been mentioned in connection with population size, because useful analysis cannot separate such factors with logical completeness. Here a first point to consider is one barely touched on previously, the effectiveness of various intensities of selection.

Selection may reach an intensity of ± 1, which represents its limit in terms of the coefficients commonly used. The range is from $+1$ for complete favorable selection through 0 for complete neutrality to -1 for complete adverse selection. Selection on a lethal dominant or a homozygous lethal recessive is -1 and of course no one has ever doubted the effectiveness of such selection. An organism must be viable, as least requirement, if it is to pass on anything to a later generation. However, the question early arose whether any degree of selection less than 1 is really effective among the viable and factually living members of natural populations. It was often pointed out that the (as he thought) one essential element of Darwin's theory was entirely inferential and rested on virtually no direct evidence.

Now finally evidence has been provided in such abundance that the reality and effectiveness of selection is not in doubt and question remains only as to the lower levels of effectiveness. The direct evidence has been reviewed by numerous authors (e.g., Dobzhansky, 1941, 1951; Huxley, 1942; Emerson, 1949) and there is no need to do more than exemplify it here. Evidence of Darwinian selection toward the mean for *Passer domesticus* in nature was among the first acquired. Bumpus (1898) found that among sparrows overcome by a storm the survivors were less variable than those that died. This was criticized as showing that selection did *not* tend to change the population, but in such obviously extremely well-adapted animals selection by a usual environmental hazard toward the existing mean (centripetal selection) is precisely what would be expected if selection does affect adaptation. Other observations on Darwinian selection both in nature and in experiments have frequently involved the selective effects of predation on cryptic or warning coloration. Dice (1947, 1949), for instance, has conducted experiments and summarized others in which extremely high selection occurred under conditions simulating those in nature. Even with several hundred prey animals, the method is not sensitive enough to demonstrate effectiveness of a selection index below about .2, but among twelve experiments the index was significant in ten and values above .5 occurred.[7] Such values are so high that they should be extremely effective even in small populations. In free populations there are various qualifying conditions, such as erratic variation in background color and gene migration between differently adapted groups, but in fact the prey animals of these experiments and their relatives are protectively colored in nature.

Still more interesting is direct evidence of genetical selection in natural and experimental populations. This includes instances of rapid adaptive change under strong environmental stress, as in the famous case of the spread of resistance to cyanide in *Aonidiella* (Quayle, 1938). Objection has been made that in this and similar examples the environmental change was introduced by man, but as already noted the obvious effects of short-range selection in nature must be expected to favor retention rather than change of existing adaptation. Only under abrupt and radical changes of environment beyond existing fluctuations

[7] This index of Darwinian selection is not the same as s, the coefficient of genetical selection previously discussed, but it tends to take comparable values. If of equal numbers of two different types of animals a and b are the numbers negatively selected (e.g., taken by predators), Dice's Darwinian index is $(a - b)/(a + b)$. Its value is positive for selection against the type for which losses are designated by a.

can we expect secular selection pressures away from the existing type strong enough for short-range detection—and such changes are rare except by human agency. Short-range, cyclic genetic adaptations to entirely natural cyclic environmental changes have also been observed (e.g., Dobzhansky, 1947; Dobzhansky and Levene, 1948).

Genetical selection between competing populations has also been repeatedly demonstrated experimentally for many different organisms (e.g., Gause, 1934). On the reality and effects of selection in intergroup competition, see also Crombie (1947). Experiments, mainly with *Drosophila,* in population cages have also demonstrated adaptive genetic changes within single populations in ways constituting direct evidence of genetical selection (e.g., Dobzhansky and Spassky, 1947).

Such observations and experiments, only a very small proportion of which has been cited, are more than sufficient to establish conclusively that selection is a real and effective evolutionary factor. It is, however, inherent in the situation that such direct evidence can be obtained only for short-range effects of exceptionally strong selection. The value of *s* in such cases, not always exactly obtainable, seems commonly to be on the order of 10^{-1} and seldom if ever below the order of 10^{-2}. Yet theoretical population genetics, as summarized in discussion of population size, suggests that *s* on the order of 10^{-4} should frequently and that much smaller values should sometimes be effective in long-range evolution in nature. The extreme precision of many cases of adaptation in nature would, moreover, require the effectiveness of such small values of *s* that these cases are regularly cited as "proof" that selection cannot account for adaptation. This, for instance, is the attitude of Goldschmidt (1945) regarding instances of mimicry so detailed that it may appear that the final steps (if small steps were involved; it is Goldschmidt's conclusion that they were not) could have had no selective advantage.

With reference to this last point and many other such claims by critics of adaptation by selection, there is a great difference between *no* advantage and an advantage so slight that the author thinks it negligible. Mention has already been made of the extreme improbability that any change in a population that is above minimal breeding strength can be, or certainly can long continue, absolutely neutral to selection. Specifically in the case of mimicry, if this is proceeding by small steps, the value of *s* will at first increase and then will start to drop when what may be called an adequate resemblance is achieved. Thereafter it will of course continue to drop more or less rapidly and will become very small, but finite, with achievement of such intricate resemblance as does sometimes occur. It will not, however, become infinitely small, i.e., reach absolute zero, until the resemblance is so complete that *no* individual predator under the *most favorable* conditions that *ever* arise can *possibly* discriminate between mimic and model. It seems certain that such a degree of mimicry is never achieved. At least it cannot be shown to exist. To say that the advantage of a given degree of increase in resemblance is negligible expresses an opinion for what it may be worth, but to say that the advantage is zero is an unscientifically loose use of terms.

As a matter of fact, the best and I think quite conclusive proof that very small values of s are often effective in nature is precisely the evidence so often cited against the effectiveness of selection. This evidence shows that adaptation may begin and proceed on the basis of very small variations and may reach a very intricate and high degree of perfection, so that its initiation, its continued trend, and the point reached must have involved very small selective values if selection was involved. Such situations are common, and the fact that they exist at all is crucial for consideration of the effectiveness of selection, but again note that this does not mean that *all* adaptation necessarily arises by steps with low selective value or that *all* changes in evolution are adaptive—such seems surely not to be the case.

The most forceful evidence is of three sorts: first, the fact that some adaptation is universal among organisms and that many of them have this to a very high degree; second, the extreme intricacy and precision of the genetic, developmental, and other physiological characters of all organisms, even the least complex; third, numerous paleontological sequences in which adaptive trends arise as minute variations and continue by slow and small changes between successive stages. The first line of evidence needs no special exemplification here. The stated fact has been noticed by everyone who ever looked at nature, from antiquity onward, and has been emphasized by every natural historian. The second point is also fairly obvious, or becomes so on acquaintance with any treatise on cellular, comparative, or human physiology, on embryology, on genetics, or on biochemistry. Some of the genetic aspects have recently been considered from quite the present point of view by Muller (1950). Examples of the paleontological evidence, the most convincing of all because it exhibits the origin and progression of adaptation and not merely the results, are legion. Here may be mentioned the rise of titanothere horns from what was at first only a slight, variable thickening of the bones, as noted by Osborn (1929), whose work contains many such examples cited as evidence against selection. Other examples will be mentioned in later discussion of trends in evolution.

The argument that such evidence demonstrates the effectiveness of very small selection pressures formerly encountered an objection not altogether insuperable but certainly very forceful. The argument sometimes seemed to take this form:

Selection produces adaptation.

Some adaptation involves changes with very slight selective value. (Some of the critics said "with no selective value," but on this see above.)

Therefore very slight selective values can be effective.

This is a perfectly valid argument in itself, but of course it is open to the objection that it takes as a premise what the critics insisted was to be proved.

Now the situation is quite different. Not only is there good, direct evidence (as above) that the premise is correct, that selection does or at the very least can produce adaptation, but also there is another conclusive point of agreement. From the known facts of genetics, entirely independent of any observations of adaptation or of how it is achieved, it is rigorously demonstrated in theory that extremely small selective pressure should be effective under certain not rare conditions, i.e., selection is effective when $s > 1/N$, a value rarely larger in order of magnitude than 10^{-2}, often as small as 10^{-4} or 10^{-5}, and probably much smaller in some instances. There is thus established a mechanism which includes and (on its own level) explains the effective action of low selection pressures. The independent observation that adaptation does frequently (but not always) involve changes of low selective value can thus no longer be cited as evidence against selection, but becomes the strongest sort of evidence that selection is, indeed, effective in producing adaptation. This same evidence also completely rules out any alternative theories, such as those of the mutationist school, that consider adaptation as the result of chance or random processes alone.[8]

(For further discussion of some of the points here touched on, see the American Philosophical Society's symposium on "Natural Selection and Adaptation," by Muller, Wright, Jepsen, Stebbins, and Mayr, 1949. Their papers are separately cited in the bibliography at the end of this book.)

The evidence indicates not only that selection is an effective factor in adaptation but also, for many factual cases of progressive, adaptive change, that it is a necessary one. The three students to whom we owe most of the theoretical structure of population genetics, Fisher, Haldane, and Wright, differ sharply on some points, but they all agree that change on the basis of genetic factors alone, mutation, random recombination, and random fixation or elimination, would be extremely slow in any but very small populations. The number of generations required for change of a single gene in the population as a whole would be of a magnitude at least that of the average number of individuals in the population and probably larger. The Equidae evidently had continuously large populations in their successful phyla. In them the time involved in specific change by genetic processes only would probably have been on the order of 10^7 years. In reality, in the line from *Hyracotherium* to *Equus* there were 9 successive genera and perhaps 30 to 40 successive species (by rough and minimal estimate; the successive species have not been adequately defined) in 6×10^7 years. This is not exceptionally rapid evolution for mammals and perhaps not greatly above the average for organisms in general, quite certainly not if we compare only with groups that were actively evolving throughout, as the horses were, rather than becoming stabilized. Some factor additional to strictly genetic processes is required to explain many observed rates of evolution, and the most reasonable probability is that that factor was selection.

In groups that are evolving adaptively (and it will later be concluded that all long- and most short-range trends consistent in direction are adaptively oriented), the control, so to speak, is by selection. If selection

[8] It is one of the commonest arguments against Darwinian or selectionist theories of evolution that they cannot be true because "they depend entirely on chance." The objection is so completely irrelevant as almost to cause despair for human reason. These theories are the *only* ones involving a factor that is *not* random with respect to adaptation and that is known factually to exist.

has a definite linear component (see the next section), the rate of evolution will tend to vary directly with intensity of selection as long as variation is adequate. The latter condition is met by most if not all long sequences, for appreciable linear selection without adequate variation is very likely to result in extinction. In most known examples it is at least a reasonable inference that rate of phylogenetic evolution has tended to vary with intensity of selection. For instance in the evolution of hypsodonty in horses (Chapter I), there was a decided acceleration in rate coincident with what must plainly have been an increase in selection pressure.

The intensity of selection depends, in turn, on the very complex relationships between the organism and its environment.[9] Selection will tend to produce modal change in the population only if something in this relationship disturbs existing adaptation. In most well-recorded sequences there is reason to believe that the rates of evolution are far from a maximum, although this is evidently not true of all evolutionary events. This is indirect evidence that in such sequences adaptation keeps up with the environment. It may then be said that ultimately it is the changes in environment that control rate of evolution, although the control is by means of the mechanism of selection. ("Changes in environment" in this connection also include the cases in which a group enters a new environment.)

When adaptation is keeping up, selection at any one time will be mainly in favor of the existing type, with centripetal direction dominating linear as discussed in the next section. In such cases, the intensity of selection tends to effect not the rate of change but the amount of variation.

If selection favored a single fixed genetic combination, Wright has shown that it would have little effect in decreasing variability unless $s > 1/(8N)$. Stronger selection would greatly reduce variability, until it would be almost eliminated at $s = 4/N$. For a population of 100,000 this figure for s is only .00004, and for $N = 1{,}000$ it is .004, values so low that they seem to prove too much. According to this, characters with appreciable selection value would be almost invariable in large populations, whereas in fact adaptive characters in such populations

[9] Here and elsewhere, unless a particular environmental factor is specified, the word "environment" is used in an unrestricted sense. It may include not only the physical and the extraspecific biotic environments but also the organism's own breeding or social group and species.

do always have a considerable amount of variability, both genetic and morphological. This is, however, readily accounted for by several factors making for variability in natural populations and not involved in this model. Mutation rates always provide some variability that cannot be wholly eliminated by the strongest selection. In a large population, spread over a considerable area, the optimum will generally tend not to be exactly the same at all points, so that even strong selection does not tend toward one point and the same sort of effect is seen in groups subject to cyclic or secular fluctuations of environment and hence similarly without one permanent optimum. Under such conditions there is an optimum amount of variability that is itself maintained by selection, as has been previously discussed. Within the common pattern of a large species divided into semi-isolated, interfertile breeding groups variation will also be maintained in the face of the strongest selection —intergroup variation by the differences in optima for the various groups, and intragroup variation by migration and marginal interbreeding. Moreover, the model assumes that *s* has time to act without a change in the optimum for the population. In fact, elimination of variation from a large population would take so much time that the optimum would be more likely than not to shift.

In spite of a paucity of adequate observational data it is clear that selection does, indeed, influence variability. In a previous section it was shown that functional, integrated structures of adaptive significance —i.e., those that are under strong centripetal selection pressure—tend to have low, although still appreciable, variability and that the variation differs surprisingly little from time to time for one variate within a group, for homologous variates in different groups, and for different analogous variates. On the other hand, vestigial and nonfunctional characters—i.e., those with low selection value—have higher average variability.

DIRECTION OF SELECTION

The characteristics of selection as a vector are intensity and direction. Direction may be centripetal, centrifugal, linear, or a resultant of combinations of these (Fig. 14). The direction is defined by the relationship of the morphology and the physiology of a population to the environmental field in which the organisms function. If the modal con-

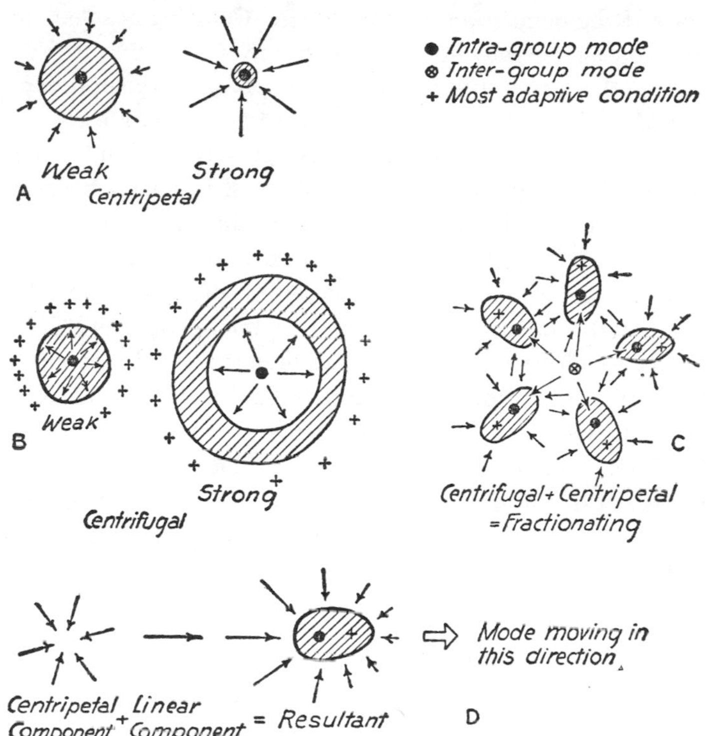

FIGURE 14. SELECTION VECTORS. Heads of arrows show direction of selection, and lengths of shafts are proportional to intensity of selection. Shaded areas show range of variation covered by population.

dition of a character in the population, that "typical" for it, has continuously greater survival (or, more strictly, reproductive) value under the given conditions than any variant or newly mutant condition, the population is well adapted as regards that character. Selection is then entirely centripetal. It acts to eliminate the variant forms, and it tends to reduce variation in any direction and to concentrate the population around a point, the optimum condition. This involves an increase of specialization, not necessarily in the usual sense of progressive adaptation of all individuals, because some are postulated as already perfectly adapted, but in the sense of reducing the number of less well-adapted individuals in the population. This is an evolutionary change, but in the more usual sense there is no evolutionary advance, because the mean characters of the population do not change. Centripetal selection is, in fact, directed against any such shifts in means.

If, on the other hand, the modal condition of a character in a population is so ill-adapted that any variant condition is more advantageous, selection is centrifugal. It favors increase of variation and progressive divergence from the population mean. As will presently be discussed, this selection can rarely be exactly realized in nature because of the other factors involved, but it will serve as a model for a theoretical analysis of the single factors underlying more complex real phenomena. In the simplest model, strong centrifugal selection would eventually result in the elimination of average individuals from the population, so that its distribution in the morphological field would be an expanding ring or a hollow circular band. The average for the character in question would remain the same, but there would be no individuals actually at the mean in this respect. Even as a model, this pattern is obviously too simple; it is made more useful if the inevitable centripetal forces within the parts of the ring are also taken into consideration. Then the ring breaks up into smaller groups, in each of which selection is actually linear, away from the intergroup mean, and the situation is evidently less abstract and closer to reality than when the unoccupied central point is considered an intragroup mean.

Pure centripetal selection occurs in nature. It may even persist for long periods of time, but in most cases it is temporary. Pure centrifugal selection probably does not occur in nature, although it is surely a component of more complex selective forces in many cases.

Centripetal selection can remain unmodified only as long as the optimum is completely static and selection is continuously effective enough to prevent any considerable deviation of population mean from the optimum. In some unusual cases (see Chapter X) this delicate equilibrium has been maintained even for millions of years, but in the great majority of cases it is continuously or continually disturbed. The simplest, or most easily understood, disturbance is some change in the external environment that causes a slow shift in the optimum. Although still centripetal, bearing toward the center from all directions and tending to concentrate the population on the optimum, selection then becomes stronger on one side than on the other, favoring variants on the side toward which the optimum is moving, and it acquires a linear component superposed upon the centripetal pattern. Progressive shift in the mean occurs, and as long as selection is effective change will be at about the same rate as the shift in the optimum. Any considerable lag is likely to leave the population behind the moving selection field and subject to a different selection pattern, resulting either in extinction or in a new direction of evolution.

In the broadest sense, environment is so complex and subtle that the optimum may shift in what appears at first sight to be a stable environment. Thus, in animals the optimum for size often shows a tendency to shift slowly in one direction, toward larger size, a tendency that may be counteracted by other influences and abruptly reversed, but is nevertheless usual in the evolution of such groups. Even though individual animals may be perfectly adapted at a particular size level, in the population as a whole there is a constant tendency to favor a size slightly above the mean. The slightly larger animals have a very small but in the long run, in large populations, decisive advantage in competition for food, also in many cases in reproductive opportunities and in escaping enemies. An abrupt increase in size in a small number of individuals at any one time is more likely to be disadvantageous, even if genetically possible, because it may not be accompanied by other necessary independent adaptations to a larger size, correlated genetic changes may be inadaptive, and in higher animals abnormally large individuals may be subjected to the hostility of competitors and potential mates. Thus, populations that are regularly evolving in this way are always well adapted as regards size in the sense that the optimum is continuously included in their normal range of variation, but a constant asymmetry in the centripetal selection favors a slow upward shift in the mean.

This is, I believe, the causal background of the empirical paleontological principle that many lineages have a steady trend toward larger size. As regards abundant groups in which selection is continuously dominant as an evolutionary factor this generalization is widely applicable—so widely that it is commonplace in the literature of paleontology to read that a certain species "cannot" be ancestral to another of later age because the earlier species was larger. There are, nevertheless, a good many exceptions that are virtually certain and more that are probable. These can invariably be explained either by special circumstances that make selection favor small size (e.g., very warm climate for homothermous animals, insular conditions, subterranean life, limitation of food supply, greater opportunities for escape from enemies because of smallness), or by other circumstances (such as small population size) that relax the control of selection. There is a large literature on this subject, which is a great favorite with paleontologists, particularly, see, e.g., Stromer (1944) or Newell (1949). Rensch (1947, 1948) has also discussed it and has suggested some other reasons why selection may favor larger size, notably the organizational advantages of having larger numbers of nerve and other cells.

I have elsewhere suggested that trends analogous to that for larger size, of which there are probably a great many, might be called "self-braking." Their characteristic is that they are controlled by selection that favors variants slightly, but only very slightly, above the mode and at the same time opposes variants in the same direction but beyond those favored. The pattern is like that of Figure 14D, but less asymmetrical, barely off circular in most cases. The resultant is a tendency for the population to change continuously but to do so at a slow rate. This helps to explain why so many evolutionary sequences do show long, slow change in one direction and to show that selection, in particular circumstances, fully explains such phenomena without recourse to some mysterious "orthogenetic" force. Self-braking trends also help to explain why evolution so rarely proceeds at rapid or maximal rates even in a direction favored by selection: selection, itself, brakes the speed. Moreover, since in such cases selection may be and probably is usually intragroup only and largely unaffected by anything except the evolving organisms themselves, this process helps to explain why change is so usual and arrested evolution (although it occurs) so relatively rare even when there are no evident environmental changes external to the population.

A change in one character may change the optimum of another even though they are genetically independent and seem to be independent as regards adaptation. Thus, an oscillating shift of both optima may operate continually and in a single direction for each. For instance, both brachydont and hypsodont teeth occur in large and in small mammals, and hypsodonty and gross size are genetically independent and, as regards extremes, adaptively independent. Nevertheless, nominally brachydont large animals almost invariably have somewhat higher-crowned teeth than their likewise brachydont smaller allies or ancestors. The comparison of the large brachydont *Hypohippus* with its small brachydont ancestor *Hyracotherium* (data in Chapter I) is a typical example. Even if no significant change in diet is involved, larger animals live longer, on an average, and require more food. They need teeth having not only larger working area but also more durability. We, ourselves, although brachydont, have relatively higher tooth crowns than do most small primates of similar omnivorous food habits. Such secondary adaptations may limit the primary adaptation. An increase in size disproportionate to the increase in durability of teeth is disadvantageous. When tooth durability has caught up, increase in size again becomes selectively valuable, making more increase in durability advantageous, and so forth, in a potentially endless cycle, while all the time the external environment may be entirely static.

Still more marked but less frequent changes in optima may occur when a mutation that was inadaptive in its original genetic surroundings happens to survive and to become fixed. Then the optima are likely to shift for all other genes the expression of which has any direct or indirect functional relationship to that of the mutation. Thus, pleiotropic genes, polygenic characters, and adaptively correlated characters form a mesh so intricate that a single change at any point may initiate a series of reactions by which selection eventually produces a change in the whole system. Such considerations help to show why (as previously noted) the differences between species involve many genes, even in cases where the differentiation might logically be traced to one "key" mutation in the ancestral population.

The most generalized abstract model of centrifugal selection can hardly be related to a real example, because it is difficult (I find it impossible) to imagine a situation in which any change whatever in a character would be advantageous and could physically occur in an unlimited number of different directions. There are, however, real and important instances in which more than one of the limited number of variant types in a population would be better adapted to available environments than the modal types, and this is a special case of centrifugal selection. Essentially such a process, one of the most common forms of evolution on low taxonomic levels, occurs when a variable and relatively unspecialized population divides into groups, each of which becomes adapted to a different ecologic niche. In longer-range phyletic studies a manifestation of similar centrifugal selection is commonly seen when branching is accompanied by evolution in two or more opposed directions from the ancestral type. Thus, the shell of shelled animals impedes motion and requires the utilization of food and vital processes that are unnecessary for reproduction and possibly detract from it. In regard to these factors thin shells or no shells will be favored by selection. On the other hand, the shell is a protection from predators and other unfavorable environmental conditions, and in regard to this factor heavier shells are favored by selection. Aside from such common developments as the strengthening of shells by ribbing instead of by thickening, the character generally evolves only in one of two directions—more shell or less shell. Under varying balances of the opposing selection factors, almost every intergrade develops; but very thin shells, not otherwise strengthened, are rare. They do not suffice for adequate protection, yet do not fully realize the advantage of maximum reduction. Here there is a balance point where variation in either direction is an advantage over the mean condition—centrifugal selection—and in fact among otherwise nearly allied molluscs some forms may become shell-less and some relatively heavy-shelled. Thus, *Chiton,* strongly shelled, and *Neomenia,* unshelled, seem to have had a common ancestry at no very remote time, and allied shelled and shell-less forms (such as slugs and land snails) occur in several different groups of gastropods.[10]

Other examples are provided by the morphologically simple character of size, which can vary only in two directions, either of which may, under different conditions, have greater selective value than an intermediate condition. This sort of dichotomy occurred at least twice in the history of the horse—when late *Miohippus* split into one line smaller than the ancestry, *Archaeohippus,* and another line larger, *Parahippus,* and one division of *Merychippus* similarly split into small *Nannippus* and large *Hipparion.*

If, through relaxation of selection or some other cause such as fractionation over a varied region, the variability of a group increases and the specificity of its total adaptation decreases, several or many of the large number of different genotypes are likely to have some advantage in particular environmental conditions, and these may be driven away from the ancestral modal condition (even if part of the population remains there and is sufficiently well adapted there) and from each other

[10] As regards this particular example there is an alternative theory that the shell-less forms are primitively so, never having had a shell. In some cases this may be true, but it seems almost certain that some, at least, represent a process of loss of shell in one branch of a phylum while the shell became stronger in another.

by a scattering effect centrifugal with regard to the ancestral configuration and linear with regard to each of the separating groups. More commonly, this situation arises when a variable group enters a new and varied environment, either by actual spread in space or by change in its own characters. Such effects are evident in the fossil record as some of the so-called "explosive" stages of evolution in various groups.

A probable example has been given by Bulman (1933), who plotted numbers of species in graptolites and found that periodic great increase in them was regularly followed by a sharp decline. His interpretation is that number of species is more or less directly proportional to adaptive value of the structures developed, and he feels that the subsequent decline is therefore inexplicable. If the opposite point of view is taken, the whole process can be easily explained. Increase in number of species represents a decline in the adaptive status of the ancestral populations, and consequent centrifugal selection and fragmentation of groups imperfectly adapted but tending more or less toward a variety of different adaptive types under the impulse of linear fractions of this total radiant selection pattern. Some of them achieve a new and perfected adaptive condition and become abundant and successful; the majority do not, and become extinct. The less adaptive phase has many species, and the more adaptive phase, few.

Wright (1931) has suggested a figure of speech and a pictorial representation that graphically portray the relationship between selection, structure, and adaptation. The field of possible structural variation is pictured as a landscape with hills and valleys, and the extent and directions of variation in a population can be represented by outlining an area and a shape on the field. Each elevation represents some particular adaptive optimum for the characters and groups under consideration, sharper and higher or broader and lower, according as the adaptation is more or less specific. The direction of positive selection is uphill, of negative selection downhill, and its intensity is proportional to the gradient.[11] The surface may be represented in two dimensions by using contour lines as in topographic maps (Figs. 15 and 16). The model of centripetal selection is a symmetrical, pointed peak and of centrifugal

[11] Some students find it easier to reverse the landscape, with adaptive optima as depressions, and to think of gravity as representing selection. I have retained Wright's representation, especially as it is useful to keep in mind that work has to be done (by selection) to carry a population to an optimum and keep it there.

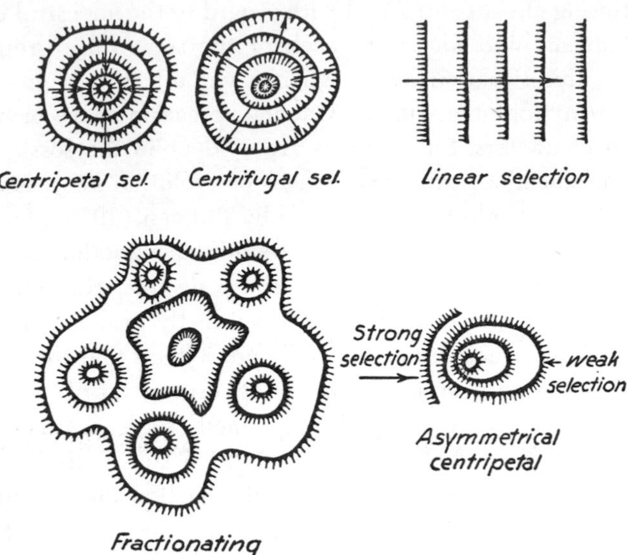

FIGURE 15. SELECTION LANDSCAPES. Contours analogous to those of topographic maps, with hachures placed on downhill side. Direction of selection is uphill, and intensity is proportional to slope.

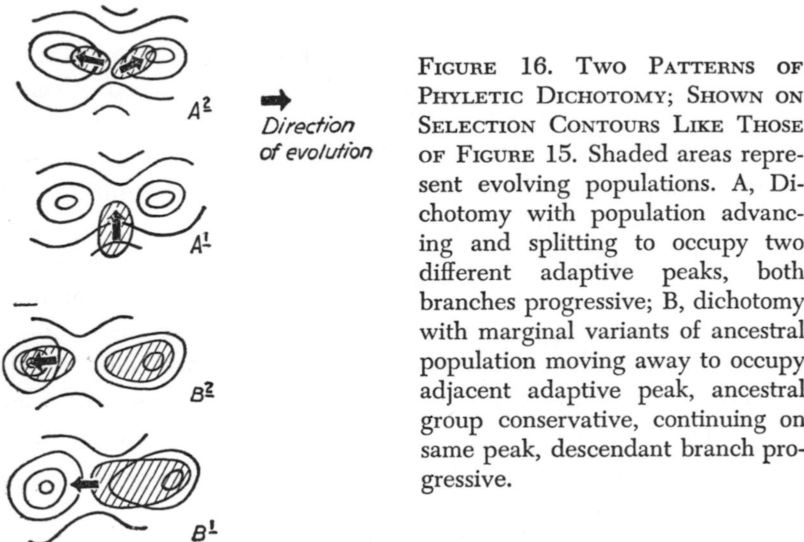

FIGURE 16. TWO PATTERNS OF PHYLETIC DICHOTOMY; SHOWN ON SELECTION CONTOURS LIKE THOSE OF FIGURE 15. Shaded areas represent evolving populations. A, Dichotomy with population advancing and splitting to occupy two different adaptive peaks, both branches progressive; B, dichotomy with marginal variants of ancestral population moving away to occupy adjacent adaptive peak, ancestral group conservative, continuing on same peak, descendant branch progressive.

selection, a complementary negative feature, a basin. Positions on uniform slopes or dip-surfaces have linear selection. The whole landscape

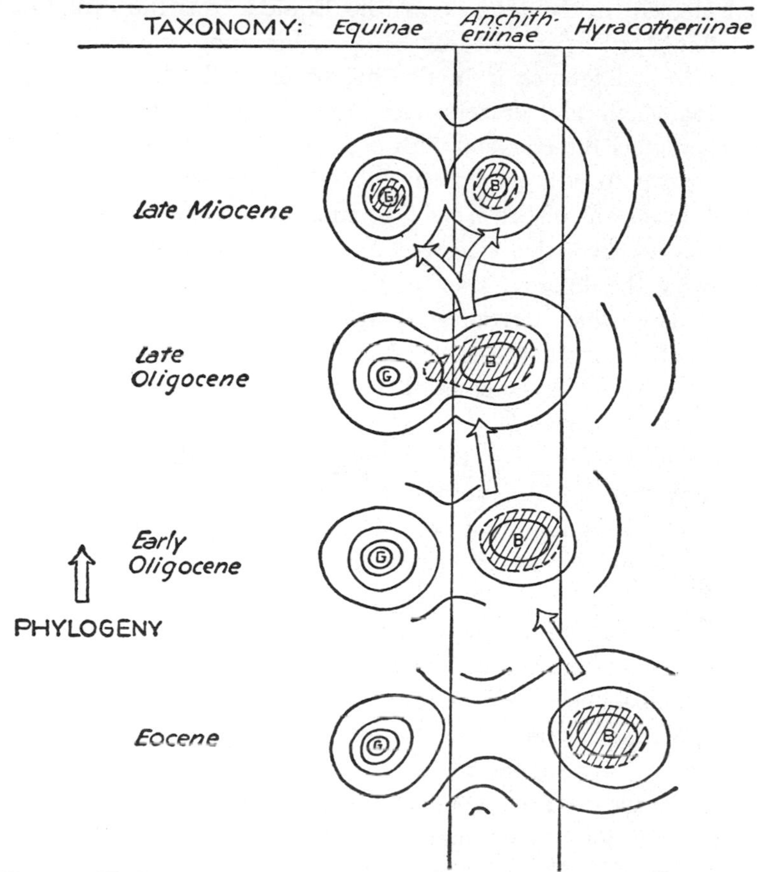

FIGURE 17. MAJOR FEATURES OF EQUID PHYLOGENY AND TAXONOMY REPRESENTED AS THE MOVEMENT OF POPULATIONS ON A DYNAMIC SELECTION LANDSCAPE. For fuller explanation see text.

is a complex of the three elements, none in entirely pure form. To complete the representation of nature, all these elements must be pictured as in almost constant motion—rising, falling, merging, separating, and moving laterally, at times more like a choppy sea than like a static landscape—but the motion is slow and might, after all, be compared with a landscape that is being eroded, rejuvenated, and so forth, rather than with a fluid surface.

One aspect of horse evolution, readily explicable in these terms, provides a well-documented, large-scale example that gives a feeling of reality to this rather theoretical discussion (see Fig. 17). An essential element in the progress of the Equidae was the evolution of food habit. There were two main types, browsing and grazing. The most obvious difference is in the height of teeth—browsers having brachydont teeth, and grazers hypsodont teeth—but there are many other characters directly and indirectly related to these habits. The tooth-crown pattern is more complicated in grazers, the occlusion is different from that of browsers, jaw musculature and movement are somewhat modified, the digestive tract is undoubtedly different, although what the digestive tract of browsing horses was is not known, since they are extinct. Less directly, there are modifications in feet and limbs and throughout the skeleton that are more or less correlated with different habitats—grasslands for the grazers and, in part, woodlands for the browsers.

In the Eocene browsing and grazing represented for the Equidae two well separated peaks, but only the browsing peak was occupied by members of this family. That peak had moderate centripetal selection, which was asymmetrical, because one kind of variation, on one side, away from the direction of the grazing peak (teeth lower than optimum, and so forth) was more strongly selected against than on the other side, in the direction of the grazing peak.

As the animals became larger—throughout the Oligocene, especially—the browsing peak moved toward the grazing peak, because some of the secondary adaptations to large size (such as higher crowns, as previously discussed) were incidentally in the direction of grazing adaptation. Although continuously well adapted in modal type, the population varied farther toward grazing than away from it, because of the asymmetry of the browsing peak. In about the late Oligocene and early Miocene the two peaks were close enough and this asymmetrical variation was great enough so that some of the variant animals were in the pass between the two peaks. These animals were relatively ill-adapted and subject to centrifugal selection in two directions. Those that gained the slope leading to grazing were, with relative suddenness, subjected to strong selection away from browsing. This slope is steeper than those of the browsing peak, and the grazing peak is higher (involves greater and more specific, less easily reversible or branching specialization to a particular mode of life). A segment of the population broke away, structurally, under this selection pressure, climbed the grazing peak with relative rapidity during the Miocene, and by the end of the Miocene occupied its summit. Variants on the browsing slope tended by slight, but in the long run effective, selection pressure to be forced back onto that peak, and the competition on both sides from the two well-adapted groups caused the intermediate, relatively inadaptive animals to become extinct. Thereafter browsing and grazing populations were quite distinct. The browsing types eventually, at about the end of the Tertiary, failed to become adapted to other shifts in environment and became extinct, while the grazing types persist today.

The structural and phyletic concomitants of this adaptive history are the basis of the usual subfamily arrangement of the Equidae. The small, primitive browsers on the early browsing peak, well isolated from the equid grazing peak, but (as shown by other data, not included in the diagram) still near other peaks occupied by their close allies, such as the early tapirs, titanotheres, and rhinoceroses, are classed as Hyracotheriinae. More advanced browsers, after the equid browsing peak had moved well away from other perissodactyl browsing peaks and nearer the equid grazing peak, are placed in a subfamily, Anchitheriinae. The grazing forms are united in the Equinae.

CHAPTER VI

Adaptation

ADAPTATION is intimately involved in almost all evolutionary processes, and particularly in those stressed in this book. It is therefore discussed under one aspect or another in every chapter, and it may not be entirely logical to set aside one chapter under this heading. There are, however, some special points regarding adaptation which can be considered separately and for which sufficient groundwork has now been laid. Some of these points are discussed here and others in the following chapter.

COMPLEXITY AND EXTENT OF ADAPTATION

On earlier pages the definition of adaptation has been taken for granted and its universal existence has been considered so obvious as hardly to require exemplification. Now it is necessary to ask more specifically what adaptation is and how widespread is its occurrence, although the subject is so intricate that even here no more than a very general and summary statement can be attempted. For our purposes, *an* adaptation is a characteristic of an organism advantageous to it or to the conspecific group in which it lives, while adaptation or the process of adaptation is the acquisition within a population of such individual adaptation. What makes a characteristic advantageous, hence adaptive, is a relationship between the organism and (in an extremely broad sense, see below) its environment. Adaptation may be on a basis of individual modification or it may have a hereditary basis. The study of evolution is of course mainly concerned with hereditary adaptation, but it has already been noted that the distinction is hard to maintain and sometimes quite artificial. The *capacity* for individual modification is generally adaptive and has a hereditary basis even though the particular form of modification in a given case does not.

It is obviously advantageous for an individual to remain alive and for a group to continue reproducing itself—or, if this does not seem obvious, we will define this as a pertinent meaning of "advantageous." While individuals live and lineages continue they are *ipso facto* adapted, and to this extent, at least, adaptation is clearly universal. There are, however, different degrees and sorts of adaptation. If one group is more successful than another under given (the same or similar) conditions of life, it is fair to conclude that it is better adapted. The surest criterion of such success is increase in relative abundance of the better adapted group.[1] Such increase in relative abundance is evidence of selection (indeed it *is* genetical selection if the advantageous characters are genetic) and again the relationship of selection to adaptation is confirmed.

Further evidence of the existence and effectiveness of adaptation is provided by adaptive correlations and facies. An animal living in a given environment has many different adaptations all correlated with particular conditions of the environment. A striking example is the lizard genus *Uma,* which lives in the Mojave and Colorado Deserts (Stebbins, 1944; Smith, 1946). These lizards have thick, serrated, overlapping eyelids with a translucent area in the lower lid, peculiar nasal passages, wedge-shaped snout with countersunk jaw, broad depressed body, scale fringes on the toes, granular coloration, and other characters all advantageous for life on and in the sand of its habitat, where it "is the only reptile which shows a decided preference for the mighty barren dunes" (Mosauer, quoted in Smith, 1946). Moreover, in a special sort of habitat different sorts of organisms tend often to have similar adaptations, producing an effect of adaptive facies. Thus in deserts small mammals are often bipedal and ricochetal and tend to have large auditory bullae; reptiles, beetles, and some other animals are usually fossorial; most animals tend to be pale in color although some are black; many plants are spiny; and so on. Adaptations to deserts, a

[1] Emerson (1949) criticized an earlier statement of this criterion on the grounds that a rare organism may be better adapted than an abundant one. That is of course true and was not at all contradicted by my statement. Emerson failed to note that the criterion was applied to changes in *relative* abundance in groups in one *certain* or *given* environment, i.e., ecologically horizontal in their relationship. Abundance of one group relative to another may increase even though absolute numbers are small or decreasing, and the criterion cannot apply to two groups adapted to quite different conditions or ecologically vertical as, for instance, parasite and host or predator and prey. Pitelka (1951) has already noted that Emerson's comment was irrelevant. I mention it here, not to reply to a criticism, but to try to clarify a point that must have been obscure.

particularly difficult sort of environment, are often striking and clear in significance so that there has grown up a large and fascinating literature on this subject (e.g., Heim de Balsac, 1936; Kachkarov and Korovine, 1942). Every geographic environment, however, has its own peculiarities which are met by special adaptations of the organisms living in it, as has been rather fully summarized by Hesse, Allee, and Schmidt (1951) for animals, and with some difference in point of view by Cain (1944) for plants, and Newbigin (1948) for both.

Concentration on such strikingly specific adaptations as those of *Uma* for life in the dunes should not make one overlook the fact that most (some students would say "all") of the other characters of such organisms are also adaptations, although they may be of a more general sort. Thus the lungs and circulatory system of *Uma* are adaptations to life in an oxygen-containing atmosphere, and so on through as long a list of characters, anatomical and physiological, as you care to compile. The adaptive significance of such characters may seldom be noted just because they are so necessary or so widely useful and occur in a multitude of different forms, but they are no less, indeed are all the more, adaptive for just those reasons.

Striking examples related to geographical environments may also tend to make us overlook the extreme complexity of the total environment and, correspondingly, of adaptations to it. The physical environment at one geographic locality not only has many different concurrent aspects, such as those of temperature, humidity, substratum, and others less obvious, but also may be radically different for different organisms, for instance for a flying bird, a burrowing mammal, and a soil microorganism. Nor is the physical environment the only one requiring adaptation. It is frequently noted that the biotic environment, the whole local community of plants and animals, has a greater or less degree of mutual adaptation and of individual and specific adaptation to it. Further, from the point of view of adaptation the conspecific breeding or social group of which an organism forms part also has its adaptations and is also part of the environment to which the individual is adapted. Even beyond this, each individual has its own internal environment to which it is also adapted and within which its various parts and functions are correlatively adapted. It seems a little paradoxical to speak of an individual's own body as part of its environment, but the distinction between inside and outside is by no means as clear as at first appears (Sinnott, 1946). The whole complexity of adaptation cannot be fully grasped without realization that what is adapted to—the "environment" in this sense—includes at least four main, strongly interacting and not sharply separable levels:

1. The physical environment
2. The extrademe biotic environment
3. The deme (conspecific local breeding or social group) environment
4. The individual (or internal) environment

Some students (Emerson, 1949) call adaptation to the first three levels "exoadaptation" and to the last "endoadaptation," but the distinction is not very clear (as Emerson notes). In fact most examples of "endoadaptation" seem merely to be anatomical or physiological correlations and integrations which are the means of achieving "exoadaptation." "Endoadaptations" under the name of "coaptations" have been stressed by Cuénot (1951 and earlier, see also Corset, 1931; Hovasse, 1950). In a broad sense the term "coaptation" is defined as an organic character the presence of which makes possible a particular function or sort of behavior. Since all functional characters do this to some extent, the term in this sense is almost meaningless or certainly unnecessary, but it has also been used more specifically for some quite extraordinary phenomena, notably "coaptations d'accrochage," in which two anatomical parts arise separately in the embryo and subsequently fit together and cooperate in a common function. They are especially numerous in insects and include, for instance, the femoral groove of the mantis into which the tibia can be folded or the many cases in which different leg segments fit together as seizing devices.[2]

It is mainly individual morphological adaptation with which this book is concerned. This is the sort most clearly visible in most of the data on long-range evolution, although with improved collecting and broadening viewpoints data for both deme or specific and interspecific

[2] I mention these here to help exemplify the variety and complexity of adaptation. Cuénot considers them evidence of finalism, and Hovasse explains them by organic selection (of Baldwin) or what he calls "parallel selection." It is unnecessary to discuss their origin here, beyond expressing the opinion that they are explicable in the same ways as any other adaptations and do not really constitute an exceptional and distinct class. That two parts should arise separately and later fit together seems to me no more—and no less—mysterious or finalistic than the fact that in epigenetic development *all* functional structures arise with apparent anticipation of their subsequent functioning.

adaptations are increasingly available. In any case, proper interpretation must be made against the broader background of adaptation in general. Much adaptation is physiological or behavioral rather than morphological although, again, the distinction is not really sharp or clear. Physiological and behavioral factors are limited by and correlated with morphology, and in a sense the adaptive characteristic of morphology is seldom strictly morphological but arises from the dynamic, i.e., physiological and behavioral, functioning of the anatomy.

That all of this applies at different levels also involves marked differences in the significance and nature of adaptation. A character adaptive in one respect may not be so in another. Adaptive alternatives may exist such that adaptation to one precludes adaptation to another. Development of armor is, for instance, a frequent adaptation for defense, but it lessens the possibility of rapid motion, which is also a frequent adaptation for defense in other animals and has generally wider usefulness than has armor. More subtle and possibly more important is the contrast between individual and group advantage in adaptation.

That contrast is, indeed, usually absent. An adaptation advantageous to the individual is also likely to be advantageous to the species. It used to be assumed rather generally that this is always true—or the question was not raised at all. This was when selection was understood and discussed in purely Darwinian terms, and Darwinian selection usually (but even it not always) acts for the advantage of the species by favoring individuals of some sorts and eliminating those of other sorts. Even selection on social aggregates generally favors the individual, his integration into the group being favorable to survival and adaptive for him, as well as the group, its social structure being favorable for continuing reproduction of the whole unit. Genetical selection as well as Darwinian selection produces no contradiction between individual and specific adaptation in such cases.

Haldane (1932) has, however, pointed out that there are "altruistic" adaptations that favor the group at the expense of the individual, that is, they shorten the lives of their individual possessors but prolong the survival of the group, as in insect societies organized as if to ensure the survival of only the small minority of reproductive members. Among others, Allee (1943) has also emphasized this situation, and Wright (1945) has suggested a genetical mechanism for it involving small, incompletely isolated populations, with intergroup selection favoring local populations in which the individually disadvantageous gene has become fixed by chance and with gene migration permitting spread of the character from the selectively increasing population. In a stimulating discussion of this whole subject, Huxley (1942, Chapter 8) has also emphasized the opposite effect, i.e., individual adaptation deleterious to the group, exemplified by development of bizarre ornamentation and overelaborate weapons by intragroup selection.

I must confess to a little skepticism regarding some of the examples on both sides, that is, as to whether "altruism" in, say, insects does develop when it is actually disadvantageous for most individuals and whether sexual characters are favored by selection if while the selection is actually operating they are disadvantageous to the group. Of course there may be in both cases a balance of advantages so that selection as a whole carries a character as far as advantages outweigh disadvantages, which may not be quite the same thing as to say that there is selection for the group and against the individual or the other way around. There is clearly also the sort of balance previously noted between narrow variation, favoring survival of more individuals but potentially fatal to the group if conditions change, and wide variation, producing more misfit individuals at the time but favoring ultimate survival of the group. But in such cases, too, one may question looking at the matter as individual *versus* group. It is rather a question of short and long range adaptation. *Both* group and individuals are favored by narrow variation while conditions are not changing, and both are favored by wide variation while conditions are changing.

Judgment as to whether characters are actually adaptive and if so in what way is quite easy in a majority of cases, which are likely to be overlooked just because they arouse no argument. This is usually because the adaptation is widespread and not specific: adaptations for respiration, reproduction, broad types of metabolism or of locomotion, and the like. It is generally only as regards characters of less basic importance, highly specific and peculiar characters, or differences between similar groups that doubt often arises and argument begins. Part of the difficulty lies in the necessarily arbitrary nature of what is called a "character" in taxonomy, which may be a purely incidental result of what is really the adaptive character. For instance, color is one of the commonest of "taxonomic characters" and we are inclined to think that if it is adaptive it is so because of its visual characteristics. Certainly this is true much of the time, but it is not necessarily true all of the time. Desert animals tend to be light in color, as mentioned, and this is generally protective coloration or, and possibly at the same time, it is protective by increasing reflection. But there is also a less common desert color facies of black animals. Are they inadaptive in color? It has been suggested that the color may protect from high-frequency radiation or that it may increase resistance to low humidity. It may actually be disadvantageous as a color, but merely the manifestation of some obscure or physiological character of outweighing advantage. That such can be the case, although with reversed value for black, has been known for many years: melano mice usually have a higher death rate than their siblings of normal color, and surely not because of their blackness, as a color. (See Haldane, 1942.) Similarly obscure correlations occur in nature. Some minor variations in scale characters of snakes, of no evident adaptive significance, are significantly associated with juvenile mortality (Dunn, 1942); the example is especially striking because this is just the sort of taxonomic character often cited as of "no conceivable survival value."

Human judgment is notoriously fallible and perhaps seldom more so than in facile decisions that a character has no adaptive significance because we do not know the use of it. The case of the European banded snails, *Cepaea nemoralis,* will probably become notorious and may cause a swing too far the other way. These snails are highly polymorphic in banding, with considerable stability in local variation and in differences between populations. The polymorphism has long been cited as selectively neutral and as firm evidence of occurrence of nonadaptive characters (e.g., still so cited in Carter, 1951). Recently, however, conclusive evidence has been produced that selection and adaptation are involved (Cain and Sheppard, 1950).

In dealing with extinct organisms, the difficulties may be even greater. Appearance of a character known in recent animals or plants and known to be adaptive in them warrants the inference that it was adaptive in the same way in the fossil group. But fossils sometimes have characters quite unknown in living animals, and evaluation of them is hazardous and often purely speculative. The complex suture lines of ammonites have no recent parallel and there is a widespread idea that they were nonadaptive, although to others the very consistency of their histories implies adaptation, and adaptive advantage is possible, at least, for example as body anchorage and leverage in varying conditions and positions or as resistance to pressures in shells of different shape and size. There are no clawed ungulates today, but such developed several times in the past. In this case few have doubted that the claws were adaptive, but there is no clear consensus as to what the adaptation was. Use for digging tubers or for clinging to trees are among the numerous, inconclusive suggestions. An even more baffling case may really be solved, although there is no possible way to establish this with certainty: extraordinary dorsal fins developed independently in several lines of Permian reptiles (pelycosaurs). Romer (1948, 1949) has shown that these have just the size and other relationships that would be expected if they were body-heat regulating mechanisms.

In view of all the difficulties and obscurities in discovering the adaptive significance of some characters, it is impossible to adduce rigid *proof* for any case that a character has no adaptive significance, either in itself or as an incidental effect of another feature that is adaptive. Since no one doubts that many characters are adaptive and since none can be proved *not* to be adaptive, some students maintain that all characters should be assumed to be adaptive (e.g., Ford, 1945b). Others maintain that the universal affirmative is equally unproven and that with so many characters that are not known to be adaptive and do not seem to be so it is illogical to suppose without proof that all are so. They therefore assign a significant role to nonadaptive change (e.g., Carter, 1951) or sometimes even the major role in evolution (e.g., Goldschmidt, 1940). Theoretical considerations from population genetics, previously reviewed, are cited on both sides of the argument. Those students favoring universal adaptation cite Fisher (with whom, in fact, Wright apparently does not disagree on this point) to the effect that no character is likely to be or, at least, long to remain neutral to the effects of selection, which therefore eventually tends to eliminate any nonadaptive character. Those assigning a role to nonadaptive change cite Wright to the effect that this is likely to occur in spite of selection in sufficiently small populations, and it was noted above that for characters of low selection value (positive or negative) the populations in question may be judged of moderate or even rather large size.

Aspects of this problem of particular importance for the present inquiry center around differences between taxonomic groups and will be somewhat further discussed in that connection.

ADAPTATION AND DIFFERENCES BETWEEN TAXONOMIC GROUPS

It is not surprising to find that opinion as to the adaptive significance of differences between taxonomic groups varies from belief that such differences have no adaptive value to the claim that they are always adaptive. Opinion often depends on whether the groups in question are of low rank, demes, subspecies, and species, or high, families and upward. It is here recognized (or perhaps I should say concluded) that there is no absolute difference of kind as regards characters and their adaptive status in the sequence of different levels in the hierarchy, but it remains possible if not probable that there is a graded quantitative difference between low and high levels. The two will therefore be discussed separately before trying to reach a general conclusion.

There seems to be a consensus that differences between subspecies and species are usually adaptive but sometimes nonadaptive. This is the opinion of both Mayr (e.g., 1942) and Rensch (e.g., 1929, 1947), who have devoted as much attention to the question as anyone.[3] Mayr tends to stress adaptive differences and Rensch nonadaptive, but it is hardly possible to take a complete census and both do agree that adaptive differences are usual but not universal. There is the strongest sort of evidence for the usually adaptive nature of low-category taxonomic differences. This evidence is especially of the following sorts:

1. Distinct but similar and related species fully established in any region have some ecological distinctions of habit, habitat, or both. In especially close cases they may overlap in some niches, but their modal populations are in different niches. For instance it has often been noted that among multiple species of rodents in any given area, unless one is clearly driving another out by competition, each has some distinction of habitat or food preference, or both.

2. There is often an observable correlation of habit and diagnostic structure among related species, e.g., in the Drepaniidae, birds with long beaks and tubular tongues regularly feed on nectar, those with short beaks and nontubular tongues do not (Amadon, 1950).

[3] Both are ornithologists, but both have extraordinarily broad grasp of neontological systematics.

3. Geographic variation generally shows graded and regular distribution of characters and this can often be correlated with environmental factors, e.g., Bergmann's rule that in warm-blooded vertebrates related forms tend to be smaller in warmer and larger in colder regions (e.g., Hesse, Allee, and Schmidt, 1951).

4. Vicarious parallel, and convergent species tend to develop similar characters in correlation with degrees of similarity of habits and habitat, e.g., the same adaptations are found independently developed in reptiles in American and African deserts (Mosauer, 1932), and in each case these characters distinguish them from their closer relatives of different habitat.

So impelling is this and related evidence that one must agree with Mayr (1949a), that "the adaptation of local populations and its causation by selection can be considered a fact" and further (1949b), that "speciation is . . . an adaptive process," but without conceding at this point that it must always and exclusively be so. The question, then, is whether nonadaptive differences also occur and, if so, whether they are common enough to have distinct significance in evolution. Evidence for nonadaptive differentiation is, in the main, the reverse of that for adaptive differentiation, hence, using the same numbers as above:

1. The rule of the ecological incompatibility of distinct groups with very similar or the same adaptive characteristics was based largely on terrestrial vertebrates. Such forms are not highly multispecific in any one habitat, and the rule seems to apply to them without known exceptions. The few apparent exceptions are explicable as involving zones of marginal contact, areas of recent invasion of a competing form, or cases of obscure but real or of periodic rather than discontinuous ecological distinction. Among other organisms, adaptive differences between related, sympatric species are certainly very common, also. I must say that it requires a measure of faith to conclude that they are universal in exuberantly multispecific communities such as, for instance, the gammarids of Lake Baikal, the molluscs or crustacea (among other groups) of a tropical reef, or the trees of a rain forest. Even in such cases faith finds support from such observations as that the gammarids of Lake Baikal do have zonal depth distribution and other adaptive differences, although the evident ecological distinctions are much less numerous than the species. The question seems still to be open,

although with some probability that ecological incompatibility will prove to be a rule for all organisms perhaps with some low-level exceptions.

2. There are innumerable examples of differences between related, sympatric species which do not seem to be correlated with existing ecological differences. Near Los Pinavetes, my home in New Mexico, *Sciurus aberti* and *Sciurus fremonti* are sympatric, similar squirrels. Food habits overlap widely. Both concentrate on available conifer seeds but also eat acorns and a wide variety of other foods as available. But, other things being equal, S. *aberti*, the larger squirrel, concentrates on the large cones of *Pinus ponderosa* and S. *fremonti* on the smaller cones of *Pinus edulis* in its lower range and of *Abies concolor* when it ranges higher. So far this is a typical example of minimal niche separation of sympatric species, but why should S. *aberti* have handsome ear tufts that are quite lacking in S. *fremonti*? Any naturalist can multiply such examples indefinitely. It is always possible, and entirely true, to say that we just do not know the adaptive value of the differences, or to guess at possible adaptive values for which there is no evidence whatever, for instance that the ear tufts are "recognition marks." Nevertheless, the myriads of such instances add up to impressive evidence that species adaptively differentiated also commonly, perhaps even usually, have other, nonadaptive differences.

3. Geographic variation is often highly irregular and with no apparent environmental correlation. Rensch (1947) has cited examples such as that parrots of the genus *Trichoglossus* in the East Indies, Southwest Pacific islands, and Australia have widely variant coloration characteristic of adjacent populations and used in taxonomic definition, but so irregularly distributed over the whole large region as to suggest no conceivable correlation with specific adaptation. Rensch has also pointed out that regularities in geographic variation do not necessarily represent adaptations. Stebbins (1950) has cited cases of plants in which some characters show definite environmental correlation and others in the same population vary quite independently of the environment. Again the negative cannot be proved; geographically irregular differences could be adaptive in some way unknown to us. But, again, their adaptive value is not proved, either, and such cases do strongly suggest nonadaptive features in geographic subspeciation and speciation.

4. The very widespread occurrence of parallelism and convergence

is the strongest sort of evidence for the efficacy of selection and for its adaptive orientation of evolution. Yet, with possible exceptions in some instances of close parallelism over geologically short periods of time, these processes do not produce *identity* even in the limited parts of structures most strongly affected by parallelism and convergence. It is beyond reasonable doubt that the parallelism and convergence as well as the concomitant divergence (on each side) from related forms usually has an adaptive basis. It does not follow (a) that the almost universal differences still present between parallel or convergent structures are adaptive, or (b) that all the differences between a branch involved in parallelism or convergence and its divergent relatives are adaptive. Nor does it follow that apparent parallelism or convergence is clearly adaptive at lower levels, at least. Complex and clear-cut examples familiar to me involve higher taxonomic categories, but one low level example may be mentioned, especially in connection with the last point. In New Mexico very closely related southern and northern populations—often distinguished as subspecies—of, respectively, *Reithrodontomys megalotus* and *Neotoma mexicana,* show parallelism in that the northern forms are larger in both cases. This follows Bergmann's rule and may be accepted as adaptive (although, in fact, it is also rather common over this short climatic cline for southern forms to be larger). But the northern forms are also paler, or a duller gray with less brown, in both cases, a fact with no evident adaptive significance and indeed contrary to the usual rule that on a cline of aridity, present in this case, the paler forms are in the more arid regions. (See Bailey, 1931, and his citations.)

No strong reason has been adduced to question the dictum of Mayr (and many other systematists before and with him) that the *prevailing* factor in most differentiation of low categories is adaptation. There is, nevertheless, strong reason to suspect that nonadaptive differences may also be quite commonly involved along with the adaptive, and even that speciation (if allopatric, as is usual or according to Mayr virtually universal) can occur in some probably rare cases without adaptive differentiation.

With regard to higher categories, the evidence is even less conclusive in some respects, but here there is at least the advantage that changes can more often be followed on the longer time scale. The evidence for the higher categories is often closely similar to that for the lower, but may be more clear-cut in accordance with greater degrees, rather than different kinds, of the changes involved. The evidence again indicates that the characters of a taxonomic category, hence of a phylogenetic unit on this time scale, are usually adaptive. Some of this evidence may be summarized in the same sequence as before:

1. Higher categories also plainly show a broad rule of ecological incompatibility, although this works out more slowly on the longer scale and has special aspects. Sympatric higher categories also regularly have ecological differences, which are usually much more obvious than for lower categories. The degree tends to be closely correlated with the categorical level. In a regional fauna including protistans, platyhelminthes, nemathelminthes, rotifers, molluscs, annelids, arthropods, chordates, in other words almost any land fauna on earth, it is obvious that each of these phyla, in spite of the tremendous variety in many cases, has a characteristic over-all adaptive type and ecological role. Within the fauna, it is also clear that, for instance, the various classes of chordates present, fishes, amphibians, reptiles, birds, and mammals, also have each a general adaptive type characteristic of it and different from any other. Apparent overlaps in type as, for instance, the possible occurrence in the same fauna of aquatic, carnivorous fish, amphibians, reptiles, and mammals are similar to overlap of species in niches not modal for more than one of them, as the stated adaptive type is not modal for any but the fish. The greater variety within higher categories (the wide adaptive radiation of each) causes such overlap, in which ecological incompatibility is actually absent because the lower categories involved (e.g., Chelydridae and Lutrinae) do have sharply different adaptive types even though both may be put under the very broad characterization of aquatic carnivores.

In the dynamics of phylogeny, ecological incompatibility is also shown in another striking way: the repeated replacement of one group by another when they are of even broadly similar adaptive type and come into contact with each other. No example is known between whole phyla of animals because, as I believe, their differentiation was adaptive in a complete and basic way, occurred early and rapidly so that it was essentially complete when the fossil record starts, and precluded any later lethal incompatibility. Radiation within phyla has produced convergence among some of their lesser categories, which has tended to reserve a certain range of adaptive types to each phylum and

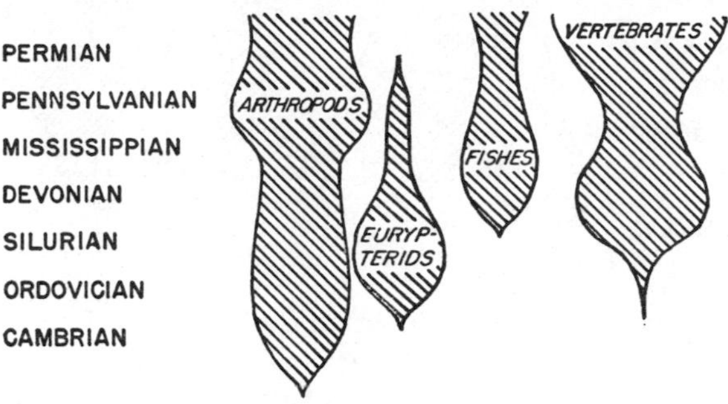

Figure 18. Early Overlap in Broad Adaptive Type between Arthropods and Vertebrates. Arthropods as a whole and vertebrates as a whole have been adaptively distinct, but eurypterids and early fishes covered some similar adaptive ranges and the eurypterids declined as fishes expanded. (The patterns of expansion and contraction are schematic and relative values for any one time are not scaled across the four curves.)

to prevent much or any ecological overlap between them. Thus Arthropoda and Chordata have decidedly different adaptive characters from the start and still have, over much larger ranges than at first. The eurypterids and early fishes covered broadly similar adaptive types and were apparently incompatible, for the eurypterids declined and became extinct as adaptively comparable fish became more abundant, with the result that this particular series of adaptive types, which might have belonged to either, came into the chordate range and remained there (Fig. 18).

Examples of ecological incompatibility are very common within phyla and within all categories below the phylum. Among chordates, "higher fishes," classes Chondrichthyes and Osteichthyes, represented essential adaptive duplication of the "lower," Agnatha and Placodermi, as attested by their incompatibility. Among ferungulates, the orders Perissodactyla and Artiodactyla were largely incompatible. The Perissodactyla, earlier much more varied and numerous, declined rapidly as the Artiodactyla increased (Fig. 19B). Among carnivores, the suborder Creodonta was replaced by the Fissipeda (Fig. 19A). And so on down to the species, with increasing narrowness of ecological specificity. Evidence that the factor is adaptive differentiation of the categories involved is completed by the fact that in supraspecific categories, ex-

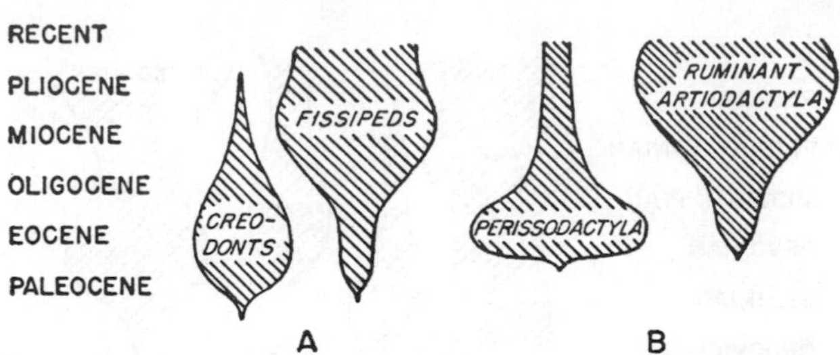

FIGURE 19. ECOLOGICAL INCOMPATIBILITY BETWEEN LARGE GROUPS OF MAMMALS. A, replacement of creodont carnivores by the fissipeds, derived from early creodonts. B, replacement of perissodactyls by ruminant artiodactyls, of independent origin. (Patterns of expansion and contraction are schematic within each group, not scaled as to relative abundance horizontally across the four groups.)

tinction of one group in the area where the two are sympatric is not always complete. When it is not complete, the evident reason is that the diminishing group included lesser categories not in the adaptive range of the other group: lampreys among Agnatha; horses, rhinoceroses, and tapirs among perissodactyls.

2. The diagnostic characters of higher categories almost always have evident adaptive significance and are related to modal differences of habit and habitat between those categories and others at the same level. In general, the adaptive significance of diagnostic characters becomes more obvious the higher the category. This fact is so evident in the thousands of systematic works with diagnoses of groups of varying rank, that it should need neither further exemplification nor argument for a systematist. That it is, nevertheless, sometimes questioned may involve a misapplication of the sort of distinction made by Gregory (1951 and earlier) between habitus and heritage. In its intended application the distinction is real and useful: habitus is the particular adaptive aspect of a group in question and heritage is the adaptive aspect general to a larger category to which the group belongs. The misunderstanding arises when heritage is somehow supposed not to be adaptive, or to be the adaptive aspect of an ancestral group.

The reason why higher categories have an adaptive aspect general or modal for their members is, directly or indirectly, because they have a unified ancestry, but this means neither that the adaptive aspect of the category is now a heritage, adaptive no longer, nor that this aspect is related to the particular adaptation of the ancestry, of its own lower-category characteristics. Indeed the diagnosis of a higher category is not a description of any ancestor or of a type of organization ever embodied in or definitive of an actual organism or population. It is an abstraction of the broader adaptive features present, usually with variations, in members of the category. The opposite opinion is the basis for the typological approach to systematics and the typostrophic (Schindewolf) theory of evolution. The origin of higher categories will be discussed later (Chapter XI). It will there be shown that the adaptive significance of the category may be quite obscure when it arises, because it actually arises as a species, or genus, or other relatively low category, and that it may not then really have any plainly significant characters that can become diagnostic of the higher category. The broad adaptive complex of the higher category is built up little by little as the category develops.

The adaptive nature and relationships of diagnostic characters of higher categories are well illustrated by the Cetacea, classic example of habitus, fishlike, and heritage, mammalian. An abbreviated but adequate working diagnosis of the order (based on Weber, 1928) is:

Obligatorily aquatic mammals with cylindrical or fusiform bodies and transverse caudal fins. Hair reduced to vestiges and dermal glands represented only by conjunctival and mammary glands. Forelimbs paddlelike, without moving joints except at the proximal end. Hind limbs not functional. Anterior end of skull produced; nares dorsal. Intervertebral movement restricted. Teeth usually simple and homodont when present, or absent with baleen present.

That whales are diagnosed as mammals, means that in addition to the above habitus features they have the following heritage characters, among others:

Lung-breathing, homothermous, viviparous vertebrates with hair and mammary glands; double occipital condyle; epiphyses on vertebrae and long bones; mandible of one bone, articulating directly with squamosal; three auditory ossicles; simple shoulder girdle; 4-chambered heart; complete diaphragm.

These characters are clearly adaptive and are as much so for the whales as for other mammals, with apparent exceptions of two sorts. In whales the presence of vestiges of hair and of hind limbs is not adaptive; it is the fact that these structures do not function that is adaptive for whales. To this extent, the diagnosis of the Mammalia represents a modal, not a general, adaptive type among included groups. "Heritage" could be defined to include only vestiges of formerly, not currently, adaptive characters, but then it would not reflect the actual diagnosis of the higher category. At first sight the persistence of functional lungs in the cetaceans seems inadaptive or at least poorly adaptive in fully aquatic animals. This is, however, integrally related with homothermy and other physiological characteristics of clear adaptive value to cetaceans, which are at least as efficient as any other aquatic animals.

3. Some of the same geographical rules apply between genera as between species and subspecies; e.g., Bergmann's rule applies reasonably well to genera of penguins. This sort of evidence from geographic variation is, however, less striking in higher than in lower categories from the mere fact that higher categories tend to cover more territory, within which the geographic variation is reflected at the lower levels of their included groups. In a broader sense, nevertheless, of general correlation of structure with environmental factors, the higher categories show this as clearly as do species and subspecies, although such evidence is little different from that under 2, above. The characteristics of the chordate classes, for instance, obviously correlate well although not fully with their typical environments, and so do those of phyla or of any categories if we get away, for higher categories, from thinking of an environment as a geographically defined place. With greater precision, characters common to all members of a higher category are frequently associated with environmental factors common to all places where they live. Thus wings in birds, excepting the few really wingless birds but not excepting penguins and probably not the so-called "ratites" with developed wings, are fully associated with the presence of a somewhat resistant but fluid medium, usually air but sometimes water, in which they move. The more probable exceptions to such associations apply to characters of ubiquitously adaptive nature, such as homothermy, which lacks evident environmental limitations and therefore associations.

4. The occurrence of parallelism and convergence is frequent at high levels as at low and may be even more striking at high levels. Convergence occurs between members of different categories up to phyla (e.g., humming birds and humming moths: Trochilidae, chor-

dates, and Sphingidae, arthropods). The evidence, again, is quite conclusive that characters distinguishing each of the groups concerned (families, in the example) from their divergent relatives are adaptive.

There is also evidence for nonadaptive differences between higher categories. With a few apparent exceptions, however, mostly related to a problem of logic and definition to be discussed shortly, the evidence seems to me less suggestive than for lower categories and to indicate that there are fewer nonadaptive differences the higher the category.

1. Related higher categories are more generally sympatric than lower categories. Subspecies of one species are almost never sympatric; species of one subgenus or species group usually are not; species of one genus are about as often as not; [4] genera of one family usually are; and so on upward. However, as higher and higher groups are considered it becomes more and more evident that the sympatric forms are adaptively different. What little weight can be given to occurrence of related sympatric groups as evidence of nonadaptive differentiation applies mainly to levels from about species to genus and becomes even less impressive for higher levels.[5] Of course this only suggests that adaptive differences do exist and do characterize the categories in question; it does not show that nonadaptive characters are not also present.

2. Evidence that differentiating characters may not be correlated with differences of habit and habitat is also decreasingly impressive for higher categories, but some evidence of the sort does exist. Most of it is of the kind provided by examples of ecologically convergent groups, such as that of the Paleozoic eurypterids and fishes, if the interpretation is correct that they competed. Since they were obviously and greatly different in structure and yet seem to have been ecologically so similar as to compete, it might appear that the structural differences had no adaptive correlation. This, however, involves a fallacy in the interpretation of adaptive development in parallel or convergent groups, as will be noticed below. The same fallacy and perhaps another (hidden

[4] If, of course, the genus is polytypic so that possibility to be sympatric exists, and so, too, for other categories.

[5] In case the original point begins to be vague on this repetition of sorts of evidence for nonadaptive differentiation, the bearing of ecological incompatibility and sympatric occurrence is as follows: If morphologically and genetically different groups live together in the same adaptive relationships and without advantage to one or the other (advantage shown by experiment and observation almost always to be eventually lethal to the less successful group), then the differences between those groups cannot be adaptive.

adaptive values, see Stebbins, 1950) seems to underlie interpretation of differences as nonadaptive in cases like the mosses, in which peristome characters diagnostic of orders, families, and genera are said to occur in species living in the same way and place (Heitz, 1944).

3. On the whole, as stated in dealing with evidence for adaptive differentiation, the bearing of geographic differentiation is less pertinent for higher categories than for lower. This is equally true of evidence for nonadaptive differences in higher categories.

4. If of two related groups, *a* and *b*, *b* converges toward an unrelated group, *c*, and *a* does not, it certainly seems justified to say that the resulting resemblances between *b* and *c* reflect similarity of adaptation and that the resulting differences between *a* and *b* reflect dissimilarity of adaptation. That is the bearing of convergence on the adaptive nature of taxonomic differences, and it has been noted that this evidence is even stronger for high than for low categories. The question of nonadaptive change in such cases arises from the fact that *b* commonly develops differences from both *a* and *c*. Then these differences from *c* are not adaptive resemblances to *a* due to common ancestry, and the differences in the same characters from *a* cannot be ascribed to adaptive resemblance to *c*. In all probability they must either be nonadaptive or represent adaptive divergence both from the ancestry and from the otherwise convergent type.

Such cases are very common, indeed universal for some characters in cases of convergence between distantly related groups. For example, in one of the most spectacular cases of convergence known, between the marsupial *Thylacosmilus* (Riggs, 1934) and the placental sabertooths, Machairodontinae, *Thylacosmilus* has numerous characters that occur neither in the ancestral and contemporary borhyaenids nor in the normal sabertooths. One of these characters, the complete postorbital bar, was later found in some rare and aberrant sabertooths. Others, like the great maxillary swelling for the canine roots, are related to extreme sabertooth adaptation even though absent in the sabertooths themselves. Still others, such as the absence of incisors, the extent and form of the jugal, and the placing and bridging of the infraorbital foramen have no obvious relationship either to the machairodontine convergence or the borhyaenid ancestry—but, even as regards these characters and the many analogous ones arising in other convergent groups, it is quite impossible to say that they are not adaptive and it is even probable that they are. There is no such thing as pure convergence. The characters of a convergent form include: adaptations or vestiges of them derived from the ancestry, convergent adaptive characters, divergent adaptive characters with lack of full identity in adaptation of convergent types, and, as a category that probably must always be doubtful in each case, perhaps some nonadaptive characters.

The evidence for nonadaptive differentiation of higher categories is found to be very weak. The general conclusion is that nonadaptive changes may be and probably are involved as a minor part of the differentiation of low categories, demes, subspecies, species, more or less up to genera, but that such a process plays a decreasing part as the taxonomic scale is ascended and is quite unimportant if not wholly absent at high levels.[6]

There remain, nevertheless, a large number of examples of differences between groups, low and high taxonomically, which seem most probably to be nonadaptive as *differences.* To conclude that the characters involved or that their evolution is nonadaptive is the fallacy to which reference has previously been made. I have elsewhere (1949a) discussed some of these cases and have called the general phenomenon that they exemplify "opportunism" in evolution. Rensch (1947) has given some of the same and other examples as showing a disorientation or "directionlessness" ("Richtungslosigkeit") in transspecific evolution.

One of the examples previously discussed by both Rensch and me is the extreme variety of species- and genus-specific forms of antelope horns, with no evident or, one might almost say, possible relation to corresponding differences in other characters or in way of life of the animals. Another is the variety of anatomical dispositions for light reception in various genera, families, classes, and phyla, for instance the marked anatomical difference between the functionally almost identical eyes of vertebrates and some cephalopods. A zoological example perhaps even clearer in significance has been provided in great detail by Sperber (1944). On exhaustive study of the kidneys of some 140 species of mammals, he found that there is a primitive type still extant and at least three derived types. The derived types arose, often in parallel

[6] One other premonitory word is required here: the actual appearance of a character may be nonadaptive as of then even though its involvement in the origin of a taxonomic category, low or high, is adaptive. This is a form of prospective adaptation, discussed in the next chapter.

and at random as regards which type would appear in any given line, in large and in aquatic mammals. Such mammals tend to have the primitive type changed into one or the other of the derived types, but there is no evidence of any adaptive correlation or significance of differences between the derived types. On the botanical side, Stebbins (1950) has shown that certain combinations of characters of flowering plants tend to occur together and to have ecological correlations, hence are almost certainly adaptive, but that some taxonomically basic differences between high categories seem to be random, as if their differentiation had not initially been adaptive.

In such cases and many like them it is probably quite correct to say that the differences between characters of various groups are non-adaptive. Nevertheless the statement is misleading and can result in a fallacy that is really quite obvious and yet that I have never seen clearly stated and that has confused me, at least, in the past. The point is whether the selective and evolutionary process involved is to be considered as operative in parallel on *changes* in characters or divergently on *differences* in characters between two groups. Intergroup selection acting by differential reproduction between two competing demes of a species or between two related and competing species may be conceived of as acting directly on differences between characters. If both groups survive, they will diverge and not only the characters developed in each but also the differences between those characters will be adaptively oriented.

On the other hand, if selection in general widely bears in the same direction within various groups but selection is not intergroup, it will not select between their characters or act on the differences in these. It will in each case and independently select for any existing variation in the direction of adaptive change. The ultimate sources, mutations, are random with respect to selection, will not be homologous in all populations, and in any case will appear in a different genetic environment in each. In this situation, every favorable variation will be selected for, but aside from one element of direction the variations may be different in the different groups and usually will be. These differences are then perpetuated as a secondary result of selection to which the differences as such are irrelevant.

There was no selection as between a cephalopod eye and a vertebrate eye producing one more adaptive for an octopus and another more adaptive for a fish. Both arose quite separately by entirely adaptive change within the group, and the difference between them arose only from the fact that different genetic materials were exposed to selection. The difference between the two adaptive characters is simply irrelevant as regards adaptation. It is opportunistic in my earlier, perhaps too anthropomorphic phrase. Similarly Sperber concluded regarding derived mammalian kidney types that all three are equally adaptive and all developed by selection; which developed in a given line depended only on what mutations happened to occur. That the differences between them are (in all probability) literally nonadaptive is irrelevant to the actual process of their evolution.

FUNCTIONS [7] OF ORGANISMS AND ENVIRONMENTS

Adaptation is a fitting together of organisms and, in the very broadest sense, environments. It is false to ascribe this action either to the organisms or to the environments. Both organisms and environments have their characteristic actions, in other words functions, in the process, and adaptation is a characteristic interaction of the two. That the physical environment has its own "fitness," that it is adapted to life as well as life to it, or that adaptation is a reaction between the two and not of one to the other, seemed a radical or even a downright subversive idea when it was promulgated by Henderson (1913). Henderson's pioneering work lacked clarity on many points, both scientific and philosophical, and it failed to find an immediate place in evolutionary theory, but through the years its basic ideas have crept into our thinking almost unperceived. This has recently been noted by Blum (1951), who has wholeheartedly revived Henderson's views and has fitted them to established modern knowledge of physics, chemistry, and biology.

The physical properties of the universe are for the most part uniquely

[7] I use the word "function" quite freely in spite of the fact that some students have criticized this use on the grounds that the word has an undesirable metaphysical or teleological tinge. A dictionary definition of "function" is "the natural or characteristic action of anything." That organs, organisms, and environments have natural, characteristic actions is obvious. Calling these "functions" is a simple and clear way of talking about them and should carry no implication as to how things came to have their functions. This whole book attests that I do not believe that functions imply prior purpose. Muller (1950) decries current leaning over backward on the subject of teleology. An eye *is* made for seeing, and this is no accident even though no one or no One planned it in advance.

"fit" for life as it actually exists here. This no more indicates that the fitness was intended than does the reciprocal and more commonly noted fact that life is fitted to these physical properties. They are the matrix in which life developed and continues to exist. If they had not evolved as they did, life would not have evolved when, where, and as it did. An alternative might have produced some other sort of self-reproducing systems, may even have done so elsewhere in the universe, but life as we exemplify it and observe it is the product of a particular set of material properties and historical events. The interaction extends from an unknown beginning in time down to now and from the most general properties of energy down to the most detailed circumstances of any single point on the earth.

It has already been noted that environment, as the complex of conditions in which an organism exists, can be artificially analyzed into physical, extrademe biotic, intrademe, and individual or "internal" components but is really an inseparable combination of all of these. It is partly alive and partly not, but even this is a transition and not a clear separation. It changes, in other words evolves, along with the organisms that are in one sense in it, in another sense part of it. There is not simply a given environment to which organisms adapt. Their own activities change the environment and are part of the environment. Grossly, it is clear that the action is integral and reciprocal when a man or a beaver purposely changes a landscape. It is scarcely less clear when a tree falls or an animal dies. More intricately, environments are seen to be created when vascular plants spread over the land and to be changed by such slight actions as the movement of an animal from one place to another.

The nature of the interaction of organism and environment, even in the limited sense of physical environment, and the fact that it is an interaction are strikingly illustrated in a highly special and concrete way by Anderson's observation (1948) that hybridization in plants depends in large measure on "hybridization" of the environment. Different species are normally adjusted each to a different and relatively uniform environment, but f_2 and later hybrids have a great variety of types potentially adaptive for environments different from those of the parental species. The hybrids are likely to persist only in "hybridized" habitats, i.e., either highly heterogeneous or without established plant associations and therefore open for development of new heterogeneity by the organisms themselves.[8]

That adaptation is not simply something imposed on the organism by the environment is evident not only from cases in which organisms change the environment but also from those in which organisms seek out an environment suitable for them or somehow preferred by them. Highly mobile animals such as birds frequently do this in a very clear way. For instance, different species may winter together at low altitudes and then return to summer each in a characteristic life zone and plant association in the mountains. It has been repeatedly suggested, first by Baldwin (1896), Morgan (1896), and Osborn (1896), that adaptation may start as a matter of social preference in what we now call a deme or as an acquired reaction in individuals and may later become genetically fixed by selection. This process, under the name of "organic selection," "stabilizing selection" (in one of its aspects), "parallel selection," "the Baldwin principle," etc., has been greatly emphasized by some recent authors (e.g., Schmalhausen, 1949; Hovasse, 1950), some of whom seem to assign it a far wider or more fundamental role in adaptation than the evidence really warrants. It is, nevertheless, an interesting facet and example of the multiplicity of adaptive processes and of the fact that all are organism-environment interactions affected ultimately by selection.[9]

It has already been concluded and will be further exemplified and argued in later chapters that most evolutionary change is adaptive. It follows that most dynamic evolutionary processes, including, as I believe, all those of long-range persistence and importance, involve changes in the organism-environment interaction. An essential factor is the entrance, figuratively or literally, of populations into environments that are in some respect new to them. The organisms concerned may

[8] Anderson does not put the matter in just this way, and it is not entirely clear to me how newly emergent land, for instance, is supposed to have provided habitats more heterogeneous than those already colonized. Offhand, it would appear that such areas would be less heterogeneous, but I take it that hybridization in environments not already organized into tightly closed biotic associations would itself produce heterogeneity of a new sort.

[9] Lutz (1948) has discussed examples of habitat selection among frogs which are interesting in this connection. It should, however, be noted that her definition of organic selection as selection of the environment by the organism does not agree with all the authorities she cites, and it is somewhat doubtful whether this is the operative factor in all her cases.

not actually go anywhere; often if not usually the new environment comes to them, so to speak. Some (only) of the ways in which such changes occur may be listed and exemplified as follows:

1. Local changes of environment:
   a. Physical, e.g., change of climate, tectonic movement, etc.
   b. Extrademe biotic, e.g., incursion of new food organisms, competitors, or predators
   c. Intrademe, e.g., by a self-braking trend toward larger size
   d. Individual, e.g., by a lesion or a change in metabolism
2. Physical movement of population:
   a. Expansion or spread, e.g., by secular or periodic increase of population
   b. Geographic movement into newly opened territory, e.g., onto emerging coastal lands, across a new land connection such as that between North and South America, or by chance dispersal such as that to the Galapagos or Hawaiian Islands
   c. Ecological movement into newly opened adaptive zones, e.g., into a niche vacated by extinction or divergent evolution of another species
3. Evolutionary movement of population:
   a. Adaptive change, most strikingly when this crosses a new threshold, e.g., tooth and correlated trends reaching a point where change of diet is possible
   b. Nonadaptive change by chance adaptive in another available environment, e.g., mutation giving salt-tolerance in fluviatile fish

It will be noticed that none of these categories is absolutely clear-cut and that they are not mutually exclusive. They grade into each other and any combination or even all at once may be involved in a given case. There is some argument as to the role of the last sort of change listed (3b), a point for special discussion below. The others certainly have all occurred and continue to occur with great frequency and all have certainly been essential elements in evolution. Their variety and ubiquity are such that an environment completely stable for any considerable period of time is inconceivable and adaptive change is the rule for all sorts of organisms. The degree and nature of the environmental change and of the adaptive response differ greatly in different cases, however, and exceptions to the rule do occur.

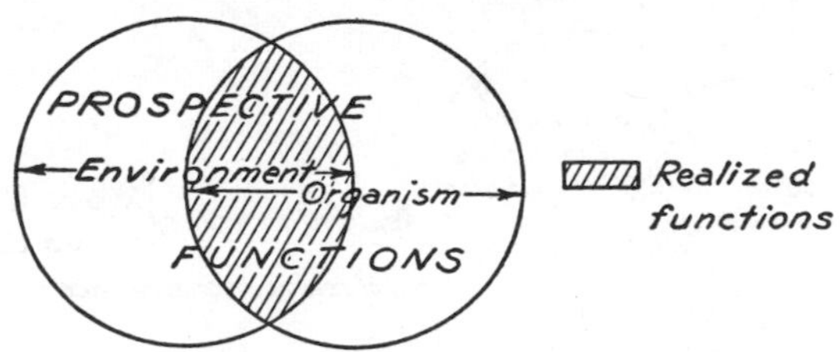

FIGURE 20. DIAGRAM OF THE REALIZATION OF FUNCTIONS BY OVERLAP OF PROSPECTIVE FUNCTIONS OF ORGANISM AND ENVIRONMENT.

In this intricate mesh of adaptive relationships as they exist momentarily and change in time, any population has distinct functions actually operative and others inoperative but possible to it. The environment also has functions some of which are operative, involved in an existing adaptive relationship with a population actually present, and others which are prospective, possible but not currently existent. We may, then, join Parr (1926) in speaking of prospective and real (or, I would prefer to say, realized) functions of organism and of environment.

Realized functions of organism and of environment represent the extent of existing overlap between the prospective functions of each (Fig. 20). This is the essence of their interaction in adaptation. The static picture is that revealed by descriptive ecology as applied to the present or to any point in past time. This is, of course, only a momentary state in a fluid historical sequence and can be understood only as part of such a sequence. The extent of realization of functions tends constantly, although usually slowly, to change. Prospective functions have a certain short-range stability, but also tend to change; in fact, change in their realization normally changes their over-all extent and nature. As of any one time, the prospective functions actually in being, so to speak, are represented on one hand by an existent, although not fully occupied, range of environmental conditions and by an existent, although not fully expressed, pool of variability in a population. Both change. Indeed a change anywhere in the system has changes everywhere in it as a concomitant.

Many important generalizations of descriptive phylogeny can be well expressed and better understood in these terms.

Sharks and whales realize some identical or closely similar (and some quite dissimilar) functions of the marine pelagic environment. The

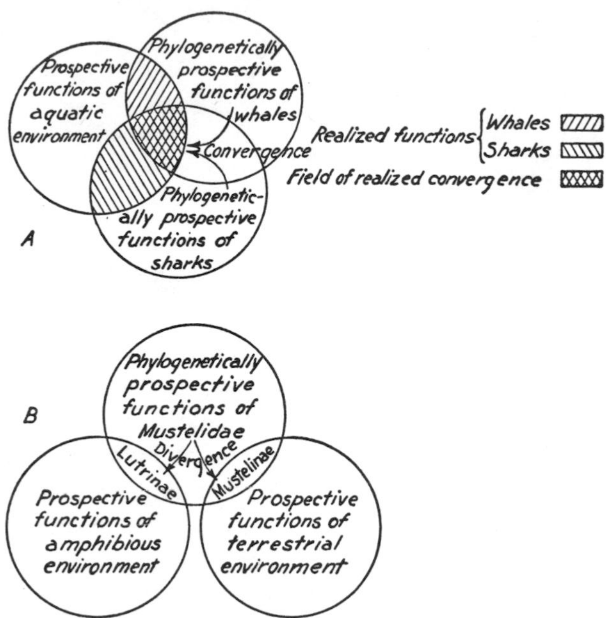

FIGURE 21. CONVERGENCE AND DIVERGENCE IN EVOLUTION OF PHYLOGENETICALLY PROSPECTIVE FUNCTIONS OF ORGANISMS THROUGH OVERLAP WITH THE PROSPECTIVE FUNCTIONS OF THEIR ENVIRONMENTS. A, Convergence in evolution through overlap of some of the same prospective functions of an environment by phylogenetically prospective functions of two different groups of organisms; B, divergence in evolution through overlap of prospective functions of two different environments by the phylogenetically prospective functions of a single group of organisms.

primitively aquatic ancestors of the sharks could and did develop in divergent ways, among which one led to a realization of particular whalelike functions, just as among the primitively terrestrial whale ancestry one line of change in prospective functions led to the realization of sharklike habitus. The triple overlap of functions in the two types of organisms and the one type of environment has an area common to all three, which is the zone of convergence in sharks and whales (Fig. 21A).

The realization of divergent functions arises from the overlap of the phylogenetically prospective functions of one group of organisms on the prospective functions of two different environments. Examples are innumerable; for example, the fission of the Mustelidae into Lutrinae and Mustelinae is typical and clear (Fig. 21B). The ancestral mustelids did not have lutrine prospective functions in a static sense but they had these in the dynamic sense that they were capable of developing them in the course of subsequent phylogeny. Development of such functions in two of the environmental conditions available to the group, amphibious and terrestrial, was the basis of divergence of the two subfamilies.

In general, it is the complex of functions of organism and of environment that determines both the direction and the rate of evolution, and the mechanism of such determination is usually natural selection. Evolutionary change in realized functions must be such as to maintain an overlap between prospective functions of environment and organism. Failure in such overlap is the sole and all-embracing cause of extinction. Again, failure results from the interrelationships and does not come from one side or the other, but in analysis the failure may be seen as it separately concerns the two prospective functions. Decrease in prospective functions of the available environments or change in their nature (figuratively, their position in the diagram) requires, for survival, increase in prospective functions of the organism, as by increased variation, or movement (evolutionary change) in these.

The wider the overlap, the greater the change that can occur either in the environment or in the organisms without extinction. This is a description of some aspects of specialization and of its effects in adaptational terms of functions. A population with wider overlap, realizing more functions of the environment, is more broadly adapted, or covers a wider ecological range, or in one usual sense of the word "specialized" is less specialized than one with narrower overlap. As a rule, the population with wider overlap will have greater evolutionary stability and will evolve more slowly. Change, even rather extensive change, in the functions of the environment may restrict some of the variation of such a population but will not necessarily cause significant change in its modal characters. More narrowly specialized populations, however, usually have an adaptive advantage and tend to usurp or at least to restrict or shift the range of realized functions of a more broadly adapted group when ranges overlap with ensuing competition.

Schmalhausen (1949) has suggested a relationship between specificity of adaptation, consequently of tendencies toward different average rates of evolution, and position in the hierarchy of nutrition and means of defense. At the bottom of the chain of nutrition and in forms without means of defense, predation is said to be nondifferential and selection to favor fecundity rather than more specific adaptation. Higher in the chain and in forms with passive defense, predation is more selective, adaptation more specific. Near the top of the chain predation is ineffective and competition produces overspecialization. The argument is too briefly summarized here, and it does suggest one among many factors involved in realization of functions and in rates of evolution, but even in Schmalhausen's full discussion the categories and claimed relationships seem rather unrealistic. To mention only one point, he considers Darwinian selection only and mainly selection by predation as bearing on relative fecundity, but it is quite clear that in many or most cases fecundity is strongly affected by kinds of genetical selection that have nothing to do with predation.

Stebbins (1949, 1950) has amplified Schmalhausen's scheme in, I would judge, a more realistic way to apply to a hierarchy of reproduction in plants. At the bottom of the scale, least specific in adaptation and with generally slower rates of evolution, are plants relying for reproduction solely on the large number of gametes and zygotes produced (many bryophytes and homosporous pteridophytes, some gymnosperms and angiosperms). In the middle of the scale are many plants, from algae to flowering plants, with large, heavily coated, resistant spores and seeds. At the top of the scale, with fast evolutionary rates and great diversity and plasticity, are angiosperms that rely on animals for pollination, fruit dispersal, or both. It is noteworthy that the scale is not taxonomic. All three reproductive levels occur in angiosperms. In grasses, wind pollination is accompanied by relative uniformity in flowers but animal dispersal of seeds in the same groups accompanies great diversity in scales and rachises, structures of the inflorescence related to seed dispersal. Such relationships underline the extreme intricacy of the processes we are trying to analyze and, no doubt, the inadequacy of any understandable analysis.

PROSPECTIVE ADAPTATION

The term "preadaptation" has been applied to a great variety of real or supposed evolutionary phenomena, from the appearance of a small mutation with selective value in the population in which it occurs to the sudden appearance of a form monstrous in its parental population but miraculously, one might almost say, adapted to some quite different way of life. A term used in so many different ways and, moreover, associated in many minds with the now usually rejected mutationist theory is of rather dubious status. Many students have discarded the term entirely. Others, apparently still a majority, continue nevertheless to use it and have described under this name many evolutionary sequences of evident importance, especially in long-range evolution and as regards higher categories. The phenomena referred to are real (whether or not their interpretation is correct in any given case) and they are important for our subject. It is therefore necessary to discuss them. The reader may call any, all, or none of them "preadaptation" as he chooses. I shall use the term in the general sense of previous authors in discussion of their work, but in order to avoid implications that still cling to it I now propose that most of these phenomena be called by the name, more neutral at present, of "prospective adaptation."

In order to realize the new functions of a changed environment, an organism must, at the moment when the change or the occupation of a new environment begins, have at least some functions prospective with regard to the new environmental functions. This is merely a more technical way of saying that organisms can continue to live only under conditions to which they are already at least minimally adapted. Prospective adaptation is not a quite distinct sort of adaptation or even a clear-cut analytical category of adaptations. It is an aspect of adaptation in general or, if you prefer, a direction of looking at adaptation: it is the possession of characteristics making possible a change in adaptation and on which the new adaptation is built. Probably all organisms except the most extremely narrowly specialized, and perhaps even those, do have characteristics that would permit some adaptive change.

In this very broad sense, which is no broader than that used by some discussants of preadaptation, preadaptation is practically universal. As prospective adaptation, it is to be recognized as a general rather than a special feature of the adaptive process. There is no real reason to discuss it unless a change in adaptation has, in fact, occurred in a given case. Then prospective adaptation involves looking backward, *ex post facto*, to the foundation and preliminary stages on and by which new adaptation arose. In the strictest sense, one to which the term is universally applied by those who use it but to which few now restrict it, "preadaptation" involves the rise of characteristics nonadaptive or inadaptive when they appear but adaptive in some different, available environment. This would be one quite special sort of prospective adaptation. The arguments about preadaptation largely concern the question whether this phenomena does occur, and misunderstanding in use of the term arises mostly from failure to define whether this or some other sort of prospective adaptation is meant.

Consideration of a large number of examples labeled "preadaptation" by one student or another, shows that most of them can be placed in the following nine categories.[10]

1. A broadly adaptive character may without change in itself permit a change or a narrowing in realization of prospective functions. For example, Cuénot (1951) cites as preadaptive polyphagous species which can change from one food to another or the descendants of which can become oligophagous or monophagous.

2. A variant arising in a population may be in the direction of an existing trend. This is simply the role of mutation and variation in all trend evolution, e.g., in increase of size among Oligocene and Miocene horses.

3. A variant arising in a population may be favored by selection in existing circumstances and may initiate a new trend. These are the so-called "key" mutations, e.g., the mutation for claws in stylinodontine ancestors postulated by Patterson (1949), see Chapter IV.

4. A specific adaptation in one ecological relationship may, without change, be requisite in another. This is a large class for which many more or less clear examples have been suggested. Pittendrigh (1948) concludes that plant adaptations to dry, impoverished desert soils are preadaptive for epiphytic habit. Stebbins (1950) cites adaptively variant plants with small leaves and hard seeds as preadaptive for increasingly dry climates. Lungs were developed in aquatic vertebrates and were, in this sense, preadaptive for land life. Carter (1951) believes that development of blood relatively insensitive to CO_2 was prerequisite to development of lungs and arose as an adaptation by gill-breathers to stagnant waters. Eigenmann (1909) maintained that evolution of nonpelagic eggs by marine fishes was a prerequisite to adaptation to

[10] Davis (1949) has recently proposed a different, less inclusive but not necessarily contradictory list.

fresh water, but whether this was adaptive in the marine fishes has been questioned. Cuénot (1951) also thinks that viviparity was nonadaptive in the terrestrial ancestors of the marine hydrophids and preadaptive to marine life, but viviparity seems clearly adaptive in some terrestrial snakes, and as a matter of fact there are oviparous hydrophids, which strongly suggests that their viviparity was secondary to marine life, not a prospective adaptation for it.

5. A specific adaptation in one situation may by intensification become adaptive to another situation more rigorously selective in the same direction. The penguin *Aptenodytes forsteri* has a whole series of adaptations to the extremely rigorous conditions of the Antarctic pack ice (Murphy, 1936). These seem in the main to be intensifications of adaptations found in its relative, *A. patagonicus,* which also lives in severe but decidedly less rigorous conditions less far south.

6. An adaptive structural modification may without essential change serve a different function in a different adaptive relationship. The broad type of digging foot may in some cases become used for swimming (*Ornithorhynchus,* some moles), and the narrow type may be used for climbing (edentates, see Simpson, 1931b). The parrot *Nestor notabilis* of New Zealand formerly fed on insects and vegetable matter and now occasionally uses the same adaptive structures to feed on live sheep (Elton, 1933).

7. Adaptive structure may by essential continuation or intensification of a former morphological trend come to serve a different function. The seal *Leptonychotes weddelli* has enlarged incisors which it uses to cut breathing holes, permitting it to winter under the Antarctic ice. (It ranges farther south than any other mammal.) Relatives farther north which do not cut holes in ice have similarly but less enlarged incisors (Bertram, 1940; the example is given at greater length and discussed in Simpson, 1944a). I would also place here Davis's examples of the origin of rays from sharks by intensification of trends connected with equilibration and resting on the bottom and of the origin of the herbivorous panda by intensification of trend from carnivorous to omnivorous diet reflected in the dentition and other characters. The change from browsing to grazing in horses (Chapter V) seems also to be in part an example of this sort, although also involving new trends.

8. Structure adaptive in one way in an ancestral group may be transformed to serve quite a different function in a descendant group. This

is one of the abundant, striking features of evolution and many examples are familiar to everyone: transformation of the reptile walking limb to the bird flying wing or of the mammalian walking limb to the bat flying wing or the cetacean swimming paddle; transformation of reptilian jaw articulating bones to mammalian auditory ossicles; transformation of breathing lung to hydrostatic bladder [11] in fishes and later incorporation of the bladder in the auditory apparatus; and many others. This sort of change has seldom been called preadaptive, but it is not sharply distinguishable from 6 and 7, usually called preadaptive and equally involving change of function, or from 3, sometimes called preadaptive and equally involving change in direction of evolution. There is in such cases as in any change of adaptation an aspect of prospective adaptation.

9. A character nonadaptive or inadaptive in an ancestral group may be adaptive in a descendant group. This is preadaptation in the strictest sense of the word. Many examples have been proposed, but in every case it is impossible to be sure that the example does not really belong in one of the preceding classes. "Mutational preadaptations," so called and exemplified by Huxley (1942) include such cases as heat-resistant mutants in *Daphnia,* cold-resistant polyploid plants,[12] or melanic mutants in mammals possibly favored by selection in damper, cooler climates. To the extent that such mutations have actually been incorporated in subsequent adaptation, I see no way to distinguish them clearly, if at all, from 2 and 3, above, or indeed from any favorable mutations, ultimately the raw material for genetic adaptation of any sort whatever. Another sort of example almost always cited in this connection and thoroughly reviewed by Hubbs (1938) is that of blind cave fish, believed to be recruited from forms already having inadaptively poor sight or perhaps sometimes to arise by inadaptive mutation causing blindness in members of a well-sighted population. Evidence of this sort, which includes also the origin of adaptively flightless island insects from supposedly inadaptive poor fliers or wingless mutants, is as impressive as any for the occurrence of "pure" preadaptation, but it is still inconclusive. Proof is lacking that the supposed "weakness" of

[11] This transformation is almost always cited as going in the opposite direction, but see Romer (1949b). In either direction it was a transformation of the sort here discussed.

[12] But the widespread impression that polyploids generally withstand colder climates is probably an error (Stebbins, 1950).

the poorly sighted or poorly flying ancestors was in fact inadaptive. If, as is entirely possible, it was adaptive, it would belong under 5, above, and would be a more usual, less controversial, sort of prospective adaptation. Most of the literature seems to involve an anthropomorphic tendency to think of blind fish as somehow disabled and of loss of sight as a degeneration caused by lack of use. It seems to me quite probable that there is selection *against* functional eyes in situations where poor sight or blindness do prevail and that the latter are definitely favored there. Even the anthropomorphic viewpoint should support this: in total darkness a blind man certainly has an advantage over a sighted one, and in dim light a man with poor vision is likely to have an advantage over either; he has visual clues denied the blind and the ability to coordinate dim visual clues with others not developed by the sighted. Goldschmidt (1940) has given a number of other examples which he thinks explicable only by single large mutations, nonadaptive or inadaptive as they occurred, preadaptive as they turned out. Typical of these is the origin of flatfish from normal, bilaterally symmetrical fish, but Norman (in Huxley, 1942) has suggested that this is quite explicable by selective, adaptive processes acting on less radical mutations.

As is so often the case in analysis of complex natural phenomena, the categorization in this list is not clear-cut. Some examples could about as well be placed in any of two or more categories and some might be placed in more than one at the same time. Even aside from disagreements as to interpretation, the phenomena intergrade. In sequence, I have followed more or less (it cannot be done exactly) the order of implied latitude in definition of "preadaptation." Practically everyone has called 9 "preadaptation," although some have denied that it occurs. No one but an extremist like Cuénot has called 1 or 2 "preadaptation." Cuénot, indeed, went to lengths that may fairly be called ridiculous, as when he called the tendency for men to go in for one sport or another "préadaptation des sportifs" (1951). Extreme latitude in use of "preadaptation" while still implying or stating that it is something different from "adaptation" has, indeed, created a supposed problem that is semantic rather than real, a "Scheinproblem."

Recognition that prospective adaptation is an aspect of any *change* in adaptation meets the objection that "preadaptation" is a "Scheinproblem," and permits clearer attention to the real problems involved in the various different ways in which prospective adaptation arises and change in adaptation occurs. In spite of the fact that really clear-cut distinctions cannot be made between all nine of the categories of the preceding list, it is clear that they do involve processes that may be quite distinct. An apparently important distinction is that in classes 1–8 the essential features seem to be adaptive throughout, that is, the characteristics that were prospective adaptations with reference to the descendant populations arose adaptively in the ancestral population. This is postulated as untrue of class 9.

Even this problem of the adaptive or nonadaptive origin of prospective adaptation reduces in some cases, at least, to a distinction without a difference. Very few students of evolution, and among them only Goldschmidt and Schindewolf (as earlier cited) with any considerable insistence and discussion of evidence, believe that marked changes in adaptation usually or normally occur by the one-step origin of a new genetic and developmental system. The only process known to have such an effect is polyploidy, and it seems to have had no great importance in major features of evolution and not to have been involved in most cases of adaptive change, nor is it what either Goldschmidt or Schindewolf has in mind.

It is generally agreed that *new* characters involved in adaptation, when they arise or later, appear by mutation of some sort and that mutation is not in general adaptively oriented. If this were all that is meant by the nonadaptive origin of prospective adaptation or "pre-adaptation," there would be no point of fact involved; nonadaptive vs. adaptive origin would be another "Scheinproblem." Whatever the mutation involved, it has not really led to or become involved in adaptation until it has become fixed in the variation pool of a population and then spread to become characteristic. There is a strong consensus that fixation and spread resulting immediately from the appearance of a "hopeful monster" do not occur, and we shall see further that this process is not required to explain anything in the record of evolution. The only ways in which fixation and spread seem even likely are (a) by positive selection, in populations of any size, or (b) by random processes in small to moderate populations, depending on a previously explained relationship between population size and selection.

If a mutation, whether adaptive, nonadaptive, or inadaptive with respect to the adaptation of an ancestral population, does become fixed and spread by selection it is adaptive *from the start* with respect to the descending populations. The prospective adaptation involved then does not really belong in 9 of the list, but in 2, 3, 5, 7, or 8, depending on just what sort of adaptive change eventuates. (1, 4, and 6 involve adaptive change without requiring genetical change, hence do not *necessarily* involve new mutation at any point.) I do not see that the "size" of the mutation or the exact way in which its results are manifested (e.g., early, middle, or late in ontogeny) makes any difference in this respect. In the process of selection, a "large" mutation is less likely to be positively selected, and hence to be involved at all in such processes, but when it is, its spread is just as adaptive as that of a small mutation and occurs on the same basis of selective value from the start. "Large" mutation is likely to require more rather than less selective action on other genetic characters, but in any case a considerable change in adaptation is almost certain to involve many genetic characters and not alone what was originally a single mutation.

The other possible process for fixation and spread of mutation, regardless of selection, is extremely unlikely ever to occur in very large populations, but may occur in moderate populations if the selective value (either positive or negative) is low and in very small populations even if the selective value is considerable. Since mutations of negative selective value are more likely to be involved when the process is random with regard to selection (there are more of them), the process is more likely to be degenerative and to lead to rapid extinction, as Wright has insistently repeated. If genetic factors fixed at random do in fact become involved in successful adaptive change, i.e., if they really are prospective adaptations in reference to descendant populations, the phase during which they are nonadaptive or inadaptive must usually if not always be very short in terms of geological time. Their selective value for the descendant populations must soon begin to be effective, and no really sharp and strong distinction from such a process as 3, above, may appear. The difference in the process of earliest fixation may, however, be important in particular cases, because the fixation of a particularly large "key" mutation or of a very strongly expressed combination of mutations may be more likely in this way than in fixation selective from the first step. It may be still more important that even with small mutations or with mutations that would have been selected from the start if selection had been effective, random fixation and spread can be more rapid than purely selective fixation and spread.

Many of the processes of sequence from prospective adaptation to new adaptation involve a threshold effect. The threshold occurs at the point where linear selection for the new adaptation begins to be effective against centripetal selection for the ancestral adaptation or against the random processes by which the prospective adaptation was fixed, in the relatively far fewer cases in which random processes were essentially involved. A threshold in this sense is not clearly involved in 1 or 2 of the list. In 3, since the variant is postulated as adaptive in the new direction from the start, a threshold is not clearly involved at that point, although one involving the whole genotype is likely to occur as modifying mutations and polygene combinations begin to come into the selective field of the shift. Patterson (1949) considers that no threshold was involved in the "key" mutation for claws in stylinodontines but that the whole shift from the ancestral conoryctine adaptive zone did come to involve a threshold in accompanying dental and other changes. (I believe, however, that even the key mutation did cross a threshold, see Chapter XII.)

Classes 4 and 6 of the preceding list do not necessarily involve a primary threshold since genetic change is not required in the initiation of such adaptive shifts. Once the shift has begun, however, there is commonly a simulation, at least, of threshold effects by relatively sudden onset of strong selection for secondary characters useful in the new environment or for radiation in it. In 5, 7, 8, and (to the extent that it is distinct and does occur) 9, a true, primary threshold must normally if not always be involved, that is, there must be a fairly well-defined point where linear selection for *new* adaptation becomes effective and the population begins definitely to evolve in that direction.

Even when the process is adaptive throughout, with respect to either ancestral or descendant populations, as I believe it usually is,[13] there is a period of instability and disequilibrium when a population is vary-

[13] This was also the opinion expressed plainly, as I thought, in Simpson (1944a), although I have since changed the way of putting it and some points of emphasis. One student criticized that work for saying that preadaptation is always adaptive and another for saying that it is usually inadaptive. It is difficult to convey a balanced opinion. Readers with strong preconceptions tend to believe that if you do not agree 100 percent with them your opinion must be the exact opposite of theirs, regardless of what you really say.

ing around a threshold; indeed this may be especially marked when selection is effective throughout. A varying population at or near a primary threshold is subjected as a unit to selection in opposite directions, back to the ancestral and forward to the descendant adaptation. (In cases where this is not true, where selection toward the new adaptation predominates from the start, no primary threshold is involved.) Possible outcomes are: (a) failure to cross the threshold, with elimination of variants in that direction and continuation of the established adaptive type, (b) failure to cross the threshold although variants in that direction are so numerous that the population has become somewhat inadaptive with respect to ancestral adaptation and extinction ensues, (c) splitting of the population, part remaining with the ancestral adaptation and part moving on to the new adaptation, or (d) movement of the whole group to the new adaptation.

Concrete examples of (a) and (b) are difficult to specify with much certainty because recognition that a threshold really existed usually depends on its having been crossed. Yet in the rather numerous cases of multiple lineages all approaching a threshold that only some of them cross, examples of both (a) and (b) certainly occur among the lines that do not cross the threshold, for instance among the choanichthyes that did not become amphibians and the therapsids that did not become mammals. Examples of (c) and (d) are common and plain enough, although which category an example is placed in is largely a verbal matter depending on the taxonomic level at which splitting is thought of as occurring. In the immediate ancestry of man, the whole population crossed the thresholds of upright posture and cerebral expansion, but the result can also be represented as splitting of the families Hominidae and Pongidae. All (long surviving) lines of *Parahippus* crossed the threshold of hypsodonty into *Merychippus*, but the result was also splitting of the subfamilies Anchitheriinae and Equinae.

If or when prospective adaptations are initially fixed by random processes, there is a nonadaptive phase, which must be relatively brief in all cases. For a still shorter time, some of the characters involved may be positively inadaptive for the organisms first having them. When the process is oriented by selection and adaptation throughout, if a primary threshold is involved there is also almost necessarily a phase *relatively* inadaptive in the sense that a population around the threshold with its opposite selection pressures cannot be as well adapted as either the ancestral population in its earlier adaptation or the descendant population in its later adaptation. This is not to say that the evolutionary movement goes contrary to a direction of some adaptation at any time or that the population must at any point be considered absolutely inadaptive, although it may become so and may become extinct if it fails to cross the threshold successfully. (In fact, the threshold may move away from it and leave it inadaptive.)

The fact that a population in this situation is relatively inadaptive —or let us say to avoid any possible misunderstanding that it is distinctly less well adapted than either of two existing alternatives—involves very high selection pressure on any population that crosses a threshold. Exceptionally high rates of evolution are therefore to be expected in any group that does successfully pass a threshold, with later deceleration as the new adaptation is more completely realized. This effect is, indeed, clearly evident in sequences with thresholds. Horse hypsodonty illustrates the occurrence of this in the midst of a sequence. It is more broadly illustrated by the general observation that phylogenetic rates of evolution are often much more rapid in early than in late phases of a group arising with a decided change in adaptive type, as previously exemplified by the lungfishes.

The rapid evolutionary movement into new adaptation after a threshold is crossed concerns not only "key" characters and other prethreshold prospective adaptations but also and often more particularly other characters of genotype and phenotype involved in integration and expansion in the new adaptive relationship. This process has been called "postadaptation," in analogy with "preadaptation," and the term perhaps may still be retained without confusion. It refers specifically to one particular phase of adaptation, readjustment to a new adaptive position after crossing a threshold. It grades into usual progressive adaptation and is not sharply distinguishable from the latter, being a phase of the whole adaptive process and not a qualitatively different sort of adaptation.

CHAPTER VII

The Evolution of Adaptation

THE PRECEDING DISCUSSIONS have led to the conclusion that most evolution involves adaptation. Absolutely or relatively inadaptive phases occur and organisms develop nonadaptive and inadaptive characteristics, but over-all patterns of evolution are predominantly adaptive and adaptation has been seen to be the usual orienting relationship even in minor details of the pattern. Adaptation, itself, evolves. We do not simply have on one side a discrete something called "environment" with a neatly fixed set of prospective functions packaged into niches and on the other side discrete things called "organisms" or "populations" the evolution of which consists of progressive occupation of the niches. That is a process that happens in the course of evolution or, at least, it is one way of stating one of the aspects of what happens. For purposes of analysis of some phases of evolution it is a valid and useful manner of speaking. For fuller understanding, however, it is equally or more useful to focus neither on environment nor on organisms but on the complex interrelationship in which they are not really separable. The present subject, then, is the evolution of this relationship, of adaptation, and the way it is reflected in the actual phylogenetic histories of organisms.

ADAPTIVE ZONES AND THE ADAPTIVE GRID

At any instant in time, the realized functions of environments and organisms define a broader or narrower field or type of adaptation. This is not precisely the same for any two organisms, but it is almost the same for individuals in the same deme, somewhat less so for demes of the same species, and decreasingly similar but still with some common ground for species of the same genus, genera of the same family, and so on. In other words, at each of these levels there is a characteristic adaptive aspect which becomes narrower and more particular in the direction from higher levels down to individuals, broader and more general in the opposite direction. Although this adaptive relationship correlates with taxonomy as suggested, it is not confined or defined by taxonomy. It is quite common for taxonomically distinct units, of different phylogenetic origin, to share an adaptive aspect with each other but not with other members of their respective taxonomic groups. Thus the thylacine, a marsupial of Australia and Tasmania long extinct in the former and possibly now extinct in the latter also, shares with the placental Canidae an easily definable adaptive type entirely distinct from that of, say, a kangaroo on one hand or a seal on the other. Nevertheless thylacines are more closely related phylogenetically to kangaroos and dogs, in about the same degree, to seals.

The definition of adaptive types has an arbitrary element. It is in part merely a matter of what analysis we care to make of an exceedingly complex and extensive continuum. Choice in such cases depends on the particular aim of the study, the nature of the relationships involved, and the taste or experience of the analyst. Thus a not literally infinite [1] but extremely large number of different definitions and arrangements of adaptive types is possible, all valid or real in the sense of corresponding to facts in nature but all arbitrary in the way those facts are analyzed. The breadth or scope assigned to such types and the number and delimitation of steps in a scale of increasing scope may also be set arbitrarily in a very large number of different ways.

Consideration of the situation with any accepted classification of adaptive types shows, however, that the distinctions of type are not merely arbitrary. They correspond with discontinuities in nature, which may tend to be more rather than less obvious the broader the scope involved. Sibling birds in the same nest do not really have absolutely identical adaptive types, but the adaptive difference between them

[1] Some readers may be interested in the somewhat abstruse but, I think, important distinction between "infinite" and "extremely" or even "inconceivably large," which has a bearing on the theory and philosophy of classification. Any realized adaptive type is a class to which one or more real organisms are referred. The greatest *possible* number of such classes (only a minute proportion of which would have biological significance, i.e., would correspond with real adaptive types) is defined by the number of combinations of individuals in groups of from one upward by integers to the total number of really existing individuals. Since this total number is finite, the number of combinations, although extremely large, is also finite and so is the number of possible adaptive types, smaller than and limited by the number of combinations.

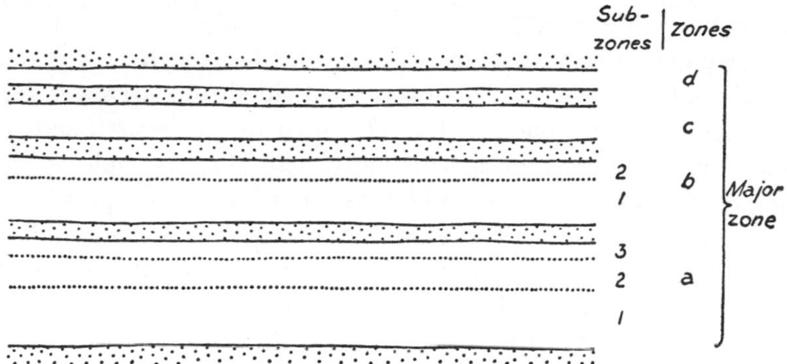

FIGURE 22. DIAGRAM SUGGESTING THE NATURE OF AN ADAPTIVE GRID. The real grid is incomparably more complex than the diagram because it may have a very large number of dimensions or number and grades of subdivision.

(which is a real discontinuity) has no practical or theoretical significance for most purposes or, at least, for our present purposes. (Even so, it does matter in some sorts of studies.) On the other hand a paramecium in a pond and a man reading this book, a fish in the sea and a bird in the air, an opossum up a tree and a bat in a cave, or a cat stalking a mouse and a dog chasing a rabbit clearly and significantly represent sharply discontinuous adaptive types with, in the sequence of examples, characteristic differences in the breadth of the discontinuity.

A visualization of some of these relationships, useful as long as one bears in mind that it involves abstraction, analogy, and oversimplification, is to represent adaptive types as zones which can be diagramed on paper as bands or pathways, together simulating in some examples a sort of grid (Fig. 22). A broader zone represents a more general adaptive type and a wider space between contiguous zones indicates greater discontinuity of type. Adaptive types may be defined in part and in some cases in geographic or physical environmental terms, but it is essential to remember that what we are talking about here is not a geographic, physical, or even in the broadest sense environmental zone but an adaptive zone, representing a characteristic reaction and mutual relationship between environment and organism, a way of life and not a place where life is led.[2]

To give even a semblance of reality to this visualization, it is necessary to think of most broad zones as subdivided into narrower zones, these again into subzones, and so on down to narrowest bands correlative with demes or even with individuals. For example, a major zone of rodentlike adaptation might, as one of numerous possible systems of subdivision, include zones one of which is squirrellike, in turn subdivided perhaps into arboreal and terrestrial subzones, the latter into nonfossorial and fossorial or maybe noncommunal and communal, and so on down to a sub-subzone representing the particular and special adaptive type of one colony of *Cynomys gunnisoni zuniensis* (such as the one visible to me as I write these lines) or of one family or individual in that colony. To be sure, somewhere along the descending scale is a point below which existing differences in adaptation are not consistently correlated with genetical or with heritable phenotypic variation and therefore do not concern us here. In the example, adaptive differences down to the deme (the colony or "prairie-dog town") probably have some correlation with heredity and those within the deme do not; this is a usual situation.

Changes in adaptation involve, figuratively, movement of phylogenetic lines within or between zones. Such movement may be an expansion or restriction, more often the latter as when a group covering a broader zone becomes specialized and more narrowly confined to one of its subzones. Movement may also be interzonal, for any level between major zones and subzones. Thus differentiation of prairie-dogs from ground squirrels was subzonal, rather far down the scale but not at its lowest levels. Change of penguin ancestors from aerial to aquatic adaptation was interzonal at a rather high level.

Discontinuities between zones as seen now or at any other point in time are generally quite clear and fixed. Canidae and Felidae are now in different, sharply discontinuous zones within a major zone of terrestrial, predaceous carnivores. Even the cheetah, a doglike animal as cats go, does not contradict the fact that animals fully intermediate in adaptive type between cats and dogs would be anomalous in the existing ecological system. Coming down the scale, there are also distinct

[2] I now find part of my earlier discussion of this subject (Simpson, 1944a) somewhat confusing in this respect.

but lesser discontinuities between, say, pantherlike and lionlike subzones, and going upward there are distinct and larger discontinuities between, for instance, terrestrial and aquatic carnivores. (Designation of zones in taxonomic terms, as "felid" or "canid," is also a convenience and a convention, as is their occasional designation in environmental terms, but is still not to be allowed to obscure the fact that it is adaptation, not organism or environment, that defines the zones.)

Since the divergence of dogs and cats in the Eocene, dogs have changed less in adaptive type. In a sense, the cats have moved from a broadly canid zone (more precisely a viverrid zone) into the felid zone. As the grid appears today, this suggests that they crossed an adaptive discontinuity. There is no doubt that such events do occur. When they do, the discontinuity corresponds with the adaptively unstable conditions around a threshold, as discussed in the last chapter. It is, however, an important additional fact that the zones themselves, and their relationship to each other, evolve. Environments change and organisms change. As they do, so do both the existing and the possible relationships between them, which are symbolized as adaptive zones. The adaptive relationship may and usually does show steady secular change. The existing canid zone is not the same as the Eocene canid zone from which the cats moved, nor was the felid zone then like the recent felid zone. The discontinuity between them was thus not the same as that now existing, and it does not follow that it was equally large or even that it was then present at all.

Many of the problems of the evolution of adaptation involve the existence and origins of discontinuities in and between adaptive zones. Some discontinuities are inherent in the ecological situation over the periods involved in an adaptive change or may even be regarded as permanently required by the environment. Thus a discontinuity between aerial and aquatic life seems inherent in the permanent physical distinction between air and water. We may be quite sure that regardless of continuous evolution of both aerial and aquatic adaptive zones, a discontinuity did have to be crossed when the penguins evolved. This example shows, nevertheless, that the inherent discontinuity may not be as great or of quite the same sort as appears at first sight. In fact there is an adaptive zone with only slight discontinuity from the strictly aerial zone and yet nearer the aquatic zone. (Fig. 23.) This intermediate zone is now occupied by oceanic birds that fly both in the

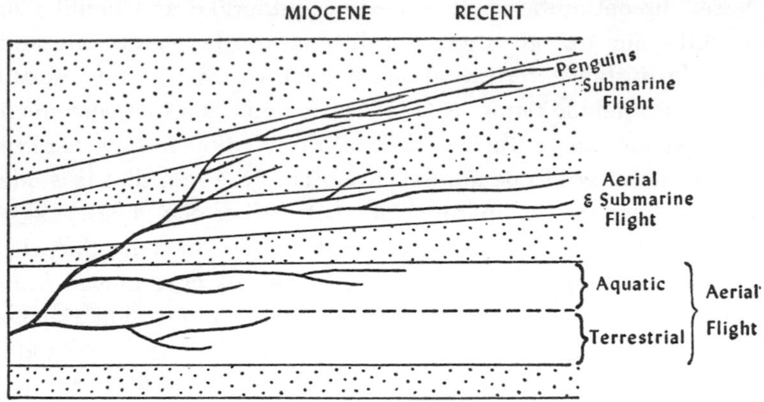

FIGURE 23. GRID DIAGRAM OF PENGUIN EVOLUTION. (From Simpson, 1946.)

air and under water, among them some of the Procellariiformes, e.g., *Pelecanoides,* diving petrels. There is still a distinct discontinuity between the adaptation of forms like this and that of the earliest true penguins. Its threshold was the point where selection for underwater flight became stronger than for aerial flight, with the ultimate effect that the former was perfected and the latter lost. Once over this threshold, the penguins had progressive and complex postadaptation to the new environmental relationship, and this, in itself, has further removed the penguin zone from that of *Pelecanoides* and still more, of exclusively aerial flight (Simpson, 1946). Incidentally, although there is now an increased discontinuity between *Pelecanoides* and the present penguin zone, that genus, perhaps with some others of similar habits, may be a rare instance of a recent animal near a major primary threshold. Study of selection on it would doubtless be extremely difficult but would also be extremely interesting.

Other inherent discontinuities include that between aquatic and land life, which was crossed, as in the somewhat reversed case of the penguins, by forms whose adaptation was alternatively to both habitats, either facultatively or at different periods in ontogeny. Discontinuities that seem to be inherent only for a particular group in the ecological situation at a given time are illustrated by such cases as change from browsing to grazing in horses. Intermediate types do occur and evidently can persist stably in other groups, but for the horses when the change occurred either alternative was advantageous over intermediate positions.

The last example, however, probably also involved and perhaps could be wholly explained by a very widespread effect that causes or increases discontinuity as and after a new zone is occupied. The effect of such occupation, in itself, tends to eliminate adaptive stability in any closely adjacent or overlapping zone. Development of specialized grazing horses placed at a selective disadvantage any less specialized grazing forms and also those in which grazing or browsing was facultative, as long as grazing was in fact continuously possible. This effect, which results in a sort of parceling out of a total ecological situation into a number of specialized and more or less sharply discontinuous adaptive zones, is very general in evolution. Examples on a grand scale will be noted toward the end of this chapter. On a small scale, the effect is a normal concomitant of speciation. After adaptive differentiation of two populations, wherever they come in contact there is a tendency to eliminate variants adaptively intermediate between them. As long as considerable gene flow continues, the contact zone is rather a line of tension than a discontinuity, but when for any reason gene flow decreases or stops (speciation, *sensu stricto,* occurs) discontinuity develops. (See, e.g., Huxley, 1939; Mayr, 1942.)

Probably almost all small and most large discontinuities between adaptive zones arise in that general way, i.e., develop along with the adaptive differentiation of the groups evolving in the two zones between which the gap arises.[3] Incidents of the opposite sort, with adaptive change initiated across a preexisting unstable zone or inherent adaptive discontinuity, seem to be relatively fewer in number. There is, however, reason to believe that they have been absolutely numerous in the course of geologic time and have been particularly important with respect to more marked changes of adaptive types and the origin of higher taxonomic categories.

In the course of the evolution of adaptation, zones not only change

[3] Although this does not represent a strong change in opinion, I formerly (1944a) failed to make this point quite clear and also stressed the less common, but frequently even more important, cases where threshold effects occur across a preexisting gap. Günther (1949) has used that treatment in support of his view that almost any change of adaptive type means crossing a discontinuity or unstable zone and has further generalized (contrary to my opinion, as he recognized) that random preadaptation is usually involved.

constantly in specificity (width) and character (direction or position) but also appear and disappear. Within the general felid zone, a machairodontine (sabertooth) subzone early developed as a consequence of the spread of numerous rather slow-moving and thick-skinned herbivores and the subzone disappeared, with extinction of its occupants, when such prey became markedly less abundant during the Pleistocene. (The example is discussed in more detail in the next section.) On a broader scale, evolution of numerous small flying insects opened a zone for aerial insectivores, still richly occupied (many birds, most bats), and more broadly still a whole sequence of wide zones successively subdivided arose progressively as vascular plants spread to the land and evolved there.

It is abundantly clear that despite periodic elimination of many adaptive zones, mostly rather minor in extent, there has been tremendous increase in the number of broad adaptive zones and in the fineness and multiplicity of subdivision of these in the course of geologic time. The broadest zones of all tended to open and be occupied early and, while such a statement depends considerably on personal definition of such zones, most of them were apparently in existence and occupied by the end of the Paleozoic. A balance toward increase in number of zones of lesser scope, at least, continued much later, probably well into the Cenozoic. Whether the balance still is tending toward increase is a moot point. There has perhaps been a decrease in the late Cenozoic, but it is impossible to say whether this indicates that the earth is, so to speak, full at last or whether much further adaptive extension and subdivision will eventually occur. If it were not for the quite unpredictable activities of man, who is eliminating many adaptive zones (or subzones) and creating many others, I would favor the latter alternative, simply as a guess subject to no verification and backed by no impelling evidence.

THE OCCUPATION OF ADAPTIVE ZONES

A basic evolutionary sequence, an element entering into most of the more complex evolutionary patterns, is the occupation by a group of organisms of an adaptive zone new to them, their subsequent diversification and phylogenetic change in that zone, and their eventual extinction in it, if this has yet occurred. For a zone to be occupied, it must first of all exist as a prospective or realized set of environmental functions, and there must be a population able to move into it. Existence of a prospective zone may involve appearance of a new environment unoccupied as far as the adaptive type in question is concerned, or new access to such an environment. Any change in the whole environmental complex (which includes the organisms in it) does produce new environments, but slow and accustomed changes usually lead not to new occupation of zones but to secular evolution of the zones (and of course of their occupants) already present. The opening of quite new zones is rather by evolutionary changes such as those indicated in the last section (e.g., spread of vascular plants on the land), by rather abrupt and widespread geographic events (rise or fall of relative sea level, advance and retreat of glaciers, etc.), or by new access (rise of a land bridge, piercing of an isthmus, waif or sweepstakes dispersal, etc.)

To enter a new zone, a group must have physical, evolutionary, and ecological access to it, all three. The need and significance of physical access are obvious. The zone must occur where the group is or where it can and does reach by dispersal. By "evolutionary access" in the present connection, I mean that the group must have at least minimal prospective adaptation for the new zone, as discussed in the last chapter. Acquisition, by any process, of such prospective adaptation may and frequently does lead to occupation of a new zone without involving physical movement of the group or ecological change around it, except such as follows its own movement into the zone. This is a relatively clear-cut event when a threshold occurs. The horses that became grazers did not go anywhere to do so, and there was no particularly relevant ecological change around them.[4]

By "ecological access" is meant that the zone must be occupied by organisms for some reason competitively inferior to the entering group or must be empty. It is highly doubtful whether a group entering an occupied zone is ever competitively superior when it is actually in the process of significant change of adaptation or when its adaptation for the new zone is prospective rather than effectively realized. Prob-

[4] It might be objected that spread of prairie grasses was a concomitant ecological change. It was a necessary preliminary in bringing the grazing adaptive zone into possible existence, but not otherwise relevant to the actual event of change in zone by the horses. Also the apparent evidence (Elias, 1942) that prairie grasses did spread in the Miocene, just when the horses took to grazing, may be a coincidence. What seem clearly to have been grazing mammals of other groups had evolved much earlier elsewhere, in the Eocene and Oligocene.

ably no group ever shifted to a distinctly new zone by progressive adaptation to it while it was already occupied by well-adapted organisms sympatric with the former group and therefore with the possibility of direct competition from the start. For success against a group already established, it seems to be a rule with few possible exceptions that the entering group must already be well-adapted to just the same zone, or to a subzone of that zone, or to an overlapping zone or one near enough to be competitive (which really means that there is at least some overlap). The first of these three situations cannot arise and the other two are unlikely to arise, if they can at all, unless the entering group evolved its adaptive type elsewhere and the "entrance" was an actual movement, geographical migration or expansion.

It follows that occupation by a different group of a zone already occupied usually, perhaps always, involves change in distribution; a geographically invading group, if the invasion is successful, ousts one already established in a region. This is a very important conclusion for, among other things, interpretation of the fossil record, which is in all cases incomplete as regards geographic sampling. It means that when one group replaces another of similar adaptive type, a common phenomenon in the record, much the most probable interpretation is that one or the other evolved elsewhere and is a rather recent migrant where found—"rather recent" because its current adaptive type has not essentially changed since it was developed elsewhere. Among many examples, that of the mammalian faunal interchange between North and South America is particularly good, because here we know with one possible exception (the opossums) where all the many groups involved came from, when the exchange began within reasonable limits, and what happened to all the groups. A fairly detailed analysis has been given elsewhere (Simpson, 1950b). In brief, there was on both continents but especially in South America rather extensive entrance into occupied faunal zones by animals *already* adapted to them on the other continent. With no exceptions, this duplication within faunal zones was temporary, one group or the other becoming extinct in a geologically short time. In most cases it was the invading type that survived, as would be expected because the ability to invade in the face of occupation implies probable competitive superiority although this might not, in exceptional cases, extend to longer competition or endure in rapidly changing conditions.

There are a number of apparent exceptions to the rule that replacement of one group by another in an adaptive zone indicates geographic invasion. I question whether any of these exceptions really are such. Some of them are clearly artifacts of classification. For instance, fissiped carnivores replaced creodonts (both suborders of Carnivora) in the late Eocene and Oligocene (Fig. 19, Chapter VI), including replacement in western United States where these groups had lived together without decided tendency toward replacement for a very long time, from the middle Paleocene. Over-all, it is entirely correct to say that fissipeds arose from creodonts and replaced the latter, but not without invasion. The fissipeds that long lived with creodonts were not competing types but a family with its own well-defined adaptive zone, Miacidae. The actually competing and replacing types probably arose from this family, in a broad sense, but not in the area of observation, where no transitional forms are found. The actual replacers are clearly invaders when they appear. Analysis at the family rather than subordinal level shows this clearly, for the invaders already belonged to other and more progressive families than the Miacidae. In fact, they replaced the native Miacidae (more similar to them in adaptive type) more quickly than they did many of the creodonts.

Another sort of apparent exception is involved in the last fact stated. A successfully invading group often tends to diversify in the invaded area, to undergo adaptive radiation as it occupies contiguous adaptive zones and specializes into subzones of its original zone. In the course of this process it may, or at least may seem to, spread progressively into zones occupied by natives but not immediately entered by the original invaders. Invading fissipeds forthwith took over the miacid zone and some creodont zones but only much later knocked out the last creodonts. Similarly in South America invading cricetid rodents almost at once replaced the natives most similar in adaptation, but since then have been expanding greatly and progressively entering more diversified zones or subzones while the natives continue to dwindle, a process probably still going on although some of the natives with distinctive adaptive zones retain these completely.

Some of this apparent expanding replacement is not really such but is the effect of repeated invasion by new groups of differing adaptive types overlapping different zones occupied by natives. Part of it seems, nevertheless, to be a real phenomenon, and yet it probably does not really represent evolution, *in situ,* of a replacing and replaced group. Some of it is due to allopatric evolution and later invasion on a smaller scale. (The areas of the examples are whole, varied continents.) Some of it is probably due to what might be called intercalary replacement. The eventually replacing groups evolve not in but between the zones of those eventually replaced. When this happens with a good many groups each of which has some, even very slight, marginal overlap on an older occupied zone, the total effect may be to make that zone untenable. The effect is not so much actual replacement in that zone as disappearance of the zone by encroachment of contiguous zones. An oversimplified model may make this clearer. Suppose a carnivore *A* is especially adapted for preying on a rodent *a,* but can supplement this diet when *a* is scarce by feeding on *b, c, d,* and *e* as available and needed. This is no serious impediment to adaptation of another carnivore to feed mainly or wholly on *b,* another on *c,* another on *d,* and another on *e.* None of these competes with *A* for *a,* its principal food, each competes for only one supplemental food of no vital importance to *A,* and they do not compete with each other. Their various adaptive subzones are intercalated around that of *A* and between it and other older carnivore adaptive types not here specified. But among them they compete with *A* for *all* its supplementary food and are severally better adapted to take each sort of this. *A* then cannot supplement its diet and it becomes extinct when *a* is in periodic low supply. The effect may, of course, apply to any environmental conditions and not to food alone. A similar effect may ensue when two groups otherwise quite differently adapted nevertheless have some one need (such as water, nesting space, etc.) for which they compete.

In the preceding model, the zone of *A* was not actually occupied by another group. It simply became empty except as different, adjacent zones might eventually cover it by combined overlap. Or, after the zone is empty, some other group may enter it by narrower specialization requiring no overlap with the surrounding zones. Then, on a small scale, there is delayed rather than competitive replacement in the zone of *A.* This is a fairly common and rather puzzling phenomena on a much grander scale. Bats (and to some extent birds) replaced pterodactyls in their adaptive zone, but not until long after the pterodactyls were extinct and the zone empty for some millions of years. Similarly cetaceans replaced ichthyosaurs long after their extinction. So, in a general way and as regards some adaptive types, did mammals replace the dinosaurs, and contrary to a popular impression this was delayed. There is no evidence of competition between mammals and dinosaurs, and mammals that most nearly occupied zones opened by dinosaur extinction did so millions of years after that extinction.

Such events are completely incomprehensible if it is supposed that a particular adaptive zone was continuously open and essentially unchanging throughout the history, for no way is known or even conceivable (to me, at least) in which a group adapted to an essentially unchanging zone can become extinct without competitive replacement (which, as a matter of fact, is a change in the zone if everything is considered). The only possible explanation is that the zone did change, but that later groups could move into the changed zone, or into one defined (more or less arbitrarily in some respects) by some of the same characteristics. After all, such analysis is always oversimple. There are very decided differences in *total* adaptation between cetaceans and ichthyosaurs, and it is an abstraction of ours when we consider the special aspects of adaptive similarity and speak of them as being in the same adaptive zone. To be more concrete, as an example of the sort of process that may occur although this particular one probably did not, a change in the total environment of the ichthyosaurs may have required thermal regulation for their survival. They did not have prospective adaptation for this change and became extinct. It was a long time before another group with prospective adaptations for the ichthyosaur zone (locomotion, aquatic life, food, etc.) plus thermal regulation developed.

In some short range processes it is possible that a transient, nonrecurrent catastrophe might cause extinction without longer alteration of an adaptive zone and that this zone would, potentially, then be open as quite the same. In general, however, it is inaccurate to speak simply of extinction as a way in which empty zones arise. The *same* zone as that formerly occupied is irretrievably gone in the complex flux of historical process. *New* zones similar in some respects or defined by the same partial specifications may arise, at once or much later, as a sequel to the extinction.

In a more or less strict sense, empty zones thus seem normally to be new zones, which are constantly arising as the whole complex of adaptations evolves. As previously suggested, such zones may arise from any sort of environmental change or any sort of evolutionary change in organisms, since any change in either obviously makes possible new relationships between them. The great majority of actually empty zones are those that never have been occupied. The most spectacular examples are provided by islands literally empty when formed and eventually reached somehow (usually by sweepstakes dispersal) by plants and animals, a sequence especially clear and well-studied in the Galápagos (Lack, 1947) and Hawaiian (Zimmerman, 1948; Amadon, 1950) Islands.

Generally less obvious but nevertheless more widely important have been cases where access to an empty zone involved more evolutionary than geographic movement. In the case of new minor zones or subzones adaptively contiguous to those already occupied, prospective adaptation for them usually occurs even when they are arising and their occupation tends to occur as soon as they appear or soon thereafter. More strikingly distinct zones may require prospective adaptations not in being when they arise and they then may remain empty for long periods of time. For instance, a broad and potentially intricately subdivisible zone for flying insectivores arose when flying insects evolved, but there were then no existing organisms with prospective adaptation for that zone and it was many millions of years before any developed. Flying insects were already common in the Pennsylvanian, and the first possibly flying insectivores (small pterodactyls) do not appear in the fossil record until the Jurassic. It seems probable (to me; some other students think it flatly impossible) that unoccupied major zones now exist. There is, for instance, no true aerial plankton although I see nothing impossible in the eventual evolution of one. If it did appear, this in turn would create other major zones. Whether prospective adaptations for this and other possible developments will appear and such zones will be occupied perhaps comes under the heading of idle speculation.[5]

EVOLUTIONARY SEQUENCES AND PRINCIPLES ON THE ADAPTIVE GRID

The same evolutionary events can be studied from many different points of view and expressed in many different terminologies. The un-

[5] Or would be a large source of material for science fiction, which has so far been singularly unimaginative and uninstructed in dealing with possible future evolutionary developments here on earth.

derlying principles, also often expressible in quite different ways with equal validity, are likely to appear more clearly and to be better understood if approaches and handling of the data are varied. With certain types of material, analogical diagrams based on the concept of the adaptive grid have proved to be enlightening for those students, at least, who habitually handle abstraction by relational symbols or visual images. For students (among them some outstanding authorities on evolutionary theory) who deal with theoretical or abstract concepts in other forms, such as purely verbal symbols, such diagrams seem to have little explanatory content. In spite of this psychological difficulty, it seems worth while briefly to pursue here the idea of grid representation for a few examples of evolutionary processes that are especially suited to this method.[6]

When a new major zone or a complex of related zones is occupied, the lower zone, so to speak, or the first one entered is commonly the widest, representing least specific adaptation or least specialization for the organisms. Successively contiguous higher zones, of which there may be a large number, are frequently (although not necessarily) narrower. They may be occupied in sequence by populations splitting off from the next lower zone in each case; this is one of several ways in which progressive[7] specialization occurs (Fig. 26A, below). Higher zones may also be occupied by populations from other origins. In either case, the result is the coexistence of groups of different but similar adaptive types some of which are more specialized than others. When the distinction between the zones is well defined, with considerable adaptive discontinuity, and when conditions affecting this relationship are fairly stable, such groups may coexist for a long time or even, so far as we yet know, as long as life persists. On the grandest scale, that is the situation of the major phyla of animals. It is also a very common situation at lower levels in all groups of animals and plants. Hence

[6] These remarks doubtless apply also to the other relatively abstract diagrams in this book. At least, I have found that some of them, such as Figure 37, give some students extensive, pleasurable insight while they merely baffle and annoy others. Perhaps every author should define his own mental processes. I habitually visualize abstractions, but I have made an attempt in this book also to verbalize them sufficiently for those who handle abstraction in words. A difficulty in this respect is keeping the words from being merely descriptions of the visual images.

[7] I am here and elsewhere in this book using "progressive" as descriptive of any sequence in which each step systematically develops from the last, i.e., in a progression. This study is not concerned with the more philosophical question as to possible progress, change for the better, in evolution.

the well-known comparative anatomical sequences from "primitive"[8] to "advanced" or "specialized" among contemporaneous recent animals: Lemuroidea, Cercopithecoidea, Hominoidea; *Squalus, Rhinobatus, Raja;* Dasyuridae, Phalangeridae, Macropidae; *Drepanis pacifica, Drepanis funerea;* and so on. Almost all of the innumerable "phylogenies" that have been based on living animals and plants are of this sort. Incidentally, with possible very rare exceptions at the level of subspecies or, at most and still more rarely, species, none of these really are phylogenies.

Although persistence, with more or less change, of populations in both lower and higher zones is thus quite common in such a situation, extinction in either lower or higher zones or, indeed, in both, is also common. One of these outcomes, survival of the less specialized, has been dignified by recognition as an evolutionary principle. Such a "principle," however, is merely a description of what sometimes happens and this one should be accompanied by all possible alternatives, all of which also frequently occur: survival of both more and less specialized (as above), survival of the more specialized, and extinction of both more and less specialized. The real point is not that one of these happens more than another or is the characteristic outcome in evolution, but that each tends to happen under certain conditions that we should seek to identify.

Survival of the more specialized, in the higher adaptive zone, is the usual outcome when the zones are not so distinct as to preclude competition, i.e., when they do overlap in some essential respect even though separated by discontinuity in all others (Fig. 24A). Many of the examples of ecological replacement of one group by another are probably of this sort, although it may be difficult to judge relative degree of specialization and to avoid the fallacy that a replacing group is *ipso facto* more specialized than the one replaced. Yet this effect is almost certainly one, at least, of the factors in such cases as the replacement of the least specialized creodonts, the Arctocyonidae, by more specialized condylarths, on one hand, and other creodonts, on the other, and then the later replacement of condylarths by still more specialized

[8] I use quotes because this is a frequent comparative anatomical designation but one that is often incorrect in its implications. No recent animal is primitive in the sense of being early or ancestral and none was ever primitive in the sense of being generalized. Most living "primitives" also have some "advanced" or strongly specialized characteristics.

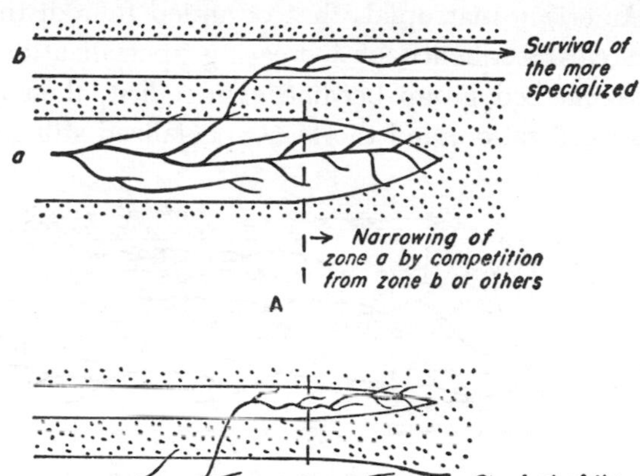

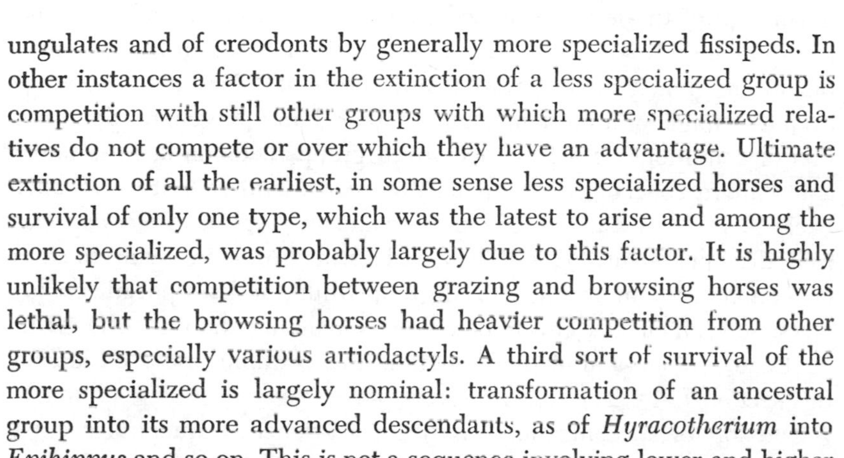

FIGURE 24. SURVIVAL OF THE MORE AND OF THE LESS SPECIALIZED.

ungulates and of creodonts by generally more specialized fissipeds. In other instances a factor in the extinction of a less specialized group is competition with still other groups with which more specialized relatives do not compete or over which they have an advantage. Ultimate extinction of all the earliest, in some sense less specialized horses and survival of only one type, which was the latest to arise and among the more specialized, was probably largely due to this factor. It is highly unlikely that competition between grazing and browsing horses was lethal, but the browsing horses had heavier competition from other groups, especially various artiodactyls. A third sort of survival of the more specialized is largely nominal: transformation of an ancestral group into its more advanced descendants, as of *Hyracotherium* into *Epihippus* and so on. This is not a sequence involving lower and higher zones but evolution of and in a single, changing zone.

Survival of the less specialized may result when conditions become generally unfavorable for all populations in related zones (Fig. 24B). Greater adaptability of the less specialized group, both in that it has, so to speak, more room for maneuvering in its wider zone and in that it often has more opportunity to change zones, then gives it greater chances for survival. Among many examples probably involving this general advantage is that of the Caenolestoidea, a group of South

American marsupials that occupied four distinct adaptive zones with a clear sequence of increasing specialization (Fig. 25).[9] The most specialized group became extinct first, then the two of intermediate specialization, and the least specialized still survives. A related factor

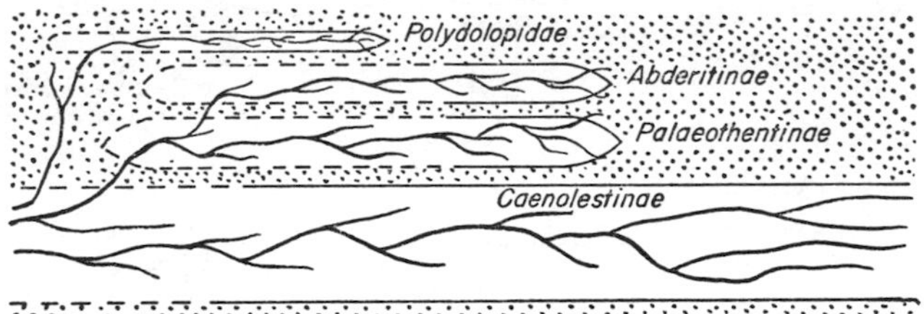

FIGURE 25. EVOLUTION OF THE CAENOLESTOIDEA. Progressive extinction of more specialized and survival of the least specialized groups. (This is a grid diagram, in which widths of bands represent relative scope of adaptive type, not abundance or variety of the animals.)

in survival of the less specialized perhaps operative in such examples and quite clearly so in others is that more restricted, narrow specialization means greater susceptibility to any pertinent environmental change, even to a single or a random change rather than any general deterioration of conditions. If a less specialized group feeds on *a, b, c,* and *d* and a more specialized group only on *a,* general scarcity of food is likely to lead to extinction of the more and survival of the less specialized; disappearance of *a,* with *b, c,* and *d* still abundant, is sure to cause extinction of the more specialized group while the less specialized survives. Such an effect seems to be involved in extinction of the Machairodontinae and survival of the Felinae in the example discussed later in this section (Fig. 28).

One sort of specialization has been mentioned, sequential occupation of zones decreasing in width (Fig. 26A). The same sort of effect, movement of populations into narrower zones, may arise in at least two other ways readily represented on an adaptive grid. One of these is for a population occupying a considerable part or even the whole

[9] My previous discussions of this example (e.g., 1944a, Fig. 21) erred in showing the most specialized group, Polydolopidae, as derived from the next lower in specialization, Abderitinae. Evidence since found suggests that the Polydolopidae arose independently from the least specialized group, Caenolestinae. This complicates the phylogeny, but it does not change the order of increasing specialization or the fact that extinction followed this order in reverse.

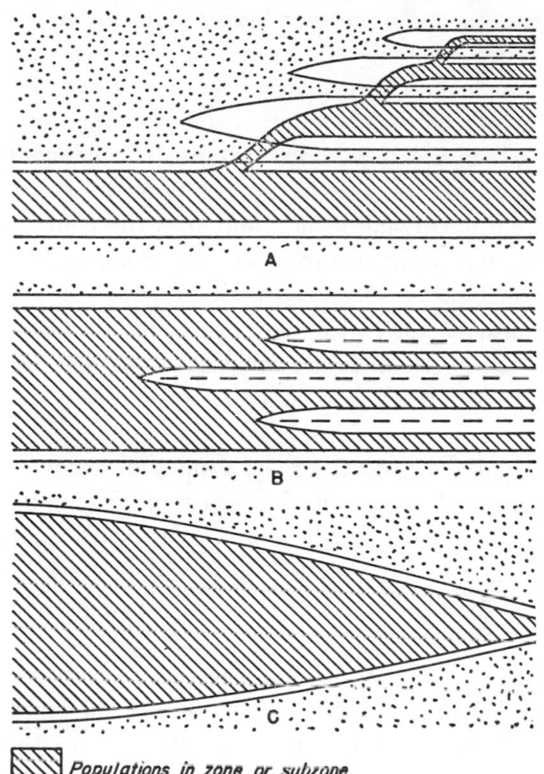

FIGURE 26. THREE PATTERNS OF INCREASING SPECIALIZATION BY DECREASING BREADTH OF ADAPTATION. A, successive occupation of narrower zones. B, segregation into subzones. C, narrowing zone.

of a broad zone to split up (by speciation) into separate populations each of which is narrower and covers only a part, of the original zone (Fig. 26B). Although this effect is not shown in the diagram, a usual concomitant and indeed, in a sense, result of this process is that the narrow zones tend to diverge from each other and eventually to cover a considerably greater total span than the broad, unified ancestral zone. This seems to be a very common relationship between speciation and the evolution of adaptation. It is perhaps seen in its purest or most clearly separable form in such examples as the differentiation among descendants of relatively unspecialized immigrants on isolated archipelagos, e.g., the Geospizidae on the Galápagos (Lack, 1947).

Another process of movement into a narrower zone involves phylogenetic progression in a zone that becomes narrower as time goes on (Fig. 26C). This, too, may be common, although one must guard against the conclusion that progressive and irreversible structural change implies in itself that adaptation is narrowing. Man, for instance, has undergone considerably more progressive change than, say, an Eocene lemuroid but obviously occupies a tremendously broader adaptive zone. Actual narrowing of adaptation in the course of phylogeny within one general zone is probably less common than change of adaptation, which has no special correlation with narrowing adaptation but is often mistaken for it. Change means loss of ancestral functions and we sometimes overlook the fact that new functions usually compensate the loss. I have the impression that progressive specialization in the sense of narrowing of adaptive zone in phyletic evolution is much less common than has generally been supposed. Yet it surely does occur, for instance in host-specific parasites and monophagous animals some of which must have arisen from less specific or polyphagous ancestors (e.g., Thorpe, 1930, 1940). But most of these and analogous cases may involve not so much the narrowing of a zone as its subdivision by speciation, as discussed above. Indeed it may be a rule that when a zone seems to narrow, and really does so for populations continuing in it, this is usually the result of occupation of subzones in that zone by other populations. At any rate, when there is no parceling out of subzones by related species or encroachment on a zone by related or unrelated populations, I see no evidence that progressive adaptation within a zone tends as a general rule toward narrower adaptation.

In passing, it should be noted that the sort of specialization just discussed is specified as narrowing of adaptive range or zone. The word "specialization" has been used loosely in many ways and only confusion results if conclusions relevant to one usage are applied to another. A wholly different but also common usage applies "specialization" to changes that limit further possibilities for change, or seem to. For instance, the specialization, in this sense, of the foot of *Equus* seems to preclude any further essential change in that structure, but I see no reason to think it gives *Equus* a narrower adaptive range than its three-toed ancestors. Some discussions seem almost to imply that any change is necessarily a specialization. The subject will again be mentioned in a later chapter (IX) in connection with the concept of overspecialization.

To turn to a different topic, it is a common evolutionary pattern for a group to expand successively into a sequence of adaptive zones which involve different sorts but not necessarily different breadths of adaptation, although narrowing in the higher or later zones is commonly present also as in the examples first discussed in this section. In this pattern, each zone normally has a considerable degree of stability when

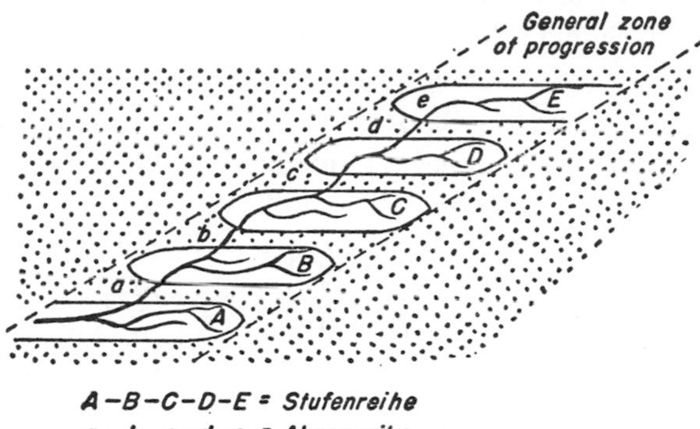

FIGURE 27. EVOLUTION INVOLVING SUCCESSIVE OCCUPATION OF ADAPTIVE ZONES AND SUCCESSIVE CLOSING OF OLDER ZONES. The Stufenreihe, A–E, indicates the general direction and nature of the progression of the Ahnenreihe, a–e, but in fact none of the populations of A–E are evolving in that direction.

occupied, and populations in each tend to survive, expand, and adapt further even while other populations are making the change to the next zone in the sequence. All or several zones may remain in existence and be occupied simultaneously, but it also frequently happens that each zone overlaps the next lower in some respect or is so related to it that occupation of a zone tends to change the next lower and to make conditions less favorable to populations in the latter. In such a case there is not only successive occupation of zones but also successive extinction of their occupants following more or less regularly at intervals after zonal occupation and in the same sequence, a situation shown diagrammatically in Figure 27.

This rather complicated pattern might be judged a priori to be so special as to be unusual in nature. Although of course the perfect regularity of the diagram is not usual, the fact is that examples clearly of this sort are very common in the fossil record. The populations evolving within a zone are frequently abundant and widespread. They are likely to be found as fossils and they provide a sequence both temporal and structural which closely simulates a phylogeny. Then the steps in the sequence are really discontinuous. The populations through which phylogenetic continuity occurred are those that moved from one zone to the next, not those that expanded within the zones. The populations in the true phylogenetic lineages that moved between zones are almost always smaller, in total numbers, and also more restricted geographically than those in each zone. Moreover, although all the interzonal transitions sometimes occur in one area, different steps may occur in quite different regions. Thus the chances of finding any interzonal line are smaller than for finding a zonal population and the chances of finding all interzonal lines in a long sequence are very slight.

The result is that paleontological sequences related to this pattern usually include the successive but discontinuous zonal populations, forming what Abel (1929 and earlier) called a "Stufenreihe," "step series." Representation of the interzonal, actually phylogenetic sequence, the "Ahnenreihe" or "ancestral series," is usually less good, and the whole of that series is seldom known for any long time span. Many paleontological "phylogenies" are really Stufenreihen or include parts that are Stufenreihen along with parts of Ahnenreihen, rather than being Ahnenreihen and hence real phylogenies throughout. The Stufenreihen are analogous to the comparative anatomical "phylogenies" mentioned above and are due to the same general sort of phenomenon, but they do usually more nearly approximate true phylogeny. They have correct and given time sequence of the stages and each is nearer to an actual phylogenetic ancestor than are the comparative anatomical stages, especially as regards the so-called "primitive" ones among the latter.

The phylogeny of the Equidae, as followed in Europe and in North America, is an example of this pattern that is unusually close to the diagram and that is open to no other reasonable interpretation because here we have not only a Stufenreihe in both regions but also very nearly, at least, the whole Ahnenreihe in one of them (see Chapter XI, Fig. 46). The lungfish phylogeny of Dollo (1896), rightly considered a triumph of the first half century of phylogenetic paleontology, was really a Stufenreihe (Abel, 1911). An example at a lower taxonomic level, of interest in showing how an Ahnenreihe can sometimes be reconstructed in almost all its detail from a Stufenreihe, is Kaufmann's *Olenus* series elsewhere discussed (Chapter VIII, Fig. 31). Other examples are legion. In fact most long paleontological series show the Stufenreihe-Ahnenreihe pattern, symbolized in Figure 27, as one element among others.

As an example of application of the adaptive grid symbolic analogy to analysis of more complex real sequences, the history of the Felidae may be summarized in this way (Fig. 28). In this diagram, innumerable complications have been omitted; all could also be analyzed in this way by using separate diagrams at successively lower taxonomic levels. The purpose here is to give an over-all picture of the evolution of a

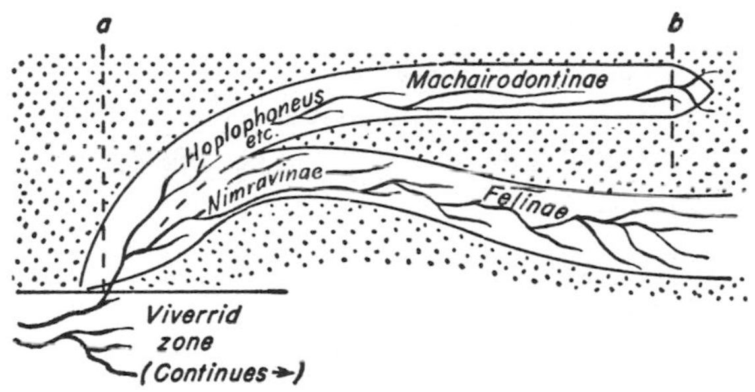

FIGURE 28. BROADEST FEATURES OF THE EVOLUTION OF THE FELIDAE. At *a* the felid zone opens by extinction of the creodonts and evolution of the herbivores. At *b* many large, slow herbivores became extinct and the machairodontine zone disappeared thereafter. For other features, see text.

large group of animals as an example only. The indicated phylogeny, which is highly schematic, is essentially that of Matthew (1910), whose views still seem to me most probable in such major points as are represented in the diagram, although Scott (e.g., 1937) has supported an interpretation radically different in some respects and Hough (1951) has suggested greater complexity and somewhat different relationships for the machairodont group. Their views could equally well be shown on such a diagram, but the purpose of exemplification of the method would be no better served.

An adaptive zone for the Felidae as a whole became available with the rise of various herbivores and the decline of creodont carnivores. It was occupied by a considerable variety of animals derived from prospectively adapted viverrids. The wide, more basal viverrid zone has persisted and is still richly occupied. The early felid major zone, which as a whole was somewhat more like the Pleistocene sabertooth than the Pleistocene or Recent cat zone, tended toward subdivision into two zones, not then very distant from each other and both rather finely subzoned among a variety of types. These two zones were occupied respectively by hoplophoneines and nimravines. Lineages of general hoplophoneine type evolved in a continuing zone which changed considerably but maintained about the same level of specialization and of differentiation from a basal, more or less viverridlike fissiped zone. Late occupants of this zone were the Machairodontinae, including the widely known genus *Smilodon*, popular exemplar of the sabertooths (or "sabertoothed tigers," a most misleading popularization). Lineages of more or less nimravine type also evolved in a continuing zone which changed more appreciably, as regards the dentition, at least, so as to become less sabertoothlike than the early Nimravinae. The Felinae arose in this zone and occupied it in later stages. Change mainly in the nimravine-feline zone tended to move the two (speaking broadly) felid zones farther apart, producing greater discontinuity between them and reducing marginal competition, which was doubtless a factor in their divergence.

In the late Cenozoic there were two quite distinct zones, the machairodontine, with adaptation particularly to preying on rather large, slow, thick-skinned herbivores,[10] and the feline, with less specific adaptation on the whole but particularly related to preying on smaller and more agile animals. By the end of the Pleistocene, the sort of prey for which machairodontines could most successfully compete had become markedly less varied and abundant all over the world and the specifically machairodontine zone effectually came to an end. Its occupants at the time, the Machairodontinae, probably had sufficient prospective adaptation to change to a contiguous zone had one been open but could not do so in the face of existing, already well adapted occupants in such zones, notably the hyaenids in a carrion-eating zone and the felines in a broad ambush and pursuit zone. The Machairodontinae became extinct.

ADAPTIVE RADIATION AND ITS SEQUELS

It is strikingly noticeable from the fossil record and from its results in the world around us that some time after a rather distinctive new

10 Bohlin (1940) thinks that they were carrion-eaters, but this seems most unlikely to me (Simpson, 1941) and in either case it is probable that this is the kind of animal they ate, whether they killed them personally or not.

adaptive type has developed it often becomes highly diversified. This may follow soon after the origin of such a type or may be long delayed (a point discussed in the next section), but it is more likely than not to occur sooner or later if the type does survive and become well established. Diversification may be brief or prolonged and may be of limited scope or may ramify into the most extraordinarily varied zones covering a breadth of total adaptation that would have been wholly unpredictable and incredible if we were aware only of the beginning of the process. (This, by the way, is one reason why I continue to be somewhat skeptical of the idea, stated in Huxley, 1942, and in work by that student not yet published at this writing, that evolution has essentially ended except in man.)

The same sort of diversification follows, and in this case begins almost immediately, when a group spreads to new and, for it, ecologically open territory. The extent of adaptive diversity eventually reached tends, although rather roughly, in the first case to be proportional to the distinctiveness of the new adaptive type and in the second to the extent and diversity of the new territory. In both cases the more direct factor may be said to be the extent of adaptive opportunity provided by the change or, in other words, the number of prospective adaptive zones opened more or less simultaneously to the groups involved. In many instances it is, however, by no means obvious when a new adaptive type is arising how distinctive it will eventually become or what zones will later open to it or develop with it. Indeed the early stages of what later becomes a strikingly novel adaptive type may have only the slightest distinction from its ancestors or collateral contemporaries, a fact which is, of course, related to delay in diversification of such groups and which does not contradict an over-all correlation of distinctiveness of type with eventual diversity.

So far as adaptive radiation can be distinguished from progressive occupation of numerous zones, a phenomenon with which it intergrades, the distinction is that adaptive radiation strictly speaking refers to more or less simultaneous divergence of numerous lines all from much the same ancestral adaptive type into different, also diverging adaptive zones. Progressive occupation of such zones is not simultaneous and usually involves in any one period of time the change of only one or a few lines from one zone to another, with each transition involving a distinctly different ancestral type. Theoretically, at least, the whole of the diversity of life is explicable by these two not sharply distinct processes plus the factor of geographic isolation which may permit essential duplication of adaptive types by different organisms in different regions.

There has been a highly irregular and yet fairly general tendency for existing diversity to increase throughout the history of life. This involves a concomitant tendency for the adaptive zones still remaining open to occupation to become narrower in scope as well as fewer in number. Because adaptive radiation, like any change of adaptive type, depends on availability of adaptive zones and because it further requires that several or many of these be available simultaneously (or nearly so on the geological scale, at least) there has also been a tendency for the scope of adaptive radiation to decrease. This has been further accentuated by the fact that major radiation requires considerable geological time. The tendency has, again, been extremely irregular but its effect is evident.

As far as environmental possibilities are concerned the very broadest possible scope for adaptive radiation existed when life first arose, even though the further evolution of life has itself from time to time opened very broad possibilities. Actual occurrence of a radiation requires, again like any adaptive change, not only this possibility in the form of prospective adaptive zones but also the existence of organisms with prospective adaptations. Protistans seem inherently limited in prospective adaptation for any but subzones of their own broad zone. After metazoan organization had become well established, there came a time when maximum prospective adaptation among them coincided with maximum scope of prospective adaptive zones in a world empty of life other than protists and a restricted range of truly primitive multicellular plants and animals. Radiation of most of the animal phyla ensued. Much of the long discussion of the rather sudden appearance of the phyla (although Cloud, 1949, has pointed out that "sudden" here involves time on the order of some tens of millions of years) seems quite beside the point because it overlooks the great probability that this was an adaptive radiation which is, to the degree required by the actual record, a "sudden" process. Each of the major phyla (i.e., those clearly warranting that grade rather than having it conferred on them because we do not know the relationships of some aberrant relict) does stem from a distinctive and basic adaptive type. When this is not obvious, the obscurity is due to later radiations within the phylum, the diversity of which overlays the relative unity of original adaptive type. No later radiation has developed the scope of this particular one, not only because there has been less time but also because the great radiation had, in a sense, preempted the broadest possibilities.

Not all origins of phyla can be assigned to the postulated great radiation. The Chordata perhaps cannot, because the first sure chordates appear rather late for a phylum (late Ordovician) and because there is on anatomical grounds some possibility that chordates arose from echinoderms (De Beer, 1951).[11]

The now existing major adaptive types of chordates do not owe their origin to early adaptive radiation in the phylum, which occurred but which involved lesser adaptive differences. The major types, corresponding to the classes in taxonomy, arose partly by radiation and partly by successive occupation of major zones: the Placodermi by succession from Agnatha, the Chondrichthyes and Osteichthyes probably as the most successful branches of a (poorly known) radiation in the Placodermi, the Amphibia by succession from Osteichthyes and the Reptilia by succession from Amphibia, the Aves and Mammalia, finally, as the most successful branches of a radiation in the Reptilia. The complexity of the process and the lack of definite distinction between radiation and succession is, however, shown by the fact that even when the relationship as between classes is successional the particular lineage that made the adaptive shift was one of many involved in lower-level radiation.

Within the Mammalia, early radiations evidently occurred but what little we know about them suggests that they were of small scope, involving few major adaptive zones and with less scope than, say, some contemporaneous single orders of reptiles. Much later occurred the combination of, on one side, primitive placental mammals occupying an adaptive zone already rather broad, the organisms with extensive prospective adaptation, and on the other side a wide range of virtually empty adaptive zones. The emptiness of these zones was due in part to extinction of many groups of Mesozoic reptiles, a series of events that no one has satisfactorily explained in separate detail for each, although the general principle back of them is clear. In part, and it may be in greater part, lack of occupation of zones for which mammals became

[11] On the other hand, there is also some possibility that the graptolites were chordates (Kozlowski, 1947) and graptolites appear early enough (late Cambrian) to have been possible results of the great radiation although there remains some slight additional possibility that they arose from the still older echinoderms.

prospectively adapted around the Mesozoic-Cenozoic transition was due to the fact that reptiles had not developed prospective adaptation for numerous zones that nevertheless became potential as evolution proceeded through the Mesozoic. Physical and ecological access to these zones existed; evolutionary access did not until mammalian evolution (a ramification of reptilian radiation) reached a certain point, a great threshold beyond which lay not only one but many diverging adaptive zones.

In the great, basic radiation of primitive placentals (nominally insectivores), which probably began in latest Cretaceous time and clearly continued through the Paleocene, various lines reached adaptive zones basic for almost the whole range later so richly developed by mammals. These most basic zones may be roughly equated with the taxonomic orders, among which at least the following rather surely arose from the great primary placental radiation: Insectivora, strictly speaking (a bundle of radiating lines that did not happen to move significantly far from the major ancestral zone), Dermoptera, Chiroptera, Primates, Tillodontia, Taeniodonta, Edentata, Lagomorpha, Rodentia, Cetacea, and Carnivora in a very broad sense (the early members were not carnivorous). For most of these, the nature and range of basic adaptation is fairly obvious, although it must be emphasized (and will be again) that when really early members are known, these are remarkably similar in all the groups named. The zones diverged and developed really clear-cut distinctions only with further evolution. For instance, the many known lineages of more or less insectivorelike and more or less primatelike mammals in the Paleocene and Eocene are all similar except in endlessly varied detail. There is no character or complex of characters, no clear-cut difference of adaptive type, that warrants dividing the lines decisively between Insectivora and Primates. Yet among these lineages were small adaptive differences that were to evolve into the tremendous adaptive discontinuity between, say, a man and a mole.

A secondary but still major, complex radiation, begun while the primary radiation was still under way, produced from primitive carnivore-condylarth or ferungulate stock a great array of carnivores, in a stricter sense, and an even greater array of ungulates and their allies. Still another radiation followed from physical access to the ecologically varied and (for mammals) empty continent of South America. There several major adaptive types, already differentiated as such, radiated simultaneously: primitive marsupials, edentates, and omnivorous to herbivorous ferungulates (condylarths and condylarthlike). In descending scale, with reference to the distinctions eventually developed between the adaptive zones occupied and the range of adaptation eventually covered by the various zones, radiation from the condylarth zone in South America gave rise to Litopterna, Notoungulata, Astrapotheria, Xenungulata, and Pyrotheria. The notoungulates, most successful of these groups, radiated early and exuberantly through late Paleocene and Eocene into what were, in the Oligocene, nine adaptive zones distinct enough to correspond with taxonomic families. Each of these zones in turn was subdivided by lesser radiation. For instance in the Notohippidae (a group with some adaptive similarity to horses, as the name implies, but no phylogenetic relationship) at least five early Oligocene genera were present simultaneously in a rather restricted area.

In such cases radiation of a minor sort, taxonomically at about the specific level, doubtless also occurred in many or most of the generic and familial zones, although geographic distribution of paleontological samples is inadequate to show the full scope of these minimal radiations. Radiation at these low levels is clearly shown in the finches and tortoises of the Galápagos or the honeycreepers, land snails, and many groups of insects of the Hawaiian Islands. It is also evident but not generally so isolable as a single event in many polytypic genera of animals and plants all over the world.

In exemplifying adaptive radiation and its scope, I have run down the scale from phyla to species. It is, however, seriously misleading as to the complexity and the true nature of the process to look at it (as some students have) only as a sequence of radiations of regularly decreasing scope. The actual phylogenetic process is not first a radiation of phyla, then of classes in each phylum, then of orders in each class, and so on down until finally a radiation of species occurs. What actually occurs when any radiation is going on is simply a divergence of populations, which are or are becoming separate species, into different adaptive zones. These zones themselves then evolve, as has been emphasized. Their characteristics, what are seen *in retrospect* to have been their potentialities, interactions between them, and the amount of time elapsed all react together to determine how distinctive they eventually become. Whether we represent the outcome as radiation of genera, orders, or phyla, depends, so to speak, on how far back we stand, how much of the whole picture we take in and how much we generalize on a pattern which is in detail no more and no less than constantly recurring radiation of populations and constantly occurring succession of adaptive changes in populations. This touches on other problems, particularly concerning the nature and origin of higher categories, that will require further discussion in this and later chapters (especially Chapter XI).

In early phases of an adaptive radiation, particularly one that does reach considerable or great scope of total adaptive range, there is usually an evident and sometimes almost a dramatic release of variability. A radiation does not occur unless prospectively adapted populations exist, and it is an aspect of their prospective adaptation that they have (as in fact do most populations, but in these cases probably to unusual degree) large pools of potential genetic variability (see Chapter III). Increase in total population, even though this is distributed among descendant species, relaxation of centripetal selection, and a centrifugal pattern of selection away from the ancestral condition and into a variety of diverging adaptive zones are concomitants of the situations in which adaptive radiation does occur. All these factors tend to release variation from existing pools of variability and also to increase the probability and rate of fixation of new variability arising by mutation.[12] One would then expect a phase in adaptive radiation in which intragroup variation was large and intergroup variation still larger and increasing rapidly. In addition to this, there would be a considerable number of adjacent and overlapping adaptive zones the boundaries of which were not yet clear and discontinuities between which were still small and fluctuating. A further expectation would therefore be development, for a time, of a great variety of adaptive types, rather poorly differentiated from each other, and partly conflicting and adaptively unstable.

These expectations are wholly fulfilled when we happen to get the appropriate early phases of an adaptive radiation. An unusually good example is provided by the Notoungulata (and some other groups) in the late Paleocene and early Eocene of South America (Riochican of Argentine Patagonia and Brazil, Casamayoran of Argentina, see Simp-

[12] There is also the possibility of selection for increased mutation rates in such a situation but it is questionable whether this is an appreciable factor or whether in any one lineage it would have time to operate before the lineage was sufficiently adjusted in its adaptive zone to make high mutation rates again disadvantageous.

son, 1948, also continuation of that work still in manuscript and studies by de Paula Couto now in progress). Intragroup variation is great, intergroup variation is unusual and is plainly increasing, the notoungulates as a whole are protean almost beyond parallel, and there are so many transitional types that the taxonomic problem seems almost hopeless.

In theory such a condition cannot persist indefinitely and in fact it does not in the known examples. It may be prolonged in a major radiation like that of the Notoungulata by the fact that the actually primary process, the splitting of populations, does not really go on all at once but now in one and now in another segment of the whole taxonomic group and in renewed bursts as new adaptive levels are reached in succession. Such a radiation, or rather, series of radiations, may thus continue in the expanding, highly variable phase for spans on the order of millions of years (see the next section of this chapter). Even in this phase, weeding out of populations and adaptive types is constantly going on; extinction rates are high although origination rates are still higher. The most prolonged radiation, however, finally reaches a time when rates of origination, averaged over the whole, fall off rapidly. In this phase rates of extinction are also usually decreasing but less rapidly, and they come to be higher than contemporaneous origination rates so that the net result is decrease in total numbers of groups present. In terms of the evolution of adaptation what is then happening is that existing adaptive zones have been occupied and appearance of new zones (a result inherent in the expansion of adaptive types) has slowed down or stopped. At the same time, overlap of zones is being eliminated, their boundaries are sharpening, discontinuities are developing between them, and intermediate or intergrading populations lose what adaptive stability they may have had and become extinct. Long range adaptive possibilities have developed and are being parceled out among a smaller number of more distinctive zones. Populations outside those zones are weeded out.

In the example of the notoungulates these effects are evident by late Eocene time and the phase, which of course is not sharply distinguishable from those preceding and following it, continued more or less throughout the Oligocene. In the Eocene there were nine families, primary products of the basic radiation. In the Oligocene there were also nine, five continuing from the Eocene and four the more highly divergent products of Eocene radiation. In the early Miocene, when the phase of parceling out and weeding out had been essentially completed, there were only six families and all of these had already been sharply distinct in the early Oligocene.

After divergence is complete and the long range main adaptive zones have become sharply defined and decidedly discontinuous from each other there follows a phase during which each main zone evolves independently with no evident further divergence or interaction between them. The radiation as such has definitely ended, both as regards its expanding phase, radiation strictly speaking, and its contracting or weeding out phase. For the notoungulates, as far as a time can be designated in a continuous, slowly shifting sequence, this change came around the end of the Oligocene. Thereafter progressive changes in the separate major zones continue and within each there are progressions between and changes within subzones, but the radiation by which the zones arose has no evident influence in these late events. There may even be renewed radiation in or from one of the zones, as a new and separate occurrence. (This did not happen among later notoungulates except on quite a minor scale within zones.)

This sequence of phases is a striking and often repeated element in the whole evolutionary pattern, but it is not the only element. It arises from time to time and on a smaller scale or a larger when a concatenation of circumstances results in the previously specified conditions for adaptive radiation. It is not constant, nor is it the only way in which stable, long range adaptive zones arise or are occupied. Their origin may be due at least as often to secular processes in the total evolution of adaptation and their occupation to successional evolution from zone to zone, and not to episodic radiation. Or both the secular and episodic elements may be involved, or again something between the two.

CYCLIC AND EPISODIC EVOLUTION

Adaptive radiation is a clear example of episodic evolution. With all its complications, prolongations, and repetitions at different levels of effects, an instance of adaptive radiation is an episode set off from preceding and following events at comparable levels and defined by a beginning and an end. It also has a characteristic course which may be closely similar in different episodes of adaptive radiation. Such observations have often suggested to students that the evolutionary process as a whole or some aspects of it might be cyclic, that is, that it might consist of episodes of the same sort recurring at more or less regular intervals. Two principal sorts of theories of cyclic evolution have been proposed: that a group or characteristic sort of organisms tends to go through a kind of racial life cycle repeated in each successive group, and that geologic history and biologic history consist over-all of a series of repeating, simultaneous physical and evolutionary cycles.

Theories of a racial life cycle usually recognize three phases described more or less as follows: an initial, relatively short phase of vigorous expansion and diversification, a longer stable phase of slow change and little increase in diversity unless among minor adaptive or geographically local types, and a final phase, again relatively short, showing more or less marked diversification into monstrous, overspecialized, degenerate, or generally inadaptive forms, followed by extinction. Numerous different terms have been proposed for these phases. Schindewolf (e.g., 1950a), who is among the few well-informed recent students who believe that such phases really do follow each other with noteworthy regularity, calls them the phases of typogenesis, typostasis, and typolysis, respectively. An analogy is often drawn with individual life and its juvenile, adult, and senile phases—an analogy obviously perilous because these individual ontogenetic phases are known to be defined and conditioned by factors that could not possibly apply to ancestral and descendant sequences of populations and that do not even have any known or likely analogy in the factors affecting such sequences.[13]

The first of these postulated phases is in general like the first phase of adaptive radiation as described above and the second suggests the phase of mainly intrazonal evolution that often follows after the radiation has quieted down and an intermediate phase of weeding out (which is not invariably present as a distinct time phase) has intervened. Adaptive radiation is common in evolution, so is stable intrazonal evolution, and to this extent the postulated phases certainly occur. That they fol-

[13] Schindewolf extends the same analogy to human cultural history also and concludes (in German) that "these manifold analogies make it probable *that all temporal-historical development follows the same sort of patterns* and that phylogenetic evolution to this extent falls into a higher category of generalization (by which, to be sure, nothing is yet expressed as to the inner determinants of these phenomena, which in themselves are only analogous)." (Italics his.)

low each other inevitably, that major radiation is the constant pattern of a "young" group, or that all of evolution can be realistically analyzed in such terms is not true.

The sequence radiation–intrazonal evolution is usual, simply because radiation does not occur unless there are diverse zones within which evolution will follow. Occasionally, nevertheless, something happens to close the zones so soon that the radiation is curtailed or the intrazonal phase is even shorter than the radiation. The camarate crinoids, for instance, seem to have been in the full swing of a radiation when they all became extinct in the Carboniferous. Occupation of a new zone and subsequent intrazonal evolution may also occur without radiation or a "youthful" phase, unless the whole scheme is reduced to mere wordage by saying that change from one zone to another *is* radiation or a "youthful" phase. No radiation seems to have been involved in the rise of the Stylinodontinae (Patterson, 1949) and none occurred among them during their whole history. Common as adaptive radiation is, such examples are also rather common, although probably less so from the mere fact that radiation may reach any extent and opportunities for radiation of small extent are likely to occur for any group that long survives.

Unless, again, the whole problem is made meaningless by calling any short-range diversification, such as the nearly universal and continual pattern of geographic speciation, "youthful," "typogenesis," or the like, it is also untrue that radiation regularly follows the appearance of a new group. It will shortly be shown that periods of most rapid diversification may occur at any time in the history of a group but seem to be less common just at its beginning than later. Note, for one example, the Mammalia, in which there was little marked or apparently basic diversification for at least 75 million years after the class arose. Nor is it literally true that basic differentiation into major adaptive zones regularly occurs within a group by one radiation (early or late), with subsidiary cycles later filling in the chinks, in a manner of speaking. Something descriptively like this can occur, as in the Paleocene origin of many of the orders of mammals, although even in such cases the divergence of the major groups, the thing that makes us call them "major groups" a posteriori, was a long and later process. In other cases not even this exiguous agreement with the racial life history theory appears. The vertebrates, for instance, certainly did not radiate into their major adaptive types in any "typogenetic" phase, but developed the major types, in large part by successive occupation of zones, over some 200 million years.

The supposed last phase of a racial life cycle, that of senility followed by death, brings up the general problem of extinction, to which this concept is only ancillary. Extinction will be discussed in Chapter IX. At this point it need only be said that the concept of senility, as applied to evolving populations, seems to be a false analogy and that a phase even analogical with old age does not appear really to occur in evolution. The whole theory of racial cycles is an overgeneralization and invalid interpretation of some phenomena that are real (buttressed by some that probably are not) but that do not really have the claimed relationships to evolution over-all.

Distinctive adaptive types and corresponding taxonomic groups are likely to be numerically small when they arise and thereafter to increase more or less in numbers and variety if they survive long. That is a necessary concomitant of any usual mode of development that can be ascribed to them. Once the group is established, it inevitably varies in size and diversity and so must have a maximum point at some time or other. In some cases the phenomena of expansion and contraction seem to have no more interest than this casual fact. In most groups there are, however, one or more well-defined maxima which clearly are not casual but represent definite and important episodes in the history of the group. Fossil sampling permits no fully reliable determination of absolute numbers of taxonomic units in a group, but at levels from genera upward, less commonly for species, it is probable that determinations of rise and fall and of times of maxima and minima are reasonably reliable for groups fairly common as fossils. Not even an approximation of figures for census population is possible except in the vaguest terms, but it may often be a reasonable (although it is not a necessary) assumption that total numbers of individuals in any one group tend to vary with numbers of taxonomic groups. (Of course this assumption is not even roughly permissible in comparison of groups of quite different adaptive types.)

Something was said about peaks of taxonomic diversity in Chapter II from the point of view of their tabulation and of the rates of evolution leading to (and from) them. The purpose here is to consider them as episodes in the evolution of adaptation, related to the subject of adaptive radiation (also of racial cycles, already discussed) and also to the theory of simultaneous geologic and biologic cycles.

Sudden rises in taxonomic diversity have frequently been discussed as examples of "explosive evolution." The term is ill chosen, as has been pointed out a number of times, for instance by Cloud (1949), although it has a certain unintended humor. These are explosions that commonly take millions of years from fuse to bang (which is silent). Cloud has suggested "eruptive evolution," which is better, surely, but still suggests an inept mental picture to me, at least. Rensch (1947) follows Wedekind in speaking of "Virenzperioden," which is better yet but might carry some undertone of the life-cycle analogy. Various other picturesque terms have been used, but I shall call these events more primly "episodes of proliferation," not a very snappy metaphor but at least a reasonable designation.

Many different factors in the complex web of evolution may underlie any given episode of proliferation. Rapid phyletic evolution, for instance, runs up the number of taxonomic units existing (although not simultaneously) in a given period of time and occurs during episodes of proliferation. The one feature probably present and determinative in all these episodes, however, is adaptive radiation. That conclusion has been questioned and it can hardly be shown to lack exceptions, but examination of many examples always reveals what seems to me clearly adaptive radiation related to each episode. This is especially convincing when more than one episode has occurred in a large group. Episodes of proliferation after the first generally (always, as far as I have found) turn out to be due to adaptive radiation in included groups. Thus in the Osteichthyes the first episode of proliferation follows an adaptive radiation in the Chondrostei, the second another such radiation in the Holostei, and the third still another in the Teleostei. In the Spiriferaceae the first peak is really a resultant of more or less independent radiations in several families and the second is mainly due to an adaptive radiation of the Athyridae.[14] (See Chapter II.)

[14] Query should hardly arise at this point as to how it is known that the radiation, which is factual, really is adaptive, which is an interpretation. Sufficient has already been said to indicate that speciation, the basic process of radiation, is normally adaptive. The general evidence for mainly adaptive control in such events is so impelling that it cannot be doubted even when, as with species or genera of brachiopods, the precise adaptive significance of each form may be dubious. In other cases, as for the teleosts, the adaptive significance is reasonably known for some if not for all of the groups involved.

It was also stated and exemplified in Chapter II that episodes of proliferation may come early, middle, or late in the history of a group. This confirms the conclusion that adaptive radiation is episodic but not cyclic. Radiations occur when certain conditions are produced by the whole concatenation of previous historical events influencing the organism and environment relationship, and not from anything regularly inherent in the organisms. The peak of an episode of proliferation follows some time (usually quite a long time) after radiation begins. This does not mean that incidence is completely random, nor would this be expected in view of the fact that there are some broad but pertinent similarities between the episodes and between the whole adaptive histories of the groups involved in them. There is, in fact, some evidence that although the peaks *may* occur at any time whatever in the span of a group they are *most likely* to occur during the first half but not the earliest part of that span.

Innumerable complications arise in attempting to compile data on such a point. Apparently objective data may merely reflect a subjective approach. For instance, as noted, the Osteichthyes had a peak of generic proliferation in the Cenozoic, near the end of their span. But this was almost entirely in the Teleostei, which originated much later than the Osteichthyes as a whole. Moreover, this peak may be considered a resultant of radiation within lines which had themselves, for the most part, arisen in an adaptive radiation in the Cretaceous. The peaks of the same episode of proliferation are different for different levels of radiation. In spite of these difficulties, an experienced paleontologist can usually designate with some assurance an epoch or period in which a group reached its generally highest point of abundance and differentiation. For instance Stromer (1944) has indicated such high points for 35 groups of about the average scope of orders, 16 invertebrate and 19 vertebrate.[15] It is an advantage for our purpose that Stromer did not select these groups to show the distribution of the high points, but for quite a different purpose (incidence of gigantism). If the spans are divided into deciles, the distribution of high points is as shown in Table 21. This is not a random distribution. The peak in the third

[15] Stromer omitted high points for two mammalian groups that he nevertheless put in his table and I have supplied these: late Paleocene—early Eocene for the "Prosimiae" and latest Cretaceous—Paleocene for the Multituberculata. I have also made some small corrections in his data, e.g., the Multituberculata as now defined are unknown before late Jurassic.

TABLE 21

RELATIVE POSITIONS OF "HIGH POINTS" IN EVOLUTION OF VARIOUS ANIMALS
(Calculated from a table by Stromer, see text)

Decile	*Invertebrates*	*Vertebrates*	*Both*
1	1	2	3
2	2	0	2
3	3	6	9
4	0	4	4
5	2	0	2
6	1	1	2
7	1	1	2
8	1	2	3
9	2	0	2
10	3	3	6
	16	19	35

decile is statistically significant and seems also to be biologically so. The peak in the last decile is also significant statistically but probably is not biologically. It is due mainly to Cenozoic high points of groups still living, hence of indeterminate span. It is practically certain that these peaks will not be in the last part of total span since the groups concerned (e.g., Rodentia) are flourishing and give no signs of approaching extinction. With allowance made for this factor, it seems probable that very late high points, in the ninth or tenth deciles, are relatively few.

Tabulation of absolute times of high points after appearances of the groups would be less significant for present purposes because of the very great differences in duration and rates of evolution in the various groups (vertebrates as a whole having decidedly shorter durations and faster rates than invertebrates). It is, however, of some interest that high points for Stromer's selection of vertebrates range from 10 or 15 (Prosimii, Mystacoceti) to more than 250 (Selachii) and average about 55 million years after appearance of the group, while for invertebrates the range is about 20 (Protodonata) to perhaps 450 (Foraminifera, Pelecypoda, Gastropoda [16]) with a mean of about 180 million years. At any rate, it is clear that considerable time elapses before a group of much scope reaches a high point and that a high or the highest point (not necessarily the first) may appear after a very long time.

[16] There is, however, little doubt that these groups now at or near maximum had earlier high points as well.

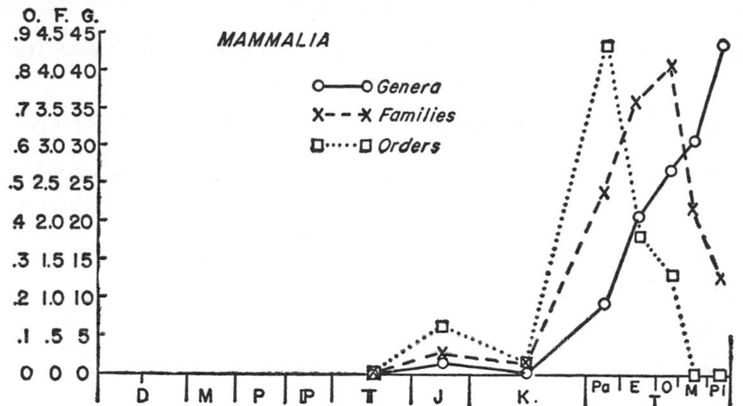

FIGURE 29. FIRST APPEARANCES PER MILLION YEARS FOR MAMMALIAN ORDERS (O), FAMILIES (F), AND GENERA (G).

Another point before now only touched on in passing, or obliquely, demands special comment. If time frequency curves are plotted for the same group in terms of different taxonomic levels, the peaks for higher categories usually appear earlier than those for lower categories. This is evident, for example, in the data for origination rates in aquatic vertebrates in Figure 9 (Chapter II), for mammals in Figure 29, and for vertebrates as a whole in Table 22. Even when, using the coarse scale of periods, peaks for different categories are in the same period, those of higher categories are earlier in the period as the data for mammals show.

Offhand, this fact suggests that orders arise first, then families, then genera, and so on, as Schindewolf has so often insisted.[17] Closer study

[17] Schindewolf's theories are so frequently criticized here that it should be pointed out that critical reference to them is worth while just because Schindewolf is among the few paleontologists (such men are few in any science) who have really broad grasp and subtle, theoretical minds. He has paid more attention than most to the really important theoretical problems of his science. His views are based, as in this case, on real evidence and must be taken seriously. The points on which I disagree naturally require more specific notice of his work than do those on which I agree.

On the present point, Schindewolf seems to mean quite literally that the higher categories of a group arise as such before the lower. Our disagreement is not entirely fundamental if this is taken as a manner of speaking, a broad and figurative view of the net result rather than a description of the process. This is the position of Wright (1949a), with which I agree in essentials. Schindewolf (1950b) has quoted two sentences from this paper of Wright's as indicating agreement with his views. The rest of that paper and, indeed, Wright's whole great body of work show that his agreement extends only as far as does mine with Schindewolf. The point actually

TABLE 22

PEAKS OF FREQUENCY OF APPEARANCE OF DIFFERENT TAXONOMIC CATEGORIES IN THE VERTEBRATE CLASSES

Period	*Peak for orders*	*Peak for families*	*Peak for genera*
Tertiary	Mammalia	Mammalia Osteichthyes 3	Mammalia Osteichthyes 3
Cretaceous	Osteichthyes 3	....	Chondrichthyes 2
Jurassic	Chondrichthyes 2	Chondrichthyes 2	....
Triassic	Osteichthyes 2	Osteichthyes 2	Osteichthyes 2
Permian	Reptilia 1	Reptilia 1	Reptilia 1
Pennsylvanian	....	Amphibia	Amphibia
Mississippian	Amphibia	Osteichthyes 1 Chondrichthyes 1	Osteichthyes 1 Chondrichthyes 1
Devonian	Osteichthyes 1 Chondrichthyes 1	Placodermi	Placodermi
Silurian	Placodermi Agnatha	Agnatha	Agnatha

shows quite clearly, I think, that this is a misleading statement as to the processes actually involved. One reason why it is misleading is evident if one considers the result of a single, isolable incident of adaptive radiation such as that of, say, the Geospizidae on the Galápagos. It would be true in a formal sense to say in this case that the family rose before most of its genera or species and before the generic peak which in turn probably came before the specific peak. But in an example so clear in most details down to the level of species (thanks largely to Lack), this is obviously only a manner of speaking or even an artifact of classification. Now that the adaptive radiation is complete, we recognize the result as a family (or subfamily). Looking backward from here, we consider that the family arose when its first species was differentiated. That species probably differed very little from its immediately ancestral species (which is, however, unknown in this example without paleontological support). The family did not arise as such but as a *species.* The family resulted from the whole radiation, and its first species is placed in it in retrospect. The peak for genera (4) obviously had to be

at issue is whether an order, say, arises and develops by processes of selection acting on ordinary mutations and in the basic form of speciation under special and necessarily infrequent circumstances (since orders are infrequent relative to species), or whether it is the immediate result of one large mutation without benefit of selection or of speciation in any ordinary sense. In the paper cited and elsewhere, Wright has been fully explicit in his opposition to the latter view.

later than the appearance of the first species now placed in the family and that for species (14) had to be later than the generic peak unless, as is improbable, the two monotypic genera arose after speciation was complete in the two polytypic ones.

In major episodes of proliferation, such as those of the vertebrate classes, analysis shows that adaptive radiation is clearly involved but that we are not dealing with a single and isolable episode of radiation corresponding with or leading to a single, short peak. There is, instead, an extremely complex series of radiations all of which began basically at the specific level but which in retrospect had different final net results. In fact it may be said that an episode of proliferation, no matter how high the level, is simply exuberant speciation irregularly continual over a long period plus a considerable amount of phyletic divergence among the lineages thus arising. Earlier radiations eventually have more far-reaching aspects not only from the simple fact that they are earlier and that their lines therefore have time to diverge farther but also because it is these lines that start the occupation of a broad new adaptive zone and that therefore have inherently more space for eventual divergence. Reviewing the whole phenomenon in retrospect and putting it into taxonomic form, we generalize by running the origins of larger subzones, corresponding with higher categories, farther back to earlier radiations than those of smaller subzones. It does not follow and when we have the crucial evidence it is found not to be a fact (as will be exemplified and further discussed in Chapter XI) that what are taxonomically the earliest representatives of, say, an order were ordinally distinct from their contemporaries in the sense of having anything like the degree of morphological or adaptive distinction and discontinuity that we associate with the concept of an order in considering the final results of the process. If, indeed, we can trace the lines back to the point where the actual biological splitting of one order from another occurred (unfortunately an impossibility in most cases), the distinction existing when the fundamental discontinuity actually arose is only specific. There is, indeed, a genus in the early Paleocene (*Protogonodon*) some species of which are *by morphological definition* members of the Order Carnivora and others members of the Order Condylarthra.

With these relationships in mind, it is seen that the spread between, for instance, ordinal and generic peaks for an episode of proliferation in a class tends to reflect the length of time during which a complex sequence of adaptive radiation continued. The ordinal peak follows, probably by quite a long interval, the beginning of important adaptive radiation in the originally relatively unified primitive representatives of the class. Other radiations then occur among the various lines from the earliest radiation, and yet others among the more and more numerous lines as radiation continues. Separate radiations at the basic level, as far as they can be separated, are quite irregularly distributed in time and also among the lineages present, but the total resultant is increase in number of *species* while the process continues. When it ends, what is taxonomically the last peak, that for species, occurs. The ending of the process in this sense does not necessarily mean that *all* adaptive radiation ceases—this is not usually the case in a very large group, at least—but that the over-all total begins to decline because extinction has overtaken origination.

The total time involved in such a major episode is large. In the vertebrate classes, the generic peaks occur 25 to 50 million years after the corresponding ordinal peaks. Since significant radiation must begin well before an ordinal peak and may not end until well after the generic peak, the whole of one of these episodes is indeed a grand event occurring over a minimum span of about 30 million years and perhaps in exceptional cases extending over 70 or more million years. Certainly we have in such episodes some of the most truly major and long-range elements of evolutionary history. Of course there are also lesser episodes of proliferation, either as parts of major episodes or as separate events, which run over much shorter spans. A lower limit cannot be fixed, but it seems unlikely that a minimal, single radiation, recognizable as such, could normally occur in much less than a million years. (Yet these are all "explosions"!)

The weeding-out effect noted in discussion of adaptive radiation is also evident in most episodes of proliferation and it begins long before the episode has run its course. On a small scale it goes on all the time, but it becomes evident on a large scale when the major subdivision, e.g., that into orders in the Mammalia, has reached a peak. It naturally affects first the results of the first radiation and so first becomes predominant over origination in the most rapidly and basically divergent groups. In a large episode of proliferation, as opposed to a single adaptive radiation, much of the weeding out is likely also to be by replacement of early groups by groups evolved in later radiations within the episode as a whole and able to reenter older zones competitively. Replacement of condylarths by perissodactyls and artiodactyls or of creodonts by fissipeds is of this sort. The result of these effects is a tendency for higher categories to decline in number while lower categories, by later radiations in the more adaptively successful, surviving higher categories, are still increasing in frequency. This, too, is striking in the Mammalia, with ordinal peak in the Eocene and generic in the Pliocene.

When one group is replaced by another in a major adaptive zone it frequently, perhaps usually, happens that both have episodes of proliferation overlapping in such a way that the earlier group is in the last expanding phases of its episode while the replacing group is beginning expansion. Thus in replacement of Agnatha by Placodermi and of Placodermi by Chondrichthyes and Osteichthyes, jointly, the ordinal peak for replacing groups coincides approximately with the generic peak of the group later replaced.

The last topic to be considered in this chapter is the theory, or rather the various theories, that involve correlated geological and biological cycles. One such theory, formerly widely held and still occasionally advanced, e.g., by Umbgrove (1942), is that there are recurrent cycles of world-wide mountain building each of which is accompanied or immediately followed by a major evolutionary outburst or "explosive" evolution on a large scale. However, even if the tectonic cycles are taken as given, there is no real correlation with major episodes of proliferation, as Umbgrove's own data show (see Simpson, 1949b). Indeed it seems rather pointless to look for a correlation when it becomes clear that such episodes of proliferation do not occur with sharp definition in time but have phases extending over as much as 50 million years or more, a figure of the same order as the average length of intervals between the supposed major tectonic episodes. It has also been assumed that the origins of major groups or the beginnings of episodes of proliferation might correlate with the tectonic episodes, but there seems to be no reliable evidence of this, either. As if to put the last nail in the coffin of this now apparently moribund group of theories, there is a growing belief among geologists (e.g., Gilluly, 1949) that the geological part of the theories was wrong, anyhow, and that the postulated tectonic phases did not occur, that is, that such phases are not really either regularly cyclic or world-wide. The least one can say is that the regularities and simple relationships demanded by the theories in question are unproven and very improbable on present evidence, but this does not quite close the subject.

There have been rather definite times in earth history when an unusual amount of evolutionary movement of one sort or another was going on in so many groups that pure coincidence seems almost out of the question and in such a variety of organisms that inherent evolutionary similarities in them as opposed to some broad environmental change also seem almost ruled out. For the vertebrates such episodes clearly occurred three times:

1. Devonian. Rapid turnover from earlier to later major groups of aquatic vertebrates with extremely high extinction rates for the former and high-category origination rates for the latter. Origin of Amphibia.

2. Permian to Triassic. Rapid turnover among both aquatic (especially Osteichthyes) and terrestrial (especially Reptilia) vertebrates with high extinction rates in the Permian or both Permian and Triassic for some groups, and high origination rates in the Triassic.

3. Cretaceous to Tertiary. Some change among aquatic vertebrates (perhaps not of crisis proportions near the era boundary). Very high turnover among terrestrial vertebrates.

Schindewolf (1950b) has made a similar list based on "turning points" (which are in general climaxes of what I have called "turnover" above) in a wider variety of organisms, including the vertebrate classes and also fourteen major groups of invertebrates and the plants as a whole. Three of his times of general evolutionary intensity are the same as those listed above for vertebrates alone. (He puts the first between Devonian and Mississippian rather than in the Devonian, but the dating is not sharp enough to give such a difference any importance.) He also lists two more:

a. Cambrian to Ordovician. Rapid turnover of marine invertebrates.

b. Triassic to Jurassic. Again considerable turnover of marine invertebrates and some of the vertebrates, but in both cases this seems less profound than the Permian-Triassic turnover.

Unquestionably still another major episode of this sort should be listed: the pre-Cambrian–Cambrian change.

For these crucial evolutionary times Schindewolf shows that, like major episodes of proliferation, they do not correlate with supposedly world-wide physical events as listed by Umbgrove and other geologists.

The wrong correlations have been sought and the situation is not as simple as was formerly believed, or perhaps hoped. The outstanding periods of fairly general evolutionary crisis are episodic and not cyclic and are not especially connected with mountain-building. Nevertheless it does seem likely that physical events influenced and helped to determine the times of these crises although they may not have been primary causes. What the record strongly suggests is that these crises followed periods of widespread hard times, that is, of deteriorating conditions in many established adaptive zones. They must almost certainly be imputed to changing environments in the broadest sense of that word. The changes involved may not have been, indeed, probably were not, the same in all crises or for all organisms concerned in one crisis. Precisely what the changes were cannot now be clearly specified in any case. Study of past environments, in the full sense of the word and on a broad scale, is one of the most difficult and one of the least advanced subjects in paleontology and geology although progress is being made. In the meantime, hints and possibilities are not lacking. For instance, the Permian-Triassic and the Cretaceous-Tertiary crises do coincide (as nearly as such very long episodes can coincide) with exceptionally great land emergence and, in the case of the earlier, with considerable evidence of abnormal climates. (Abnormal climate is also frequently claimed around the time of the Cretaceous-Tertiary crisis, but the concrete evidence is still very unconvincing.)

Given the way adaptation evolves, it is probable that most or all of the turnovers that occurred in these crises would have occurred sooner or later in any case. Their tendency to group together suggests that their occurring just when they did was influenced by widespread environmental factors, quite possibly physical in part, at least.

To avoid any misunderstanding it may be well to restate the obvious fact that the physical environment and the history of the earth have certainly and strongly influenced the history of life. The influence is always there, as so much of the history of life is the evolution of adaptation and physical environment is always an important factor in adaptation. Sometimes the influence of geologic history is quite obvious, for instance in the shifting relationships of land and sea connections. Often its influence is inferred but the exact bearing is obscure. As of now, it does not seem to be true that rhythms or cycles of the earth are simply and directly reflected in cycles of evolution. It does still seem probable that episodic earth changes have had an effect on the timing, at least, of episodic crises in evolution.

CHAPTER VIII

Trends and Orientation

ALL THAT HAS GONE BEFORE involves the proposition that evolution has orientation. It has, certainly, random elements and even phases, but as a whole it obviously is not random. Since this is obvious, the fact does not need belaboring, but merely stating the fact does not settle many questions that it raises. Since evolution is to some evidently large degree nonrandom and oriented, phylogenetic sequences or many of them have some element of sustained direction; they show a prevailing tendency; or, in other words, they have trends. It need not be questioned that trends are common, but are they usual or universal? Are they more characteristic of one phase of evolution than of another? What is their general nature (if they have one)? To what extent do related or simultaneous groups have similar trends? Why and how do different characteristics in the same phyletic progressions show correlated trends? And, especially, what are the factors involved in the orientation of evolution and the limiting of trends?

Some of these points have been touched on before in different contexts, and answers to some of these questions are implicit in previous discussion. Now the phenomena of trends are to be examined in their own terms and explicit answers to the questions sought.

TRENDS AND THEIR PART IN EVOLUTION

An evolutionary trend is a sustained, prevailing tendency in a phylogenetic progression.[1] Almost all fossil sequences long enough to be called "sustained" show prevailing tendencies in some characters and over a part, at least, of the sequence. Trends are thus extremely com-

[1] Some students (e.g., Swinnerton, 1947) define the word in a more restricted way, applying it to a tendency for the same changes to occur in different lines of descent, i.e., to parallel trends. Here, "trend" is used for a general phenomenon of which parallel trends are one case. It is, further, intended to be a neutral, descriptive term without implication as to causes.

mon in paleontological data, as far as these do include good ancestral-descendant series. The most frequently cited example is the known sequence of Equidae, to be reviewed later in this chapter. A classic invertebrate example is that of *Micraster,* a sea-urchin, from the English Chalk (late Cretaceous), first satisfactorily worked out by Rowe (1899). In this complex, well-preserved organism, Rowe found definite trends in eleven apparently independent characters, although the trends are not uniform throughout, which is a different point. For instance, a continuous sequence of overlapping populations from low-zonal *M. corbovis* to high-zonal *M. coranguinam* shows, among others, trends for increasing relative breadth and height, forward migration of mouth, and development of a tip posteroventrally to the mouth.

It would be pointless here to multiply examples simply to show that trends do occur. The literature of paleontology is full of them. Two generalizations are clear from the start: (1) trends are usual in paleontological sequences, and (2) they may involve any visible characters of organisms.

A third generalization is that diverse descendant lines of common ancestry often show similar, or more technically, parallel trends. This is what some English students call "programme evolution." Examples of this, too, are very common. A good vertebrate example is the sequence of the late Miocene to Recent Myospalacini (rodents, "mole mice" or sokhors) of China described by Teilhard (1942, 1950).[2] Early Pliocene speciation produced three lines that can be followed in much detail. Thereafter all three simultaneously showed trends (1) to larger size, (2) to fusion of the cervical vertebrae, and (3) to increasing hypsodonty. Another vertebrate example on a grand scale is that of the therapsid reptiles, all of which in their various lines and major groups became more mammallike in numerous characters (Olson, 1944). Among invertebrates, graptolites show striking "programme evolution" (e.g., Bulman, 1933; Thomas, 1940; George, 1948). Particularly interesting is the clear evidence that different lineages of *Dimorphograptus* each independently and progressively reduced one of the two rows of thecae and so polyphyletically gave rise to the "genus" *Monograptus* and indeed the "family" Monograptidae of many authors (Fig. 30). (Of course the "genus" and "family" thus originated are not genetical units

[2] He uses the antedated name *Siphneus* for *Myospalax.*

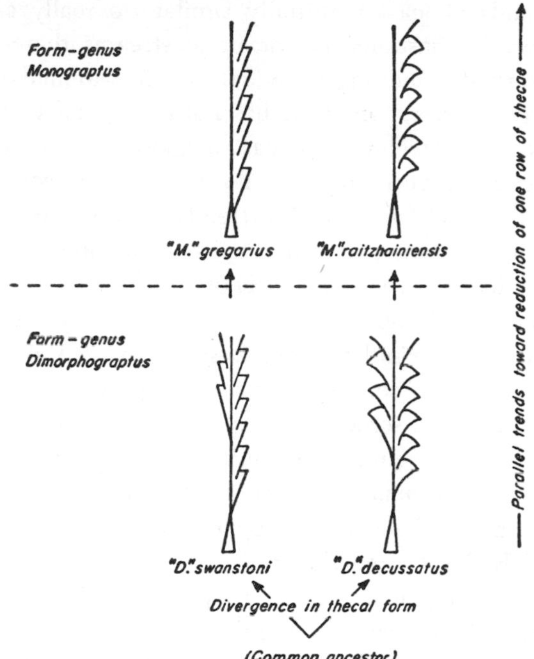

FIGURE 30. PARALLEL TRENDS IN GRAPTOLITES. Divergent evolution from a common ancestor produced several lines (only two of which are shown) with different thecal forms. Parallel trends in several of these lines resulted in reduction of one row of proximal thecae, without noticeable effect on thecal form. (Based on a diagram by George, 1948.)

but are conventional form-units of paleontological taxonomy, a point that need not bother us here.)

It has been supposed more broadly that at certain times similar trends commonly appear in different and unrelated lines, so that there are "Zeitcharaktere" (Beurlen, 1929), "Zeitbaustile, Zeitsignaturen" (Dacqué, 1935), or "Trachtenmoden" (Brinkmann, 1929). Examination of the proposed examples, however, shows that in most cases the lines are in fact related, so that they merely constitute more or less extended instances of parallel trends, and that other cases are so general that they amount to little more than a verbalization of the obvious fact that each geologic period does have characteristic fossils. (See further discussion and references in Haas and Simpson, 1946, also Stromer, 1944.)

Sometimes trends at least nominally similar do really affect wholly distinct groups. For instance in torrential streams there is a similar tendency toward streamlining in widely diverse animals. This is obviously adaptive convergence that has nothing to do with contemporaneity of the animals (but may cast a good deal of light on "programme evolution"). All in all, I do not see that the concept of "time signatures," as distinct from parallel trends, on one hand, or adaptive convergence, on the other, designates a real phenomenon or has any value and meaning for evolutionary theory.

A related series of phenomena that are undoubtedly real involves the repeated occurrence of similar trends in successive offshoots of a group, often from a continuing "conservative stem," a process called "iteration" by Beurlen (1929) and some students following him. (His term goes back to the "iterative Artenbildung" of Koken.) Repeated trends toward coiling in a plane (*Gryphaea*) or spiral (*Exogyra;* both are collective form-genera) appeared in oysters separately arising from a long persistent, relatively flat, *Ostrea*-like stock.

Another well-documented example may be given in somewhat greater detail because it is of interest not only in this connection but also as illuminating a number of different points. The data were provided by Kaufmann (1933, 1935) in an extraordinary study that was slow to receive the attention it deserved.[3] He studied a late Cambrian sequence of the trilobite *Olenus* in a limited part of southern Sweden. Four distinct lineages could be recognized:

Oldest: 1. *Olenus gibbosus anterior–O. g. media–O. g. posterior.*

Middle: 2. *Olenus transversus–O. truncatus–O. wahlenbergi.*

Youngest: Two overlapping lineages, 3 appearing somewhat earlier and also becoming extinct earlier than 4–

3. *Olenus attenuatus anterior–O. a. media–O. a. posterior.*
4. *Olenus dentatus anterior–O. d. media–O. d. posterior.*

Between the disappearance of 1 and appearance of 2 and again be-

[3] It should be stated that it is possible to combine Kaufmann's brilliant observation and analysis with later advances in theoretical paleontology to reach conclusions additional to his and in some respects different from his. This is done here without pointing out the differences in detail. Only one is of possibly essential importance. Kaufmann recognized gradual progressive change, "Artabwandlung," within each of his four lines but thought that each arose by a different process, "Artumbildung," probably involving saltation. His data better fit the hypothesis that the lines arose by ordinary splitting, speciation, and "Artabwandlung" from a continuing stem, itself evolving by slower "Artabwandlung," with no saltation evident or required at any point.

tween the disappearance of 2 and appearance of 3 are beds barren of trilobites. They apparently represent times of local environments unfavorable to these animals, with resulting local extinction of older lines followed by re-invasion when the environment again became favorable. Extinction of line 3 may have resulted from competition of the similar, invading line 4.

Throughout lines 1, 3, and 4 there is essentially complete morphological intergradation in successive populations with gradually changing means. This is essentially true also of line 2, but there the samples are not quite continuous as each is from a well-defined bed, apparently representing mass mortality.

The earliest members of lines 1 and 2 are practically indistinguishable. They clearly represent offshoots from the same ancestral population which persisted elsewhere without much change while line 1 was evolving here. The earliest specimens of line 4 are also similar, but a little more distinctive. This can be imputed in the main to evolution in the ancestral population elsewhere. (More time had elapsed than between appearances of 1 and 2.) From these earliest members of lines 1, 2, and 4, the character of the conservative ancestral stock and its evolution can be reconstructed in some detail and with considerable certainty (see Fig. 31). The earliest members of line 3 are more distinctive and must have characters not present in the line ancestral to 1, 2, and 4. Line 3 may nevertheless have arisen from that stock and have evolved more rapidly than the others while migrating, or it may be from a different, closely related stock. The difference between the two possibilities is not very important, because both involve geographic differentiation of species of *Olenus* and a double migration resulting in sympatric occurrence of closely similar species. Significantly, the two did not long persist together but the earlier migrant became extinct here soon after the later invader appeared.

All four lines independently showed certain trends after reaching this area. In the pygidium (tail piece), which I have used in my summary figure, the most obvious trend or "iteration" is in the fact that this part became relatively longer and narrower in all four lines.

There are also differences of two sorts between the lines:

a. Those due to evolution elsewhere; for 1, 2, and 4, at least, this represents slow progression in the ancestral stock between the times when these lines split off.

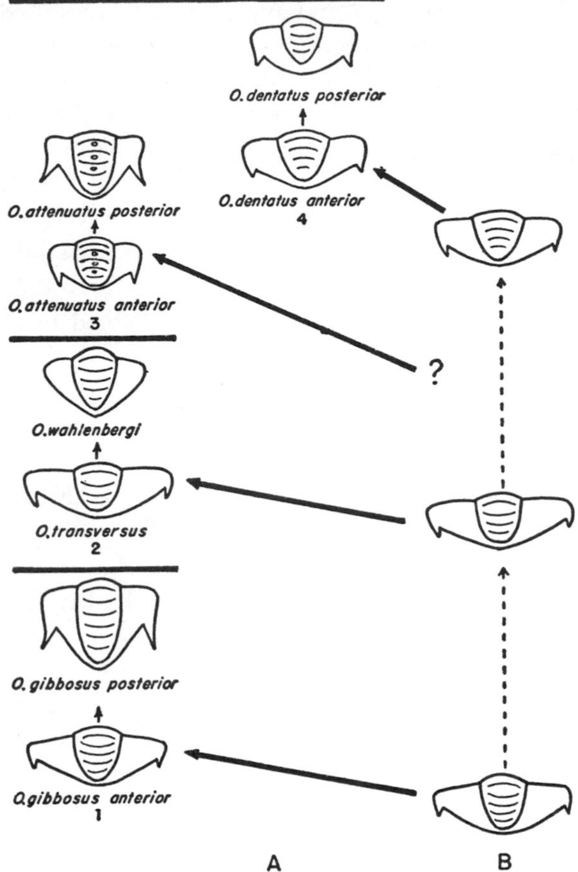

FIGURE 31. PARALLEL TRENDS IN THE TRILOBITE GENUS *Olenus*. A, four separate lineages as represented by their pygidia; early and late members of each are shown—in all cases intermediate forms are also known. B, reconstruction of conservative ancestral stem from early members of lineages. Relative time sequence is correctly shown, oldest at bottom. Heavy transverse lines indicate local conditions lethal to trilobites. Short arrows indicate phylogeny *in situ;* broken arrows, phylogeny of allopatric ancestry. Long, heavy arrows indicate migration into region where specimens were collected. (Data from Kaufmann, 1933.)

b. Those due to noniterative evolution within the lines (e.g., on the pygidium, loss of lateral spines in line 2).

Iteration has exactly the same aspect as have parallel trends, and in both cases the lines concerned are more or less closely related in about the degree that the common trends are more or less similar. The conclusion is inescapable that these are, in fact, the same phenomenon except for the distinction, unimportant for interpretation of the trends, that in one case (parallel trends) the splitting of lines from a common ancestry was roughly simultaneous and in the other case (iteration) it was appreciably successive but from a continuing ancestral group that did not change much during the intervals.

In the example of *Olenus* it is important that the iteration involves certain rather general characteristics and neither obliterates differences inherited from the ancestry nor precludes divergence within the lines. This is true as another generalization for both parallel and iterative trends. In *Monograptus*, for instance, the parallel trends involve a single feature, which happens to be obvious and convenient for the practical taxonomist, but which very likely was a simple change genetically and ontogenetically. At the same time, the forms of the thecae, which are likely to reflect more complex genetic and developmental factors, have characteristics distinctive of each line and clearly derived from ancestors in *Dimorphograptus* (Fig. 30). (In passing, note that this has a bearing on the fact that distinction between diagnostic characters of lower and higher categories is quite arbitrary: it would be a better representation of the genetics of the situation to consider the thecal characters as generic and the change from biserial to uniserial arrangement as specific, but the taxonomists have found it more convenient to make the latter generic and the former specific.)

Since both parallel and iterative trends involve definite genetic resemblance among the lineages affected, it would be expected that common directional tendencies due to any cause would work out rather similarly in them. Similarity in genetic systems would reinforce similarity in whatever factors make for the change. The question arises whether the similarity is not, in fact, wholly in the genetic systems, whether the change, too, is not entirely inherent in them rather than an interaction between them and something else, such as the environment-selection-adaptation complex. Among many reasons for believing that this is not true is the fact that similar trends often appear under similar circumstances in organisms so distantly related that the common trends cannot reasonably be ascribed to common "internal" or genetic factors. These trends in unrelated groups also frequently have incidences that have nothing to do with whether the organisms are contemporaneous or successive but show that the relative ages of the organisms are irrelevant. Thus parallel and iterative trends are seen to be only special cases of a much more general phenomenon, cases that are special to the extent that genetic similarity exists and strengthens the similarity in trends. Genetic similarity to a significant degree exists only if the organisms are of roughly the same geologic age, not far from their common ancestry, or are successive branches from a conservative stem. There is, of course, complete gradation in all these respects, so that the special cases are, after all, just those more toward one end of a complete quantitative continuum and not qualitatively different from the general phenomenon (Fig. 32).

This general phenomenon involves many of the principles, in fact descriptive generalizations, that have been recognized by paleontologists and other evolutionists and that are sometimes misleadingly labeled as "laws." Among these, perhaps the one that has received most attention is the previously mentioned frequent tendency for organisms to become larger as their lineages evolve, sometimes called "Cope's law." (But this is also "Depéret's law" and other "laws" have been called "Cope's.") This is, indeed, a very widespread trend although by no means universal or irreversible, as previously noted. For general reviews of the subject, with many examples and citations, see Stromer (1944), Rensch (1947), and Newell (1949), and for a particularly good example and its interpretation, published too recently for inclusion in those reviews, see Stenzel (1949). Although just at this point in the discussion concern is more with descriptive generalization about trends than with theories regarding their determination, it may be added that all these students [4] reflect a clear consensus that phyletic increase in size of individuals occurs when and while such increase is adaptive, and at no other times. The generality of the trend is held to coincide exactly with the generality of selective advantage in (usually slow) increase of size.

Another example of generalized trends is that toward hypsodonty in many herbivorous mammals. This is also common, although its special circumstances make it far less so than a trend like that toward larger size. Incidentally, the definition of these circumstances make it still

[4] Stromer, not concerned with explanatory theories, does not explicitly commit himself, but his remarks are consistent with this point of view.

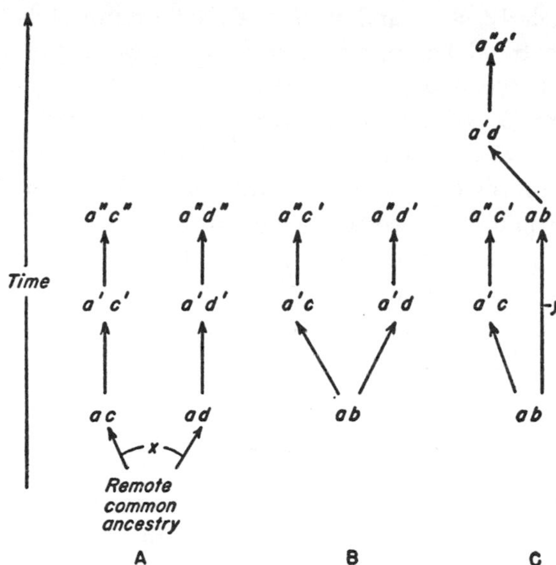

FIGURE 32. TRENDS IN GENERAL, PARALLEL TRENDS, AND ITERATION. A, a general trend in character or character complex *a,* with trend *a-a′-a″*, shown as happening to occur in two groups of remote common ancestry; in general the existence of a common ancestry and the ages of the two lines relative to each other are here irrelevant; the lines are distinguished by the different character complexes *c* and *d,* which also evolve but without parallelism of trends. B, parallel trends in two lines from an immediate common ancestry; *a-a′-a″*, common trend; *b, c, d,* differentiating characters; the fact that the trends are shown as simultaneous is here relevant. C, iteration; symbols as in B; relative time relationships also relevant. B and C are shown to be fundamentally similar special cases of A. If lines *x* in A are short and equal (i.e., common ancestry is not remote and lines arose at same time), the situation is formally identical with B. If line *y* in C is insignificantly long or shortened to zero, this situation is also formally identical with B. All intermediate and combined types occur.

clearer that the trend is always adaptive.[5] This trend was noted above in *Myospalax* (as discussed by Teilhard, 1950), and Stirton (1947) has

[5] As usual, dissenters can be found, but it seems to me that their dissent is in some cases merely formal and in others unsupported. Thus Stirton makes a strong point of his opinion that hypsodonty *alone* was not the determining factor in survival of the various hypsodont horses, but I am not aware that anyone ever thought it was, and Stirton agrees that hypsodonty was adaptive in horses. Teilhard implies that hypsodonty in *Myospalax* is not explained by selection and adaptation, but his alternative is only to say, "Reste à trouver la bonne explication."

given particularly clear and well analyzed examples in horses and beavers. The trend has occurred in marsupials (e.g., *Phascolomis*), taeniodonts (*Stylinodontinae*), edentates (all that retained teeth), lagomorphs (all), rodents (a very large number in practically all main groups of rodents), notoungulates (almost all groups except the earliest), proboscideans (Elephantidae), hyracoids (Myohyracidae), perissodactyla (Equinae and some rhinoceroses), and artiodactyls (many different groups, e.g., some pigs, camels, antelopes, etc., in varying degree). In all cases where food habits are known, the trend toward hypsodonty occurs in animals that eat especially fibrous, siliceous, or gritty food. (The edentates are a special case; enamel was early irretrievably lost and in all groups that did take to even moderately abrasive food the softness of the teeth was compensated by hypsodonty.)

Still another example is of further interest because it involves trends observable among both animals and plants, a most unusual situation in view of the complete lack of homology in gross anatomy. Two complementary sorts of trends occur rather widely: (1) a series of numerous similar parts tends to be reduced in number while the remaining parts become more differentiated from each other, and (2) a series of relatively few similar parts tends to increase in number and the differences between the parts tend to become less. Examples among vertebrates include: (1) in many fishes, and in the whole sequence from fish to man, a usual trend to reduce the number of dermal bones in the skull roof with increasing individuality of those remaining; and (2) increase in number of vertebrae in most snakes, with loss of sharp distinction between regions of the series. Gregory (e.g., 1935a, b, and c, 1951) has been especially interested in these trends in animals and has repeatedly discussed them and related phenomena in various aspects and under various names such as "Williston's law,"[6] "anisomerism," or "emphasis"

[6] Stromer (1944), who has also repeatedly discussed trends of the general sort of (1) above, took Gregory to task for not knowing that he, Stromer, had stated this "law" before Williston. Most of these generalized trends have been noticed over and over again from the early days of paleontology down to now. Perhaps some day historians may find it possible or even worth the trouble to designate who first mentioned each of them. In the meantime, few "laws" are named after the student who has real priority. "Dollo's law," for instance, of the irreversibility of evolution, perhaps the most widely known of personally named evolutionary principles, was plainly stated by Scott before Dollo, and for all I know by others before Scott. Incidentally, Gregory (1951) has now duly acknowledged Stromer's priority in statement of "Williston's law."

in relation to (1) and "polyisomerism" or "reduplication" in relation to (2).

Similar trends occur in plants. For instance, (1) simple flowers have arisen from shoots with reduction in number (potential, at least) of parts and increase in their differentiation, and (2) compound flowers represent great multiplication of similar and simplified structures. Stebbins (1950) has summarized these relationships. He adds the important observation, which also applies to many animal examples, that increase in size and complexity of structures often accompanies reduction in number (related to trend 1, above), while increase in number often accompanies decrease in size and complexity (related to 2). It might further be noted that both sorts of trends in all the many kinds of organisms that show them involve a widely general functional relationship: a given function (e.g., locomotion of arthropods, gamete production of plants) may be performed by a series of similar parts acting as a whole or it may be performed by one or more parts acting as separate units or limited sets (although generally in coordination with each other or still other parts). Adaptive change may go in either direction and in so doing it involves one or the other of the complementary trends here considered.

When a trend of any sort occurs, it is unusual for it to involve only a single characteristic or sharply separable phenotypic "unit character." Even when it seems to do so, as in the graptolite trend from biserial to uniserial arrangement of thecae, there are probably trends in other characters that we cannot or do not observe. Both organic structures and genetic systems are so integrated that a change in one respect almost necessarily involves changes in others if integration is to be maintained—and if integration is long or seriously upset, extinction ensues. The occurrence of trends correlated within organisms is so usual and raises so many interesting points that a special section will be devoted to it later in this chapter.

An important question regarding trends in general is whether they are always, usually, or often undeviating or rectilinear,[7] that is, to what extent is it true that once change starts the same sort of change con-

[7] Objection has been made to this term on the grounds that it is a misplaced metaphor with no literal meaning as applied to evolution. The objection seems to me ill taken. I cannot believe that many students fail to grasp the visual abstraction of a line as symbolic of evolutionary change and a straight line as symbolic of continued change of the same sort.

tinues indefinitely or to its possible limit. All trends do have rigid limits of various sorts. For instance, Lull and Gray (1949) found that when skull changes in the ceratopsian dinosaurs are studied by d'Arcy Thompson's method of deformation of a Cartesian grid, some trends cannot be extrapolated very far without encountering a geometric impossibility, such as a triangle with the sum of two sides equal to the third side. There are also definite mechanical limits to morphologically possible changes; for instance, any group of mobile, terrestrial organisms has an upper limit of size above which it cannot continue to be mobile. There are many trends that are inherently self-limited. For instance, hypsodonty can go no farther than continuous growth of teeth throughout life, reduction of toes cannot go beyond one, decrease of surface/volume ratio cannot go beyond a sphere, etc. There are also many more subtle biological and ecological limits that belong in quite a different category. For instance, increase in size may also be limited by a balance between individual food requirements, total available food, and population size adequate for continuous reproduction.

It is not, then, surprising to find that all trends do stop either in continuing phyla or with extinction. (Whether in any sense the trend, itself, causes extinction is a separate question to be discussed in the next chapter.) Often when a trend stops it has reached an inherent or mechanical limit. Such cases of course would not constitute real exceptions to a possible (but in fact untrue) generalization that trends are undeviating. There are, however, innumerable cases in which a trend has stopped before reaching an inherent or mechanical limit. For instance, a trend toward hypsodonty does often go on to the limit of continuous growth, but in even more cases (including, for example, cheek teeth in the numerous trends toward hypsodonty in perissodactyls and artiodactyls) it definitely stopped before that limit was reached. Horses have certainly not reached a mechanical limit of gross size, but their trend toward size increase has clearly stopped. Indeed, the great majority of trends toward size increase can be shown, without reasonable doubt, to have stopped before a mechanical limit was reached.

There is therefore nothing intrinsic in trends that causes them to go on as long as possible. Certainly *something* stops trends when an inherent or mechanical limit does not. To think that trends simply have various predestined spans or different amounts of momentum would be idle metaphysics. There is no reasonable alternative to the conclusion that when a trend stops short of other limits, it has reached an adaptive limit, in other words, it has ceased to be advantageous to the organisms involved. Often we do not know enough about the functional aspects of a trend to know why it ceased to be advantageous, but in many examples this is quite clear. For instance, under normal conditions a wild horse's grinding teeth last until the animal is senile from general organic aging and past breeding age. There would be no advantage to the species in greater hypsodonty so that senile horses died with unused grinding capacity.

It would be interesting to know whether little *Nannippus phlegon*, most hypsodont of known Equidae, did not have an exceptionally long life span or live on exceptionally abrasive food. Or it may merely be that a small animal has relatively greater tooth wear. Brachydont horses averaged significantly larger than contemporaneous hypsodont forms. Most mammals with continuously growing teeth are small.[8] That brings up another point that may be discussed here parenthetically: whether, since failure of the trend to go on to continuous growth was due to an adaptive limit, the cases in which growth does become continuous may not represent momentum carrying a trend beyond an adaptive limit. I think this is not the case. When animals progressively adapt to abrasive food there comes a point when increasing hypsodonty means that the whole tooth cannot be accommodated in the jaw at once. The most hypsodont horses are just beyond this point; the basal part of the tooth has not yet formed when the top begins to wear, but it forms not long after. With small animals adapted to a high rate of wear, the total height of tooth required for a normal reproductive lifetime is several times greater than can be accommodated in the jaw at any one time. Somewhere between these two conditions is a threshold where the mechanism of continuous growth becomes more effective than that of delayed root closure.

In cases like hypsodonty in Equidae limitation of the trend is clearly adaptive rather than intrinsic and this may be advanced as a generaliza-

[8] The thermodynamic efficiency of small mammals is less than for large mammals, that is, the smaller forms require relatively more food in order to maintain equivalent activity. That point has been emphasized by Watson (1949). (His recognition of its evolutionary significance is, incidentally, an interesting result of assigning a paleontologist to study of food production during the war.) Of course small size may have advantages outweighing thermodynamic disadvantage. The shrews, for instance, are a successful group in spite of their extreme thermodynamic inefficiency.

tion true of all limitation of trends. It would obviously be inadaptive for a trend to go beyond a mechanical limit (claims that some trends do so will be discussed and discarded in the next chapter), and since it is physically impossible to go beyond inherent limits of the sort exemplified above those are irrelevant to adaptive control.

That trends are not intrinsically undeviating is sufficiently evident from the plain fact that a great many of them do, in fact, deviate. This will be abundantly exemplified in the following summary of trends in the Equidae. Many other examples are known. For instance, there was a long over-all trend in the shelled cephalopods, and on a smaller scale in many separate lineages of these, toward increased and tighter plane coiling of the shell. But in many different lines from the Triassic onwards this trend changed its course (e.g., to spiral rather than plane coiling) or was quite reversed, with a new trend for the shells to unroll. It is also common and indeed the rule for long trends that they fluctuate in rate and, when their nature permits of alternatives, in direction. Variation in the opposite direction to the trend always persists and may even be more marked than in the direction of the trend. For instance, in populations of *Gryphaea,* with a trend toward coiling, less coiled variants often are more numerous and extend farther from the mode than more coiled.

When there are alternative directions, variants toward these commonly also occur. Such alternatives often exist even when our way of studying changes obscures them. For instance, form is often studied by a series of linear dimensions which can only increase or decrease, but the form as such has more, sometimes a great many, possible directions of change. Even in a simple consideration of form by three measurements, as in the trend toward relatively broader and deeper test in *Micraster,* the possible directions of change and the actually observed directions of variation in the population are not two, as for each dimension, but eight. Added to this are variations in relative position of maximum dimensions, in outline of sections, etc., so that possible directions of variation in the functionally single character of over-all shape of test are really very numerous.

The answer to the question whether trends are inherently or universally undeviating is certainly "No!" It is nevertheless a fact of observation that short trends and segments of long trends are, as a rule, essentially rectilinear. Of course with any sort of change whatever, segments may be isolated that are so short as to have essentially a single predominant direction and so to be essentially rectilinear, but that makes the whole problem merely verbal. There can be no clear-cut definition as to how long a sequence must be in order to be designated a trend, but most students would probably agree that trends on the order of $10^6 - 10^7$ years should be so called and are common in the paleontological record. Essentially undeviating trends on the order of 10^6 years also seem to be common, although far from universal. Essentially undeviating trends as long as 10^7 years also occur, but are less common. Trends much longer than this, say on the order of 5×10^7 years or more, seem rarely if ever to be essentially undeviating. At least, all claimed examples of such trends seem to me to show essential deviation if the data are reexamined. For instance, increase in size of Equidae over about 6×10^7 years or of size of sabertooth canines over about 4×10^7 years are classic examples of supposedly long-sustained rectilinear trends. In fact and beyond any doubt neither trend was either rectilinear or sustained for the stated length of time.

It is significant that most of the known cases of moderately sustained, essentially rectilinear evolution come from what seem to have been abundant populations evolving at moderate rates in rather stable environmental (and adaptive) conditions. Trends are especially characteristic of a particular phase of evolution: progression within an adaptive zone after this has been occupied by shift from another zone or by splitting and radiation and after new adaptive equilibrium and new genetic integration have been achieved.[9]

TRENDS IN THE EQUIDAE

The evolution of the Equidae is one example always cited in any discussion of trends. It is extraordinary and disheartening that outside of students who have actually worked extensively on fossil horses it is still commonly cited in support of opinions which in plain fact it conclusively disproves. As Jepsen (1949) has noted, many recent comments on equid history reflect no essential progress in knowledge since the first outline of the horse family tree was put together by Marsh and

[9] This means the same thing *descriptively* as Schindewolf (e.g., 1950a) means when he says that orthogenesis is characteristic of the typostatic phase of evolution, although I reject the theoretical implications of his statement as decisively as he rejects mine.

Cope 65 or 70 years ago. More recent illustrations are sometimes copied, but texts as recent as, say, Matthew and Chubb (1913) seem not to have been read or, at least, understood by the many who still write about "orthogenesis" in the Equidae. As Jepsen also quoted, "There is a tendency to put the chart before the horse."

Under these circumstances, it seems advisable again briefly to summarize what can be said to be really known today about trends in the Equidae. Most of these trends are more fully but less technically discussed in Simpson (1951a), and of course most of the older literature, as well as thousands of specimens representative of virtually all known groups of horses, have been considered but need not be specially cited except on one or two specific points. There were changes in every single part of equid anatomy, but only a few of these need to be followed here. The others would not modify conclusions based on the examples here selected.

Before examining particular phenotypic trends, a remark is needed on equid phylogeny. Even in some of the most recent works on evolutionary theory (e.g., Cuénot, 1951)[10] and general texts (e.g., Hegner and Stiles, 1951), this phylogeny is presented as a single line of gradual transformation of *Hyracotherium* into *Equus*. It has been well known to the better informed for more than two generations that the phylogeny includes considerable branching, and for the last ten or fifteen years it has been increasingly evident that the really striking and characteristic part of the pattern is precisely its repeated and intricately radiating splitting. Its botanical analogue would be more like a bush than like a tree, and even if the tree figure of speech were used, *Equus* would not correctly represent the tip of the trunk but one of the last bundles of twigs on a side branch from a main branch sharply divergent from the trunk. A summary of present authoritative views on horse phylogeny is given in Figure 33. *This figure is still grossly oversimplified.* The genera are drawn very broadly, much more so than by some specialists, and some of the groups named as genera represent structural stages rather than genera by a truly phylogenetic definition. Moreover, if the actually evolving units could be shown, the lineages in a strict sense,

[10] A remarkable example because Cuénot has actually compiled a phylogeny correct in showing branching of the Equidae (although highly incorrect in many other respects) and then has completely ignored this in his other illustrations and in the text discussion of supposed orthogenesis in the family.

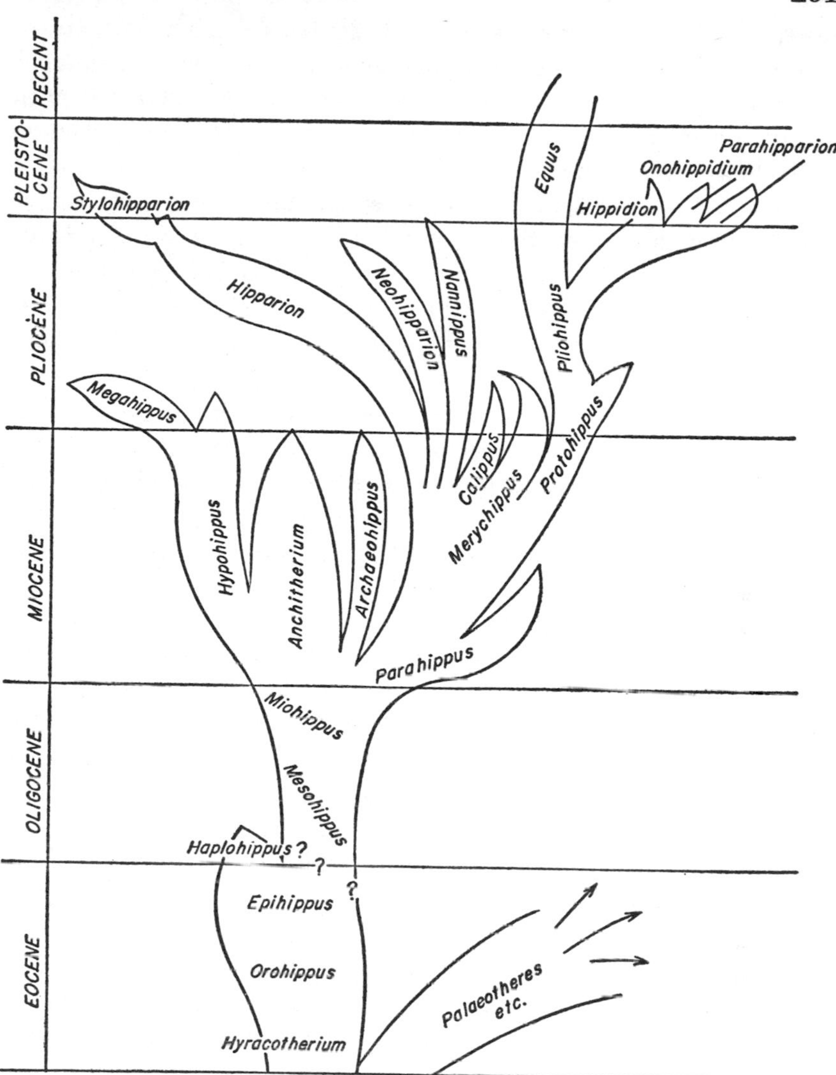

FIGURE 33. GREATLY SIMPLIFIED REPRESENTATION OF THE PHYLOGENY OF THE EQUIDAE. Only generic branches are indicated, and even at this level many students would recognize more branches. Each generic area should be thought of as made up of from several to thousands of different strands, the real lineages. Time scale not proportional to absolute lengths of epochs.

there would be certainly thousands and probably tens of thousands of these, each one in some respects divergent from all others. For instance the simplicity of the Eocene sequence is fictitious in the sense that it is known that at all times in the Eocene there were more or less numerous different lineages of horses. Those for any one time just do not happen to have diverged far enough for taxonomists to find it convenient to call them different genera.

Size trends: The Equidae show no significant trend in size, either toward larger or smaller, throughout the Eocene, at least a fourth of the whole history of the family. From early Oligocene into the Pleistocene, many lines of horses showed a quite irregular and fluctuating but still rather general tendency to increase in average size. At all times there were large differences in mean size between different species, even within one genus. From Pleistocene to Recent there has been some decrease in mean size in *Equus*. At least three lineages at various times developed strong, significant trends toward *smaller* size (*Archaeohippus* through the Miocene, *Calippus* in early Pliocene, *Nannippus* through the Pliocene). (See Romer's graph, 1949a, on length of cheek-tooth series, which is fairly but not entirely correlated with gross size.)

Trends in the skull: Pre-optic dominance, relative elongation of the front part of the skull, also showed no defined trend in the Eocene but thereafter and in most lines there was a trend toward increase. As discussed in Chapter I, this tends to have an allometric relationship with gross size and part, at least, of the change is dependent on established growth pattern and not on genetic change. There may have been a genetic change in growth pattern in *Merychippus*. For most branches of the family there are no good data on this point. Most other skull characters show quite erratic trends and none show any over-all or very long continued trend. For instance, most lines of Miocene horses developed peculiar pre-orbital pits, partly distinctive of different lines. These were lost in the ancestry of *Equus*. The evolution of the brain (Edinger, 1948) is known only in the lines leading to *Equus*, and there only in part. It is known that the rate of evolution varied markedly in different lines and at different times.

Trends in the dentition: Molarization of the last premolars progressed steadily through the Eocene and was essentially complete in the early Oligocene. A trend toward a high degree of lophiodonty continued slowly and somewhat irregularly through the Eocene and Oligocene. A trend to complicate the cheek-tooth pattern by secondary spurs and folds is practically absent in early forms and in all the browsing lines, but began to appear in the common ancestry of the grazing forms toward the end of the Oligocene and was rather rapidly progressive in some lines during the Miocene. Thereafter it was extremely irregular, with some loose correlation with size, and perhaps with reversal in some lines. Recent *Equus* has simpler patterns than some Pleistocene species. Early forms and browsers became slightly more hypsodont in loose correlation with their size. Accelerated hypsodonty and development of cement began in a single group, perhaps even a single lineage, in *Parahippus,* and was general in the numerous different lineages of *Merychippus,* derived from *Parahippus,* but occurred somewhat irregularly and at different rates in the various lines. Thereafter each lineage or group of related lineages carried hypsodonty to a characteristic point and after this was reached the trend stopped and hypsodonty simply fluctuated around a mean value.

Trends in limb ratios: There is a widespread impression that the so-called "speed ratios," proportions of lower to upper limb segments, showed a constant trend to increase in the Equidae. Romer (1949a) has demonstrated that this is decidedly false. The tibia/femur ratio did probably rise significantly somewhere before *Mesohippus* but then shows what looks like merely erratic variation, without defined trends, for most later horses. It seems, however, to have had a strong trend toward *decrease* in mean value from early *Pliohippus* to recent *Equus.* The radius/humerus ratio had no established trend in the Eocene, was higher in *Mesohippus,* rose rather steadily in average value from *Mesohippus* to *Merychippus,* and thereafter showed a fairly definite reversal of trend from mid-Pliocene *Pliohippus* to recent *Equus.* Data are not available for other branches.

Changes in foot mechanism: The most famous of all equid trends, "gradual reduction of the side toes," is flatly fictitious. There was no such trend in any line of Equidae. Instead there was a sequence of rather rapid transitions from one adaptive type of foot mechanism to another. Once established, each type fluctuated in the various lines or showed certain changes of proportion related to the sizes of the animals, but had no defined trend. Eocene horses all had digitigrade, padded, doglike feet, with four functional toes in front and three behind. In a rapid transition (not actually represented by fossils), early Oligocene horses lost one functional front toe and concentrated weight a little more on the middle hoof as a step-off point, but retained a padded, unspringing foot. This type persisted without essential change in all browsing horses. The *Parahippus-Merychippus* lineages in a rapid but gradational transition (which *is* represented by fossils) lost the pad as a functional element, became truly unguligrade, developed a complex springing mechanism in the middle toe, which normally carried all the weight, and reduced the side toes which were, however, retained and functional as buffers and stops. This type of foot was retained, again without essential change or evident trends, in all lineages derived from *Parahippus* through *Merychippus* except *Pliohippus*. In early *Pliohippus* another rapid transition (also known from fossils) involved loss of the side-toes, strengthening and simplification of the springing mechanism, and development of check ligaments. This type persisted without essential change or trends in all the lineages descended from *Pliohippus*. (See Camp and Smith, 1942.)

The Equidae had no trends that: (1) continued throughout the history of the family in any line, (2) affected all lines at any one time, (3) occurred in all lines at some time in their history, or (4) were even approximately constant in direction and rate in any line for periods longer than on the order of 15 to 20 million years at most (usually much less). Figure 34, diagrammatically summarizing some of the changes as they really did occur, further shows that trends were often quite different in rate and direction in different lines and in the same lines at different times (e.g., Fig. 34A); that in a single functional system different trends occurred at different times and rates (Fig. 34B); that a character with marked trends either earlier or later may show no evident trend at all over long periods of time (Fig. 34A and B); and that a character that nevertheless has strongly marked progression within the history may at no point have what can strictly be called a trend but instead may progress from one stable adaptive level to another by a sequence of short, steplike transitions (Fig. 34C).

In addition to the characters that do show some definite trends or progression in the Equidae, there are many that do not. Some merely show fluctuation or relative constancy throughout (all basic mammalian and placental characters), and others show speciational segregation in various lines (diagnostic tooth patterns in the *Hipparion* group), or sporadic, episodic incidence in single lines (nasal notch in *Hippidion*).

The whole picture is more complex, but also more instructive, than

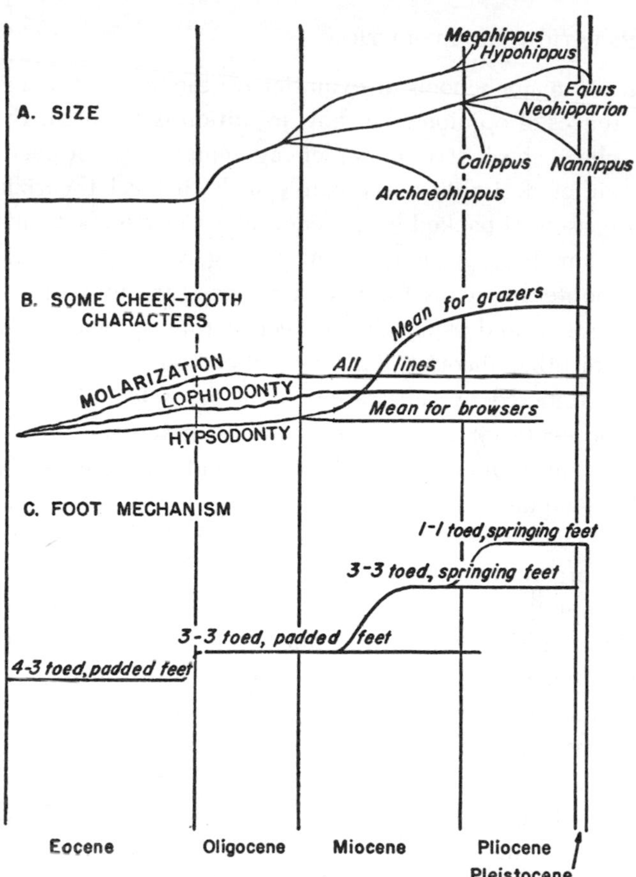

FIGURE 34. DIAGRAM OF THE EVOLUTION OF SOME CHARACTERS IN EQUIDAE. A, size; only a few lines are shown, labeled by terminal genera. B, some characters of cheek teeth, without attempt to distinguish lines except broadly the grazers and browsers on the curve for hypsodonty. C, foot mechanism; broadly speaking, all lines are on some level or branch shown. The vertical scales are not exactly proportional to the changes involved. The curves show in a relative way, only, quite approximate rates, times, and directions of change. The time scale is in proportion to estimated absolute lengths of the epochs.

the orthogenetic progression that is still being taught to students as the history of the Equidae. It is a picture of a great group of real animals living their history in nature, not of robots on a one-way road to a predestined end.

THE ORIENTATION OF EVOLUTION

The various major schools of evolutionary theory have arisen mainly from differences of opinion as to how evolution is oriented. Every possibility has been considered. Complete agreement has not been reached (that is neither desirable nor possible in high level theorizing), but there is a consensus backed by an enormous body of facts and logical induction from these. As long as everything is not known about life, which undoubtedly means forever, it is necessary to keep other possibilities in mind, and especially to keep before us evidence adduced in support of other theories. The conviction that evolution is usually and mainly oriented by adaptation involving selection does not exclude the necessity for considering some other factors.

To be sure, little time now needs to be spent on the neo-Lamarckian theory that hereditary adaptation arises by *direct* interaction of organism and environment. This theory was, I think, surely correct in designating organism-environment interaction as the cause of adaptation and adaptation as the main orienting force in evolution (which, incidentally, Lamarck himself did not believe). The neo-Lamarckians did great service to evolutionary theory by insisting on this relationship and by producing a great deal of evidence for it. There is, however, no longer any real doubt that they mistook the nature of the interaction, which does not produce adaptation directly but through the mediation of genetical selection. The postulated neo-Lamarckian mechanism for direct transfer of modification to a genetical system does not exist, and all the evidence for its action as well as much evidence it could not possibly explain can be wholly accounted for by the indirect mechanism of selection.

Some of those who still raise doubts about selection as the main orienting force in evolution seem to be wandering in a sort of verbal maze rather than discussing a real problem. Thus Arambourg (1950) says (in French) that "detailed analysis of proposed mechanisms shows us in all cases the inadequacy of mutationist or Darwinian concepts and always takes us back to envisaging as final determinant an action external to the organisms. I am well aware that Lamarckism is no longer in style at the moment and perhaps I am going to appear behind the times, but it seems to me impossible to separate the living world from its base and to make the biosphere independent of the vicissitudes of the toposphere." But in this last sentence Arambourg has stated precisely what the modern synthesis of mutationist and Darwinian concepts shows and insists on; indeed this is a main support for that theory. Arambourg's explanation of extinction, the main subject of the cited paper, is also in every essential that advanced on the basis of this synthesis. His implication that recognition of interaction between "biosphere" and "toposphere" is *ipso facto* Lamarckian thus raises a problem that is merely semantic.

That the issue raised is semantic and not objective is even truer of the numerous statements that the modern synthesis, or the so-called "neo-Darwinism" of today, relies on chance alone. (This is at present the obligatory official view in the U.S.S.R., but it used to be aired often enough in free countries and still is occasionally.) It is quite true that the theory embodies factors which are stochastic or are unoriented with respect to some other factors (not quite the same thing as saying that they depend on chance alone), but it is also true that it embodies a factor, i.e., selection, which is *the only nonrandom or antichance evolutionary factor that is objectively demonstrated to exist.*

Search for a factor of antichance and failure to appreciate that selection supplies this exactly to the degree demanded by the facts lie back of the numerous theories involving vitalism and finalism. The vitalists insist that antichance or the orienting factor of evolution is inherent in organisms. As to what there is about or in organisms that orients their evolution, the vitalist solution is to propose names (e.g., "entelechy") but not to specify just what is being named. The finalists postulate a goal or a literal purpose in evolution. Some of them explicitly ascribe goal and purpose to God (Du Noüy, 1947) or to spirits (Broom, 1933). Others merely evade such ascription (Cuénot, 1941) or fall back on the vitalist naming fallacy ("aristogenesis," Osborn, 1934).[11] None

[11] I think it unfortunate that Carter (1951), in his excellent exposition of modern, decidedly nonvitalist and even more nonfinalist views on evolution has adopted several of Osborn's terms, including "aristogenesis." Carter does caution in a footnote that "Osborn associated with the term theoretical views that are not implied here," but in fact Osborn *defined* his terms as finalistic. He said explicitly that "definite variation," "rectigradation," and "aristogenesis" are the same thing and that they refer to a creative process of change which is "definite in the direction of future adaptation." "Aristogenesis" has been repeatedly discussed by other authors (e.g., Rensch, 1947), and always with recognition that it means a perfectionistic finalistic factor. Indeed, this meaning is inherent in the root *aristo-*, deliberately used by Osborn on that account. An attempt to revive the word in a very different sense is sure to cause serious confusion.

drops any hint as to how, in the actual process of reproductive continuity that is before our eyes, the postulated future goal is supposed to orient change.

The propositions that evolution is oriented because that is its nature (vitalism) or because it or something knows where it is going (finalism) claim evidential support in the supposed existence, or even universality, of rigidly undeviating trends that are not oriented by current adaptation. This is the supposed phenomenon often implied by the term "orthogenesis." A great deal of time and argumentative fury have been wasted because two students seldom seem to mean the same thing when they say "orthogenesis." Some use the word merely to designate the fact, so obvious as to be entirely banal, that evolution, like any change, goes in some direction or other. Others apply it to the also obvious but more interesting fact that evolution does not go in just any direction, that it does have orientation. Still others use it to express their opinion that evolution must go, and keep on going, in just one direction. These and many other usages have been reviewed by me (1950c) and especially by Jepsen (1949).

A word for which everyone has a different definition, usually unstated, ceases to serve the function of communication and its use results in futile arguments about nothing. There is also a sort of Gresham's Law for words; redefine them as we will, their worst or most extreme meaning is almost certain to remain current and to tend to drive out the meaning we might prefer. Thus it is futile for a paleontologist to say, as in effect many have, "By 'orthogenesis' I mean to describe cases in which a trend has gone on for some time without significant deviation; I intend no implication as to why it did so." An odor of finalism still clings.

The question pertinent at this point and usually associated with the ambiguous term "orthogenesis" is whether there is something intrinsic in organisms or in evolution which causes trends to continue indefinitely, undeviatingly, in one direction irrespective of current environmental factors or of current adaptation. It will be necessary later to discuss some other real and supposed phenomena that also have some bearing, but enough has already been said and exemplified about trends to warrant a flat "no" to this question. This conclusion is supported by a strong consensus and by a vast literature, further review of which is not needed here (again, see Jepsen, 1949; the subject is also discussed well and at some length by Rensch, 1947). Jepsen has pointed out the interesting fact that nonpaleontologists commonly assume that the concept of such trends was developed by paleontologists and that their reality is generally upheld or has even been "proved" by paleontologists. In fact the concept was first clearly stated and has often been supported by neontologists who arranged various contemporaneous recent forms in a sequence and called the result "orthogenesis." Of course *any* group of objects can be arranged in graded series according to some criterion. Unless the criterion is correctly and solely phylogenetic, the result has no relevance whatever as to how the characteristics of the objects arose.

Even paleontologists have been guilty of this fallacious procedure. For example, Schindewolf (1950a) gives an illustration purporting to show "orthogenetic enlargement and overspecialization of the upper canine in the family Felidae." In the sequence demanded by this interpretation, the ages of the forms shown are early Pliocene–late Oligocene–middle Oligocene–Pleistocene. They belong to at least two and almost certainly three sharply different lines of descent. *Dinictis* may, indeed, be somewhere near the ancestry of *Metailurus*. If so, in the real temporal sequence these two forms show *reduction* in size of the canine. In the other line, *Hoplophoneus* is almost certainly not ancestral to *Smilodon*. Oligocene sabertooths as a group do suggest what the ancestors of *Smilodon* were like, and some of them have canines relatively at least as large as in that genus. The indication is that there was no trend for canine enlargement in sabertooths after the beginning of the Oligocene and that the relative size of the canines thereafter simply fluctuated or became now larger, now smaller, in accordance with temporary or particular adaptive advantage. If, as claimed, the large sabers made it very difficult to eat, the animals took 40 million years to starve to death. The fact is that this famous example of "orthogenesis," presented as gospel over and over again in the literature, does not exist. It is pure fiction. Paleontologists familiar with the specimens have been pointing this out repeatedly for more than forty years now, but it still has not penetrated the consciousness even of all their fellow specialists. Such is the stubbornness of error, and such some of the "evidence" for nonadaptive orientation of trends.

It is true that there have been many and still are some paleontologists who believe in the existence of sustained, rigidly undeviating, nonadaptive trends. Nevertheless there are more paleontologists, an increasing number in recent years, who point out that the supposed evidence for trends of this sort is worthless or misinterpreted and who conclude that such trends are entirely absent in the known fossil record.

Another sort of supposed evidence for the occurrence of nonadaptive trends is the claimed generality or universality of the same trends throughout otherwise varied groups or occurrence of the same trend in quite different environments. The latter point can conveniently be discussed at the end of this chapter in connection with a particularly striking claimed example that happens also to involve an unusually interesting correlation of trends. The claimed generality of trends can be briefly dismissed here with little more than the statement that trends are not really more general than the *adaptive* relationships involved in them. One reason why similar trends sometimes seem to be so common among members of a given group is that if organisms do develop a distinctly different trend, taxonomists remove them from the group. Thus early amphibians show considerable (but decidely relative) similarity of trends because the amphibians that had very different trends are called "reptiles." Another frequent reason for apparent uniformity of trends in a group is ignorance of the fact that other trends had occurred but had turned out in the long run to be, precisely, less adaptive. Thus until 1931 a trend toward depressed skull (in part allometric) was considered universal in Crocodilia, but then I found one skull of a fossil crocodilian with exactly the opposite trend (*Sebecus,* see Colbert, 1946; several other members of this peculiar group are now known).

The very gradual appearance of new characters at the beginning of a trend is common. Disbelief that the earliest stages can be sufficiently marked for the action of selection, even though later stages are admittedly so, has been a frequent argument against adaptive initiation of trends, for instance by Osborn (almost all his evolutionary studies), Robson and Richards (1936), Willis (1940), and many others. This long seemed an extremely forceful argument, but now it can be dismissed with little serious discussion. If a trend is advantageous at any point, even its earliest stages have *some* advantage. Thus if an animal butts others with its head, as titanotheres surely did, the slightest thickening as presage of later horns already reduced danger of fractures by however small an amount. That the amount in a given case is too small for selection to act is an opinion now quite unjustified in view of the demonstrated great effectiveness of genetical selection in most situations (see Chapter V). It is certain that if we can see any advantage whatever in a small variation (and sometimes even if we cannot), selection sees more.

Opposing evidence is quite inadequate to upset the previous conclusion that trends are largely oriented by adaptation, the mechanism of which is genetical selection. By this is meant that adaptation is the driving force, so to speak, that keeps a trend going in a given direction as long as it does and that then stops the trend or changes its direction. In the whole complex intricacy of the organism-environment system there are a number of interrelated but more or less analyzable factors that help to determine the particular direction taken by evolutionary change, whether in trends or otherwise. The possibilities for change are always limited and sometimes have quite narrow limits.

The existence and nature of environmental and adaptive limitations have been shown in Chapters VI and VII. In terms of the descriptive symbolism there developed, every group of organisms lives in an adaptive zone. Some changes may occur without altering the adaptive relationship. Others cannot, and that is one basic limitation on evolutionary change. Change of the adaptive zone calls for some change in the populations occupying it and determines a required direction of change if they are to continue in it. Change to another adaptive zone can occur only if that zone actually exists, is physically accessible, and is either empty or occupied by competitively inferior organisms. In broadest outline, these are the factors in adaptive orientation of evolution.

Other important factors are involved in the genotype as an element in the system: environmental[12] change–selection–genotypic change–phenotypic change–adaptation. Any living population has some degree of balance and integration of the genetic system of the population as a whole and in the genotypes of its constituent individuals. If the population is already well-adapted, the integration tends to be very extensive and intricate, although it may involve a mobile rather than a fixed equilibrium. In any case, changes in the genetic system are admissible (that is, can occur without ensuing extinction) only when they are such in nature and extent as not to upset prior balance too severely and not to preclude reaching a new adequate balance quickly. A poorly or broadly

[12] It should be remembered without specification each time the word is used that "environment," unless qualified, is here used in the broadest sense, including the organism itself as well as everything exterior to it and affecting it in any way.

adapted population can tolerate more genetic change of some sorts than a well or narrowly adapted one, but there are always limits, broad or narrow. This is another factor, not really additional to adaptive control but as a particular sort of adaptive control, that certainly orients evolution.

This factor applies broadly to whole populations, but it has concomitants in individual structure. Any individual is a working organism with parts that exist together and operate with a high degree of harmony. No change can occur unless it maintains a working relationship with parts unchanged or less changed—and even in cases of "large" mutations the great majority of characteristics of the mutant are those inherited with little or no change from its parents. In terms of the genetical background, Waddington (1939) has pointed out that, "Granted that a character is dependent on the interaction of many genes, it will be easier to continue a line of evolutionary change, for which many of the modifiers are already present, than to start off on a completely new line. Thus the uni-directional nature of trend evolution is not particularly surprising."

Another aspect of the same sort of limitation is seen in the fact that a population always has a certain range of realized or expressed phenotypic and genotypic variation, another much larger range of existing potential variation stored in the genetic pool of variability, and a third prospective range involved in mutations not yet fixed in the pool. Selection can act directly only on expressed variation. Change in a direction represented by such variation is, so to speak, easiest, certainly most likely and most rapid. Change in a direction covered only in pooled, potential variability is less likely and generally slower because it cannot occur until variation is released or realized. Change in the direction of variation that is purely prospective is least likely and slowest because it must await occurrence, fixation, and realization of mutations. Change in a direction not represented in any of these ranges of variation cannot occur.

Since all change depends ultimately on mutation, either current or in the past and working down through the various pools (Fig. 12, Chapter IV), possible directions of change are ultimately and absolutely limited by the mutations that have occurred and are occurring in a population. Even though trends in different lines may have the same adaptive direction, they may in detail follow different directions of morphological change because they are dependent on different mutations which, moreover, are occurring in different genetic systems. For instance, in the rhinoceroses horns developed, either in short trends or by transition between adaptive subzones, at least four and perhaps more times quite independently. The adaptive orientation is clear and was of nearly the same sort in all instances, but the horns, morphological expressions of the same adaptive change, are quite different. Two horns transversely paired (*Diceratherium*), one nasal horn (*Rhinoceros*), one frontal horn (*Elasmotherium*), and two tandem horns (*Dicerorhinus, Ceratotherium, Diceros*) severally evolved. This is, of course, another case of the opportunistic aspect of evolution, previously mentioned. Similar examples are legion, and they constitute extremely convincing evidence that the direction of evolution is largely determined by adaptively oriented selection acting on adaptively unoriented materials that limit possible avenues of change. That some rhinoceroses did not develop horns reinforces more than it contradicts the conclusion. In some cases, as among the many early, small, running rhinoceroses, horns would probably have been inadaptive even if materials for their development existed. Among some of the later hornless types horns might have been adaptive, but then it seems clear that mutations leading to them did not occur, or at least did not become expressed sufficiently to be acted on by selection.

In such examples and, it may be said, in general, mutation is a permissive and limiting factor. Horns do not develop without mutation. With appropriate and sufficient mutation, they develop if they are selectively advantageous. Mutation is orienting to the extent of determining the particular sort of horns. It has frequently been suggested that mutation may be an orienting factor in a fuller sense, that it may determine the whole direction of evolution, as regards given characters, with no significant involvement of selection or adaptation. This was the viewpoint of early mutationists, more or less Devriesian in outlook, a viewpoint that we have already found good reason to abandon. With some differences and as regards major changes in evolution, it is also the viewpoint of Schindewolf and of Goldschmidt, likewise discussed adversely heretofore.

Within the firm framework of modern population genetics, it seems that mutation pressure, alone, could not orient evolution unless mutations were strongly directional and occurred with such high frequencies as to overcome omnipresent selection. As noted in Chapter IV, mutation is, indeed, strongly directional in some cases. Incidence of some mutations is much higher than that of others and much higher than the rate of back mutation. This very fact, however, becomes an argument against mutational orientation of evolution when it is noted that in all the (admittedly few) cases in which relative rates of mutation have been studied, their dominant direction is not the one in which past evolution has occurred or in which changes in recent populations are noted. It seems to be the virtually unanimous decision of geneticists that the predominant direction of mutation is degenerative and that the actual utilization of mutation in evolution depends either on happy accidents that make rather rare large mutations progressive (Goldschmidt) or on a factor that overrides relative mutation rates and literally selects the more favorable mutations even if their rates happen to be relatively low (most other geneticists). The latter point of view is strongly supported by the great mass of evidence, reviewed throughout this book, that evolution is in fact mainly adaptive and does involve almost exclusively gradational change in populations rather than saltation.

It remains true that extremely little is known about mutation rates in wild populations. The possibility is by no means excluded that in some instances and phases of evolution a predominant direction of mutation might have been an effective factor. It does, however, seem warranted to conclude from evidence already available that this is not the usual or even a common way in which orientation occurs. If direction of mutation and of selection did somehow and sometimes happen to coincide, this would unquestionably accelerate change. If very high mutation against selection occurred, this would probably be inadaptive and presumably might cause extinction. There seems to be no convincing evidence that it has ever done so, in fact. Some supposed examples that have been or could be interpreted in this way are discussed in the next chapter.

CORRELATED CHANGE

It has been remarked that trends or evolutionary changes in general rarely, if ever, occur singly. Change in one character is almost always accompanied by changes in others, a fact which is both interesting in itself and significant for interpretation of many evolutionary phenomena.

The gross fact that some phenotypic characters show a high degree of association was noted long ago; it is the so-called Cuvierian principle, which even antedated Cuvier. A certain disillusionment resulted when it was found, for instance, that claws may be associated with ungulate teeth, as in the chalicotheres and several other extinct groups. Nevertheless it is a fact that certain associations, even between such disparate parts as teeth and feet, are much more frequent in nature than others. Beyond this fact and in moving evolutionary series it is common for different characters to tend to evolve together.

There is a variety of known mechanisms at different levels that can produce correlated change in different characters of the same organisms. A partial list of these includes:

1. Genetic mechanisms, among them:
   a. Pleiotropy
   b. Gene association, not only linkage and the factors reducing or eliminating crossing over, but also factors influencing gene frequencies and through them the frequencies of gene combinations in populations
2. Ontogenetic mechanisms, especially:
   a. Allometry, and growth fields and gradients in general
3. Functional and adaptive mechanisms, among them:
   a. Coaction or interaction (e.g., occlusion of teeth)
   b. Congruence of functions (e.g., of teeth and digestive system)
   c. Adaptive correlation, involving different characteristics both or all useful in a given adaptive zone (e.g., adaptations for grazing and for rapid locomotion)

It is also probable that characters frequently become associated by sheer coincidence, especially at the level of unit mutations subject to sampling effects. Although casual associations may thus be random, it is very unlikely that any continuing, simultaneous trends occur without correlation by one of the mechanisms listed above, or something like them. The genetic mechanisms for correlated change and to less extent the ontogenetic mechanisms have the peculiar feature that it is possible for them to produce nonadaptive or inadaptive changes under the influence of selection for a correlated adaptive change. This can hardly be true of the functional or adaptive mechanisms, in which the correlation itself is produced by selection, except to the extent that genetic and ontogenetic mechanisms may also be or become involved in such cases.

In some cases there is a reasonably clear sense in which one change is primary or independent and another secondary or dependent on the first. With genetic correlations in which one or more characters are selected for and others are not, changes in the latter clearly are dependent, secondary, or satellite. Thus there is reason to believe that in some birds and mammals darker and lighter pigmentation reflects physiological characters that are subject to selection when the colors, as such, are not (Dementiev and Larionov, 1945). Adaptive correlation may also primarily involve one change to which others are adjusted. Thus increase in gross size may be a primary trend with increased thickness of limb bones proportionate to their length as an accompanying secondary trend. (Weight increases approximately as the cube of linear dimensions, strength of bones or muscles approximately with the square.) This correlation can, indeed, be followed in most groups of land vertebrates that show marked increase in size.

A more complex example of correlated change in which genetic, ontogenetic, and functional correlations are all involved is provided by simultaneous changes in gross size, brain size, skull form, mandibular musculature, and other features in the evolution of man and the apes (e.g., Weidenreich, 1941). It seems reasonable to say that the primary change was that in effective brain size, that this is rather intricately influenced and associated with a separate trend toward larger gross size, and that the other changes noted are secondary to the change in relative brain size. This is essentially Weidenreich's view (although his equation of three sorts of changes with distinct taxonomic levels seems to me fundamentally incorrect). The actual way in which these characters are tied together is by the growth pattern and it might be said that what actually changed is just this one thing, the growth pattern. But the growth pattern changes because of genetical changes, which occur because of selection, which changes because of changes in the adaptive zones, and so on. It is largely an artifact to isolate features and factors in this complex.

Watson (especially 1919, 1926, 1949, 1950) has devoted particular attention to primary and secondary trends and has discussed them in some detail, especially in two examples, horses and early amphibians. With respect to the horses, he points out that increase in size entails a complex series of mechanical and physiological changes which are quite clearly secondary to primary size changes.[13] The increase in size itself (when it occurred) is attributed to natural selection, the advantage involved being an increase in thermodynamic efficiency. (This is undoubtedly a factor in the horses and other mammals, but there are also other probable factors, such as intraspecific competition for food and increased complexity of neurological and other systems, that are also involved in these cases and that apply to others, such as the oysters, in which thermodynamic efficiency is not likely to be an important factor.) Watson reaches similar conclusions regarding the proboscideans and a number of other groups: they show one or a few primary trends, adaptive in nature and induced by natural selection, and a variety of secondary trends, also oriented by selection, adaptive to the primary trends.

In the evolution of certain early amphibians, especially several lineages of Permian to Triassic labyrinthodonts, Watson finds two primary trends: progressive loss of replacing cartilage bone, notably in the brain case, and progressive flattening of the skull and eventually of the whole body. Both trends appear in a number of different lines, although with considerable differences of incidence and rate. Correlated with them are numerous other changes, most of which are evidently secondary to the primary trends. For instance, posterior prolongation of the retroarticular process and changes in structure and position of the occipital condyle are mechanical concomitants of the fact that an animal flattened against the substrate opens its mouth by raising the skull rather than by lowering the mandible.

This is an extraordinarily clear and interesting example of correlated primary and secondary trends, and to this extent it is indisputable. Interpretation in terms of orienting factors does raise some further important and disputable points. Watson considers the secondary trends adaptations under the influence of selection, but he thinks that this ex-

[13] It is nevertheless not only gross oversimplification but also actual misstatement to conclude that, "It seems probable that the whole story of horse evolution, which is orthogenetic in the actual etymological sense of the word, may be accounted for entirely by the bringing about of an increase in size by natural selection." It has been shown above that the whole story of horse evolution is not orthogenetic in *any* sense even of that word to which so many different senses have been given. Nor is it true that the whole story, even as to broad essentials, can be accounted for by increase in size, although many of its details can. For instance, this cannot be reconciled with the fact that three-toed browsers, three-toed grazers, and one-toed grazers of the *same size* occurred, and long persisted, not to mention many lesser features that are not correlated with size.

planation cannot apply to the primary trends, which must therefore have some (unknown) internal cause in the organisms themselves. The argument that the primary trends are not adaptive is: (a) that Watson has been able to find no advantage in them; (b) that they occur in animals with different physical environments (i.e., mainly terrestrial or mainly aquatic) and persist through changes of environment; and (c), as regards the reduction of replacing bone, that a biochemical explanation is ruled out because a thickening of membrane bones often occurred simultaneously with a reduction of cartilage bones. It is worth while to devote a little space to this argument, because the example is similar to but more extreme and clear-cut and better worked out than a large number of others in which nonadaptive trends are claimed to exist.

In the first place, as a matter of scientific method, when there are two sets of similar phenomena, such as primary and secondary trends in Perissodactyla and Labyrinthodontia, and when the explanation is known in one case (as Watson submits that it is in the Perissodactyla and I agree), the minimal and most likely hypothesis is that the same explanation applies to the second case. Rejection of the most likely hypothesis requires *positive* evidence that it cannot or probably does not apply. The fact that we do not know whether it applies or not is irrelevant unless there is strong reason to believe that we would know if it did. Here there is no such reason. The fallibility of personal judgment as to the adaptive value of particular characters, most especially when these occur in animals quite unlike any now living, is notorious.

Thus our ignorance as to whether the trends in question were adaptive or not certainly fails to suggest that they were not. In fact they involve sorts of characters that are often adaptive, and it is quite easy to see that they could have been adaptive even though it cannot be proved that a given possibility is indeed the right one. There is reason to believe that cartilage is an embryonic adaptation (Romer, 1942). Increasing retention of cartilage in adults is an example of neoteny affecting a caenogenetic character (De Beer, 1951). Neoteny is a frequent ontogenetic mechanism in adaptation and, as it happens, strikingly so among amphibians, perhaps because they do so commonly have different juvenile and adult adaptations. It is thus entirely possible or even probable that this apparently primary trend in labyrinthodonts was really a secondary trend accompanying a primary trend toward adaptive neoteny. Bystrow (1938) has adduced strong evidence that neoteny is involved in labyrinthodont evolution. Alternatively, the fact that membrane bone increased while cartilage bone decreased could well indicate not that a biochemical factor was absent but just the opposite, that the trend was biochemical. If calcium and phosphate were limited or (more likely) if the organisms' capacity to utilize and deposit them were limited, and if there were strong selection pressure to increase their deposition in the protective membrane bone armor, it seems more likely than not that the organisms would, so to speak, increasingly spend their supply and utilization of these substances on the membrane bone and skimp on the replacing bone where cartilage was just as good and quite possibly better.

As to the other primary trend, dorsoventral flattening is quite common among recent poikilothermous vertebrates and is often clearly adaptive. (It also occurs in a few mammals, notably the extraordinary Australian bat *Tylonycteris,* in which it is an adaptation for getting through narrow cracks in stalks of bamboo.) It is further interesting and highly pertinent that this adaptation occurs repeatedly in such (apparently!) completely different physical environments as deserts (in lizards) and the bottom of shallow seas (in skates, etc.). The adaptations are, of course, different in many respects in these two greatly disparate cases, yet the flattening is clearly related to one feature in which these two environments are alike: presence of a barren substrate, closely applied to which the flattened animals live or spend much of their time, and which, moreover, is in both cases commonly composed of loose sand or silt. Such facts demonstrate not only that a trend toward dorsoventral flattening is quite likely to be adaptive but also that occurrence of the trend in "different" environments is not a valid argument against its being similarly adaptive in both.

In fact there are many examples of similar or effectively the same adaptation in decidedly different environments, either because the adaptation is broader than the physical environment (e.g., large, complex brains in whales and in elephants) or because the environments actually are the same in some respect that may have been overlooked (as in the example above). (In Australia there is also a pouched "mouse" that is dorsoventrally flattened but that lives in arid country and has nothing to do with bamboo; nevertheless the adaptation is really the same as in the bat: the "mouse" hides in narrow cracks in the ground.) The fact that a characteristic appears in different physical environments is irrelevant with respect to its adaptive value unless in the environments (within the whole of them, not just their physical aspects) there are different aspects directly affecting that particular characteristic. Abel (1928) has argued in a way similar to Watson that various trends in the Equidae were nonadaptive because they appeared in both forest and grassland horses. In that case it is abundantly clear that the argument is irrelevant. Trends that really did occur in different environments, for instance increase in size in some phases and lineages, are certainly not involved in adaptation to forest or grassland but in adaptation to biotic factors that are the same in both. Characters that are involved in specific adaptation to forest or grassland do not show parallel trends in the two.

It seems proper to conclude that the labyrinthodonts provide another example, and an unusually good one, of primary trends adaptive to the environment and oriented by selection and of secondary trends similarly oriented and adaptive to the primary trends.

A great many trends, probably a majority of them, do involve secondary effects. Almost any change is likely to have others carried along with it, more or less incidentally, by genetic correlation, by effects of an established growth pattern, or by secondary adaptation to the change, itself. It is, nevertheless, rather unusual to be able to say definitely that a given trend is really primary in an absolute sense or that one of two simultaneous trends is fully independent and the other fully dependent. Interdependency in the total adaptive situation is usually too complex for such simple descriptions to be really accurate. Thus in some of the horses there were simultaneous trends toward larger size and toward hypsodonty. Surely these trends were not independent; the more hypsodont teeth were part of the means of obtaining nourishment for attaining and maintaining larger size. But the fact that brachydont horses were simultaneously attaining still larger size shows that it is inadequate to represent hypsodonty as dependent on size or size as dependent on hypsodonty. Both are, in a truer sense, dependent on the whole pattern of adaptive change, which is in the last analysis the true and primary trend to which all others contribute and within which all others are intricately interdependent.[14]

[14] In statistics it is a required convention to designate one variate in a regression as "independent" and the other as "dependent." Of course this has nothing to do with independence and dependence of characters in an evolutionary trend.

CHAPTER IX

Extinction, Relicts, and Irreversibility

IN THE GRAND PATTERN of evolution nothing is more dramatic than the prevalence of extinction. The earth is a charnel house for species as for individuals. If this were not so, progressive evolution would have slowed or stopped long since. The total energy available for life is limited, and in the long run some groups must become extinct if others are to arise. Romer (1949) remarks that "extinction is the common lot, survival the exception," and he estimates that perhaps not more than one percent of middle Mesozoic tetrapods have living descendants. The problems of extinction are as pertinent to the whole course of evolution as the problems of origination. Extinction has repeatedly been discussed as it has affected given groups or in general, for instance recently by Colbert (1949a) and Arambourg (1950), whose reviews of the subject have been constantly useful in preparing this chapter. In spite of much attention to the subject, there is a lack of decisiveness and explicitness in most discussions of it. This treatment will not be an exception, but it will attempt to point out why we can seldom be decisive and explicit.

Four sorts of fully general theories of extinction stand out among more detailed suggestions. One sort maintains that extinction is a kind of momentum effect, that organisms evolve themselves out of existence. Another draws from the racial life cycle analogy the idea that races have a fixed span, as have individuals, and finally die of old age. The third sees the environment as executioner, slaughtering hapless populations. The fourth considers extinction as a derangement of coordination between organism and environment, a failure of adaptive response. The third sort of theory need not be discussed apart from the last, which, as well as the first two, will receive separate consideration. There are also some special topics which have aspects, at least, related to extinction and which will be briefly reviewed in this chapter, particularly relicts, blind alleys, and the irreversibility of evolution.

HYPERTELY

Evolution has some characteristics that resemble inertia and momentum. A few students have gone so far as to take these indications literally, naively applying to evolutionary sequences the classical laws of mechanical motion. Abel (1928) did this rather elaborately and justified it by saying that organisms function, function is activity, activity is motion, therefore—! This is such obvious semantic double-talk that it need not detain us. Nevertheless, the thought that somehow evolutionary momentum carries trends to inadaptive lengths and causes extinction has been expressed so often and with such a wealth of supposed examples that it cannot be ignored. Such an effect produces hypertely, a term that literally means that change has been carried too far. It is noteworthy that critics of selectionist theory in general and of the modern synthetic theory in particular use hypertely as their crushing argument. Some leading proponents of this theory have also frankly admitted that hypertely is a major stumbling block. Thus Haldane (1932) despaired that such phenomena "are not obviously explicable on any theory of evolution whatever," Waddington (1937) emphasized that no explanation in terms of selection had been offered, and Huxley (1942) essentially repeated Waddington's observation. On the other hand, I (1944a) tried to show that the synthetic theory (which involves a great deal more than selection) can well account for claimed cases of hypertely, and I believe there is a growing consensus to this effect.

Almost any case in which size, structure, or habit (although this is less often mentioned) is carried to extremes may be cited as hypertely, especially if it is bizarre to our human eyes. (To a lemur, man must be strongly hypertelic.) The assumption is that in true hypertely, the extreme is inadaptive and did or will lead to extinction. Cuénot (1925, 1951) mentions as hypertelic the displays of male pheasants, peacocks, and birds of paradise, the horns of various beetles, peculiar appendages in Hemiptera, the hornlike canines of *Babirussa*, and numerous other oddities.

At the outset, the fact that the organisms bearing the hypertelic structures just mentioned are alive and doing well (except as molested by man) casts some doubt on the idea that hypertely is inherently inadaptive and is a or *the* cause of extinction. Moreover, most or all of these hypertelic structures do have definite uses to the organisms, whether we approve the uses (e.g., to stimulate females) or think them performed as well as possible (e.g., by the pseudo-horns of *Babirussa*). As to size, the most hypertelic of all organisms, whales, are also living and their size is apparently useful to them. As to habit, the courtship procedures of some birds or the social organizations of some insects seem about as hypertelic as possible and yet are clearly useful to those concerned. Such observations enjoin a certain caution in assuming that if extinct animals had some peculiar feature, it caused their extinction by hypertely.

Classic examples of hypertely in extinct animals, cited over and over again in almost all discussions either of hypertely or of extinction, are the coiling of *Gryphaea,* the antlers of *Megaloceros,* the elongated dorsal spines of some pelycosaurs, and the canines of the sabertooths. It has already been noted (Chapter VIII) that the sabertooth example is fictitious, and also (Chapter VI) that the pelycosaur dorsal protuberance had a probable adaptive significance. Romer (1949a) also adds that the "spinescent" pelycosaurs were very flourishing in their time and became extinct no more rapidly than their spineless relatives. The cases of *Gryphaea* and *Megaloceros* are sufficiently interesting and well analyzed to treat them as key examples of hypertely and as evidence on its nature and causes.

Gryphaea was a Mesozoic pelecypod, a form-genus derived with iteration from a conservative oysterlike stock. In it, one valve became progressively more tightly coiled until "in some individuals the umbo of the left valve actually pressed against the outer surface of the right valve, so that this could be opened only slightly if at all. Such a state of affairs could only lead to the death of the individual and the extinction of the race" (Swinnerton, 1923; I do not find the statement in the last edition of that book). In a long series of studies, British students have analyzed this trend and the populations showing it in much detail (especially, Trueman, Maclennan and Trueman, Swinnerton, and Arkell, see references in Westoll, 1950) and Westoll (1950) has lately given a new interpretation. The complex trend involved shortening of the area of attachment, earlier onset of arching, increase in degree of arching or spiral angle, thickening of left valve, etc. The net effect (which in this case of "hypermorphosis" follows more or less the same sequence in ontogeny and phylogeny) is a series starting with a normal oyster, left valve broadly attached and shell aperture not far from

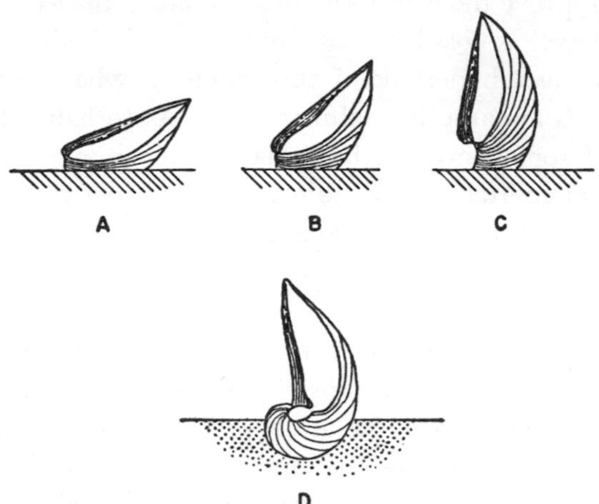

FIGURE 35. EVOLUTION OF *Gryphaea*. A–C, Sections through progressive stages in a lineage of *Gryphaea*, growing on a solid substrate, showing effect of shortening of the area of attachment and increase in the spiral angle. D, section of an adult in the final stage of the *Gryphaea incurva* lineage; the shell is shown as broken free from its juvenile attachment and lying partly embedded in a soft substrate. (Slightly altered from Westoll, 1950.)

parallel to the attachment (Fig. 35). The attachment becomes progressively smaller and the aperture increasingly vertical until its center is straight up. Finally, the now heavy shell breaks loose from the small attachment and is anchored and oriented on the substrate by its own weight and shape.

Westoll interprets the history in terms of two trends. One, rather steady, was toward larger size. The other, more rapid and accelerating, was toward tighter coiling (and associated characters). The latter trend can be represented by the size at which the coiling, if continued at the same rate, would cause closure of the two valves (Fig. 36). In early forms with slight coiling, the size at closure would have been very large, about 50 centimeters, a size never reached or even approached in the actual populations. With increased coiling, however, this critical value fell rapidly. At the same time, size increased, and in a final stage, *Gryphaea incurva*, the two trend lines began to intersect so that the

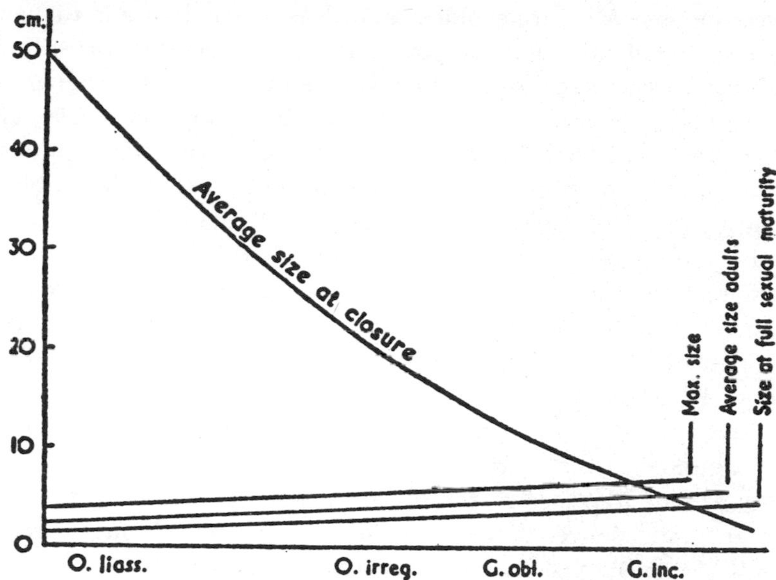

FIGURE 36. TRENDS OF SIZE AND CLOSURE IN A LINEAGE OF *Gryphaea*. The "closure" trend reflects tightness of coiling and shows the size at which the valves would have clamped shut if the animals had grown to that size. The three straight lines show maximum size, mean adult size, and size at sexual maturity in observed and (to the right) extrapolated populations. At the bottom are labeled stages in the sequence conventionalized as species, from *Ostrea liassica* to *Gryphaea incurva*. A population would necessarily become extinct from closure near the point where the bottom size line intersects the closure line. No known population reaches that point. (After Westoll, 1950.)

large members of the population did reach closure, which entailed their death.

Both of these trends were adaptive. Increase in size, as previously discussed, is a very common trend and in such a case as this is almost certainly favored by selection because of the advantages of larger size and of quicker growth for obtaining food. The trend toward coiling, which raised the aperture and later led to unattached life on the bottom, was adaptive for living on an unconsolidated or muddy bottom. It is, in fact, known from the sediments in which *Gryphaea* is found that it flourished in such environments. Nevertheless, the two adaptive trends have a limit imposed by intersection of the size trend and closure trend.

In the actual case, this limit was reached but not passed. Only the largest and hence, as a rule, oldest animals reached closure in the most advanced populations. For the population as a whole this was still not inadaptive. Death of nonsocial animals once they are past breeding age or even in late breeding age is not disadvantageous to the species, and it may be advantageous in increasing the space for younger, vigorously breeding individuals. Further selection on the two trends favorable to younger and younger animals, might, as Westoll suggests, have brought closure so early in the breeding period as really to interfere with reproduction of the population. This point does not seem to me to have been reached, but the somewhat delicate equilibrium would certainly make the population especially sensitive to competition or to environmental change. In either case, the whole phenomenon is most reasonably explained as occurring under the influence of selection and no mysterious momentum effect is required or really suggested.

The case of *Megaloceros,* the so-called Irish elk (which was a large stag and ranged over a wide territory in Europe), is more widely familiar. In a trend that can be followed approximately from the Pliocene through the Pleistocene, these animals became very large and the antlers of the males became proportionately still larger until they were truly enormous. Many students have considered them so large that they "must have been" disadvantageous. I can only feel awe for anyone who *knows* that structures were disadvantageous in animals that were very abundant for tens of thousands of years and more, but let us suppose that they were, or were becoming so at their largest. (They were certainly not lethally so before that stage, if ever.) There still are several possible explanations relying on known mechanisms and excluding an unknown momentum mechanism. One is selection for youth, mentioned in connection with *Ostrea* and previously suggested for this case by Haldane. Rapid antler growth would favor younger stags in reproduction and would not reduce population size and reproduction even if it produced disadvantageously large antlers in old stags. Only when the disadvantage reduced the total capacity of the whole group of stags, which would be quite late in a polygamous species, would the trend for accelerated growth really become disadvantageous to the population.

Another possibility, or indeed probability, is that this was an example of selection in correlated trends. Body size and antler size were probably allometric in *Megaloceros,* as they are known to be in its ally

Cervus elaphus (Huxley, 1932), with $k > 1$ for antler on body regression (see Chapter I). In early stages, selection was for both larger body and larger antlers, the allometric relation then accelerating the trend. When the point was reached where antler size ceased to be advantageous, selection against further increase in antlers was weaker than that for further increase in body size. The latter trend then continued, and therefore allometric increase in antler size continued, until the opposite selection pressures became equal. Body size was then somewhat under its optimum and antler size somewhat over its. That so specialized a creature might then be especially susceptible to extinction with environmental change is a different point, invoking no momentum effect.

This sort of balanced effect is probably very general in cases of correlated trends, which are themselves very frequent (Chapter VIII) and could involve any of the sorts of correlation previously discussed with the possible exception of adaptive correlation. When changes are correlated in any way through the genetic system (which also determines allometry, of course), the net effect of selection will be the algebraic sum for *all* characters involved. In such a case the net effect may quite frequently be some change against selection for some one character. It is again the total situation and not an artificially isolated element that determines the outcome.

Many of the characters commonly designated as hypertelic are striking secondary sexual characters. These are in all analyzed cases either neutral secondary effects of sex or, especially in those called hypertelic, they are advantageous to the individual in sexual selection. They are then adaptive, not, to be sure, to some other environmental factors but to factors in the intrademe environment. Intrademe selection may be dysgenic for the species (although this is seldom clear), but it is adaptive if the total situation is kept in mind. The same may be said of "altruistic" selection (about which I also have misgivings, see Chapter V): it may seem inadaptive for the individual but it is adaptive for the species and a balance is struck.

The nearest thing to a true momentum effect that is at all likely really to occur in nature would result if mutation overcame selection pressure and directed a change or trend. It has heretofore been pointed out that this is extremely improbable. Among the very few examples that might just possibly involve such an effect is that of the brachiopods described by Fenton (1935). He found that in *Atrypa* the most consistent trends were degenerative, mostly a loss of regularity in details of shell ornamentation. The adaptive significance, if any, of these details is unknown and can hardly have been very important. This may have been, in a manner of speaking, permissible nonadaptive change oriented by mutation. The fact (as I see it) that over-all control of evolutionary change is usually adaptive by no means rules out the possibility of a considerable amount of such permissible, nonadaptive change.

In a case like the progressive disruption of striations in *Atrypa*, calling changes "degenerative" may well be a human aesthetic judgment that has no real bearing. There are occasional instances, however, that do seem degenerative in a pathological sense. Fenton also noted a slowness in repair of lesions in late populations of *Atrypa*. More striking is the high incidence of bone diseases in the latest European cave bears (*Ursus spelaeus*) studied by Breuer, Ehrenberg, and others (see summary in Abel, 1935). It is a reasonable interpretation that the amount of disease was increased by life in damp, dark caves which were, nevertheless, advantageous as lairs. Again there was a balance of selection pressures on two different aspects of adaptation. One might say that the bears paid for their homes with a certain amount of disease. Again, too, this was a precarious balance, which in this case was upset by man. *Ursus spelaeus* was one of the earliest species to be helped to extinction by his destructive cave-mate *Homo sapiens*.

Nopcsa (1923) and others have noted that certain evolutionary changes characterizing whole populations and normal in that sense closely resemble the results of disease in other groups. Thus bone-thickening resembling pathological pachyostosis is common in early stages of aquatic adaptation. Pituitary enlargement and gigantism may be associated as in acromegaly (e.g., Edinger, 1942). Amynodont rhinoceroses have a short-legged, big-headed facies similar to that of achondroplasia (Wood, 1949). These and all the examples of what Nopcsa called "arrhostia" seem clearly to be adaptations definitely advantageous to the groups in which they were normal and they are not really either pathological or hypertelic in any special sense of the word. In some cases the resemblance to pathological conditions in other groups is merely a coincidence. In others, it may reflect the same mechanism (perhaps so in racial and pathological gigantism). This raises the interesting point that one group's pathology may be another's health.

The growth mechanism is doubtless much the same in both, but it is normal and adaptive for some lizards to grow to one foot in length and for some dinosaurs to grow to fifty feet. It would be equally pathological for the same mechanism to produce a dinosaur a foot long or a lizard fifty feet long. "Arrhostia" involves no evidence for racial pathology or inadaptive momentum or hypertely.

Cases in which there really is an over-all inadaptive balance for the whole organism in the total environmental situation apparently usually have the same general cause: what was a favorable balance has become unfavorable because of environmental change more rapid than possible adaptive response by the populations affected. This lag (Darlington, 1939) is the usual cause of extinction and it is discussed as such below. It is mentioned here in order to point out that it has nothing to do with hypertely or momentum although it can be readily mistaken for the supposed effects of those dubious factors.

The following list summarizes various possible ways in which there may arise characters that *seem* to be extreme and inadaptive and that have been called hypertelic by some students.

1. The character in question is really adaptive (e.g., the pelycosaur "fin")
2. The character may be more or less inadaptive in degree, at least, but is linked with adaptive characters and involved in a balance the net effect of which is adaptive:
   a. Allometry (e.g., antlers of *Megaloceros*)
   b. Other forms of genetic correlation (this is a possible or probable factor in many cases, e.g., in *Atrypa,* but it cannot be clearly designated in examples known to me)
   c. Altruistic and individual (e.g., juvenile, sexual) selection (probably a factor in *Gryphaea,* perhaps in *Megaloceros*)
   d. Ecological, biochemical, etc., balance (e.g., cave bears)
3. The characters are actually inadaptive and not balanced with adaptive characters:
   a. Drift (possible in many cases, but not clearly distinguishable from other possibilities in given examples)
   b. Mutation pressure (improbable; possible in a few cases, e.g., *Atrypa*)
   c. Lag (extremely common, the usual or universal cause of extinction)

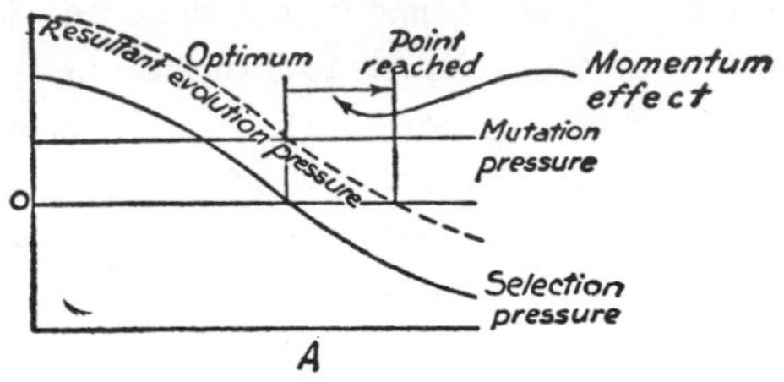

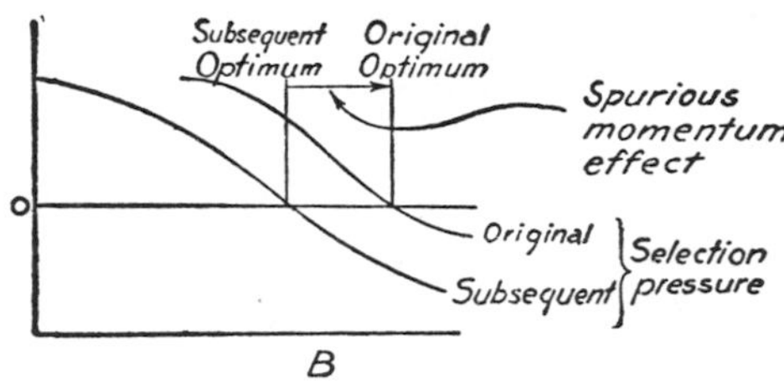

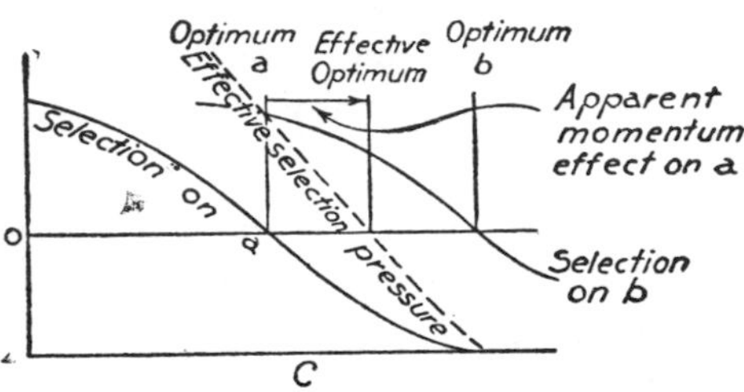

FIGURE 37. ANALYSIS OF SOME SUPPOSED MOMENTUM EFFECTS IN EVOLUTION. The tendency of a given evolutionary factor is to carry the character to the point where the line graphing intensity of that factor intersects line O. A, constant mutation pressure plus selection pressure about the optimum produces a resultant evolution pressure that intersects O beyond the selective optimum; B, backward shift of optimum makes original optimum inadaptive and produces spurious aspect of momentum for group that had reached the original optimum; C, correlation of two characters with different optima makes effective selection pressure different from selection for either character alone and places the effective optimum beyond the optimum for one of the characters.

Figure 37 symbolizes three general cases in which characters really inadaptive may occur. A fourth possible case, involving drift, is not shown as its bearing seems theoretically clear enough without symbolization.

The conclusion is that true momentum or hypertely defined as the result of momentum does not occur in evolution, except for the slight possibility of evolution dominated by mutation pressure, if that be considered momentum.

RACIAL SENILITY

The idea that groups of organisms become old and die of senile degeneration was mentioned in Chapter VIII in connection with the theory of racial cycles. It has also been advanced as a general cause of extinction. Little space need here be devoted to this hypothesis, which merely seems absurd to me (e.g., 1949a) and to many other students (e.g., Arambourg, 1950; Romer, 1949a; Rensch, 1947; Carter, 1951—to mention only a few).

On the face of it, the term "senility" applied to evolving groups is a misused metaphor where not even analogy exists. Nothing in a continuously reproducing population does or can possibly correspond with the process of aging in an individual. Moreover, all lineages existing at any one time are of precisely the same age, so how can some be "young" and some "old"? Unless life has arisen in more than one period of earth history, which is extremely improbable, all must necessarily have undergone the same span of continuous reproduction. Why, too, should related lineages or groups even at the same taxonomic level have such very different spans? To reply that some are cut off before they reached old age is to deny what is claimed, i.e., that extinction is caused by old age. To reply that some lines age more rapidly than others is a circular argument that assumes what is to be proved.

Supposed evidence for racial senility is of two intergrading sorts: it is pointed out that the last members of a lineage are often extreme or hypertelic, and it is claimed that an outburst of variation (mainly intergroup), of degeneration, and of bizarre and inadaptive forms often precedes extinction. As to the first point, it is obvious that if a group is changing at all, its last members will have changed the most, regardless of when extinction occurs. That they have usually, or ever, changed too much at the time of extinction is either a statement of the obvious (they became extinct because something was wrong) or an assumption of evolutionary momentum, a concept that we have just seen reason to discard.

The second sort of argument for racial senility, which is at least more interesting than the first, dates especially from Hyatt in the 1890's. He pointed out that ammonites toward the end of their history developed remarkable complication and variety of sutures and ornamentation and that they took on numerous (to our eyes) bizarre forms, showing spiral rather than plane coiling, unrolling in various ways, and even becoming irregularly serpentine. This is still the standard example of racial senility. It has, however, repeatedly been pointed out that: (a) "senile" forms began to appear 100 million years and more before extinction of the ammonites; (b) particular groups in which "senile" forms are most common are among the longest-lived groups of ammonites; (c) "senile" forms were often very abundant and obviously successful over long periods of time; (d) progressively adaptive significance can be assigned to some supposedly degenerative "senile" trends; and (e) normal or "youthful" forms persisted to the very end of ammonite history and then became just as extinct as the "senile" forms.

Many or all of the same insuperable objections apply to every other supposed case of racial senility. Some of the hadrosaurian and ceratopsian dinosaurs developed grotesque skull characters toward the end of their evolution, but lineages with normal skulls also persisted and it was the extinction of these that ended the history. Among dinosaurs in general, some of the last were as small and relatively unspecialized as the first.

Outbursts of peculiar forms that would certainly be considered senile if the group happened to be extinct occur among organisms that are now living and thriving. Many Recent beetles are as "hypertelic" and "senescent" as ever were the ammonites or any other extinct group, but beetles are obviously among the most successful forms of life in the world today and to consider them as on the verge of extinction would be ridiculous.

The appearance toward the end of a group's history of anything that could reasonably and even descriptively be called hypertely or senile degeneration is, moreover, the exception rather than the rule. I think any unbiased review would support Rensch's (1947) statement (in German) that, "In innumerable cases lineages become extinct without there being recognizable in the last forms any sort of morphological or pathological degenerative phenomena." For instance in the radiation of the notoungulates (Chapter VII), all the groups became extinct one by one and in no case did the last forms look like anything but perfectly normal members of the group. They did not become senile; they merely became extinct.

Quite aside from the absurd idea of racial senility, it would be expected theoretically that a group on the verge of extinction would show increased variation and degeneration. As an adaptive zone is becoming untenable, which in broad terms is what happens in any case of extinction, centripetal selection relaxes and centrifugal selection comes into play. Survival depends on getting out of the zone and effective selection will finally all be away from the ancestral type. In dwindling populations dysgenic inbreeding and genetic drift are likely to occur. In some recent animals approaching extinction there is some evidence that such effects do indeed appear. There are some possible examples among fossils (Fenton's *Atrypa* could be one), but instances that can reasonably and clearly be so interpreted are remarkably rare in view of the theoretical expectation. The reason for their rarity is probably that by the time these effects begin to occur extinction is usually so near that it follows instantly, as far as the fossil record could show, and morphological degeneration does not have time to become really appreciable.

In a broader sense, something related to this expected phenomenon, but without clearly degenerative small population effects as far as the record usually shows, may be involved in groups with a multiplicity of phyla up to their extinction. Sometimes this multiplicity seems to represent, in a sense, a sort of exploration of possibilities in an environment that is degenerating with respect to ancestral adaptations (Fig. 38A). In that case it would be a more reasonable figure of speech to say that the group shows virility in the face of adversity rather than to call it senile. Sometimes one or more of the exploring lines do find a new and successful adaptation (Fig. 38B). This diagram quite closely resembles the history of late therapsid reptiles, which split into a large number of lines most of which became extinct but three or more of which became "mammals."

LOSS OF ADAPTATION

Adaptation is an extremely complex two-way fit between population and environment (Chapters VI, VII). Changes in environment are incessant; no environment is really and completely the same for two seconds in succession. In all cases adaptation necessarily includes a cer-

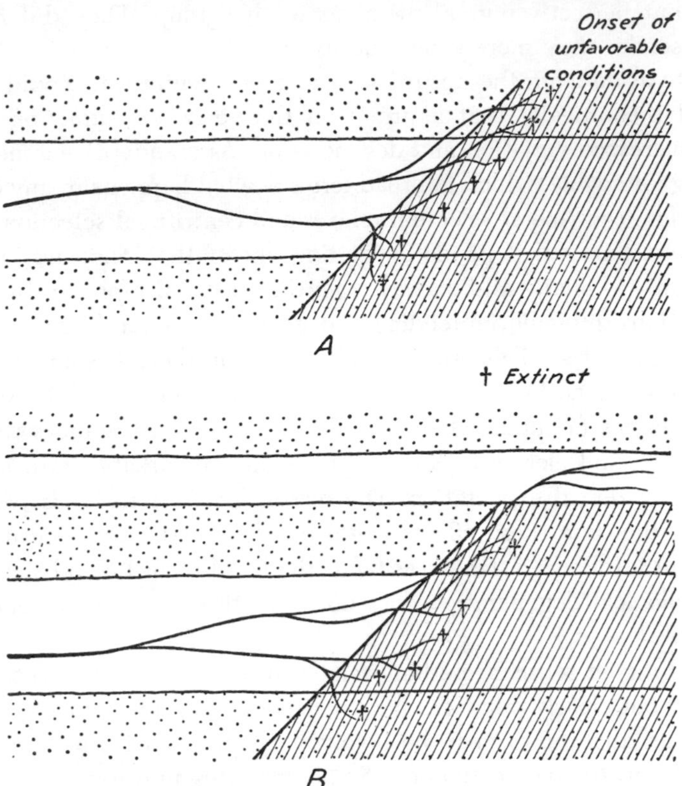

FIGURE 38. "EXPLORATORY" LINES IN A DEGENERATING ADAPTIVE ZONE. A, no other adaptive zone is reached and the whole group becomes extinct. B, another adaptive zone is reached by one (or more) lines; a new adaptive type arises and the old type becomes extinct.

tain range of tolerance for environmental change. When, as *almost* always happens sooner or later, environmental change exceeds a former range of tolerance one of two things must happen. Populations involved may change so as to maintain adaptation or to achieve new adaptation: that is the usual, but it would appear not universal, reason for evolutionary change. Or the populations involved may not change sufficiently and adaptation may be lost: that, quite simply, is the usual cause of extinction, the only other possible cause being loss of existing adaptation by genetic drift.

Since some degree of adaptation is, almost by definition, the requisite for continued existence, saying that loss of adaptation is the cause of extinction comes perilously near to saying that loss of existence causes extinction. The statement is not quite as trivial as that, however. By designating adaptation as the key to extinction, as well as the key to existence, an end can be put to the old, long, and futile argument about whether "internal" or "external" factors cause extinction. It becomes clear that the cause is neither in the organisms nor in their environment but in the relationship between the two.

The problem is thus broken down into two aspects. It seems to be true, as many students have said (e.g., Arambourg, 1950, with quite unnecessary apologies for being "Lamarckian" because he took the environment into consideration), that extinction rarely if ever occurs without change of environment. It is, however, equally true that extinction never occurs without a failure of response in the populations affected. The change in environment is not the "cause" of the characters of populations that make them fail and therefore is not, as often stated, the "cause" of extinction. Extinction is not the invariable or even the usual outcome of change in environment. Of two populations undergoing comparable changes of environment, one often fails while the other succeeds. To explain any particular case of extinction it is, then, necessary to specify two things: what pertinent change occurred in the environment, and what factors in the population prevented sufficient adaptive change.

Some environmental changes are so rapid and brutal that they are obviously beyond the capacity for adaptive change in any population. Then it is unnecessary to specify the population factors, not because they are any less involved, but because we know that the whole mechanism of population change is simply incapable of rising to such a rate of action. This is true of practically all cases of extinction that have had any study while they were actually occurring, most of which were due to man, either because his destructive power was itself the precipitating change in the environment or because he changed other aspects of the environment (see the interesting popular summary by Williams, 1951).

It has been seen (Chapters I and II) that really marked changes in natural populations, even under strong selection pressure, commonly take on the order of 10^5 years and upward, even though relatively trivial changes as in frequencies of existing alleles or chromosome arrangements may be very rapid. Any environmental factor capable of killing off the whole population of a species in 10^5 years or less is more likely than not to cause extinction. Man obviously can do this for almost any species to which he devotes his attention. (Perhaps there are exceptions among organisms that take refuge in an environment man would prefer not to destroy, i.e., himself, and that are, moreover, capable of exceptionally rapid change: parasitic and pathogenic protistans.)

It is also evident that such events may occur without man. Volcanic eruptions have wiped out whole species in a day, but only insular forms or others very limited in distribution. Human introduction of competitors (e.g., rabbits in Australia) or predators (e.g., mongooses in the West Indies) have also led to extinction so rapid that adaptive response was simply out of the question, and such invasions have repeatedly occurred in the past without human aid. There is a possibility that a single extreme fluctuation in a normal environmental variant, in weather, for instance, would cause immediate extinction, and incidentally such a fluctuation would in almost all cases be absolutely impossible to observe in the stratigraphic record. We are, however, protected from the probability of being thus baffled by the fact that almost any group old enough to have a distinctive adaptive type must already have persisted through the whole possible range of weather and could only be seriously affected by slower change in climate.

The point about such intense, quick environmental changes is that they are beyond the capacity for change of the evolutionary mechanism as a whole. *Any* group subjected to them becomes extinct, so that there is no point in discussing differences that determine differential extinction. Rapid and nondifferential extinctions probably have been very common. For instance local (at least) extinctions of three of the four iterative lines of *Olenus* (Chapter VIII) were probably of this sort: all trilobites disappeared from the area from one stratigraphic plane to the next. Nevertheless, slower extinction and clearly differential extinction are also common. The notoungulates took a long time to die out, even after the irruption of competitors from North America. Horses became extinct in North America while bison, living with the horses and with considerable adaptive similarity, did not.

The ultimate genetical causes for failure of adaptive response in a population are simple and clear, although more explicit analysis is often complex and obscure to the point of impossibility. Materials for change are expressed variation, pooled potential variability, and new mutations. Obviously, loss of adaptation will follow environmental change if expressed variation is not adequate or appropriate, if, further, appropriate potential variability either does not exist or cannot be released fast enough, and if, still further, appropriate new mutations do not occur or are not fixed or do not achieve phenotypic realization at a sufficient rate. Faced by a given degree of environmental change, groups with lower phenotypic variation, lower genetic variability, lower mutation rates, or any combination of these are more likely to become extinct.

For various reasons, small populations and panmictic populations of any size are more likely to become extinct than moderate to large populations and populations subdivided into partly isolated breeding groups or demes. Small populations have limited variability at any one time and low absolute incidence of mutation, and they may be subject to genetic drift. They are also likely to be narrowly localized and so more subject to rapid extinction by a regional catastrophe. Panmictic populations have relatively low expressed and potential variability. Other things being equal, the larger the population the more potential variability, at least, it is likely to have and the larger its absolute rate of mutation will be. Size for size, the more subdivided a population is, the higher will be both expressed and potential variability, and also the less likely is any temporary or local disturbance to upset the balance of the whole.[1]

Most organisms have optimal and minimal population sizes and densities (see review in Allee, Emerson, Park, Park, and Schmidt, 1949). Under normal conditions, the optimum is a stable equilibrium to which the population tends to return after fluctuations either above or below. The minimum (for a self-reproducing population) is, however, an unstable point. Fluctuation above the minimum tends to go on to the optimum and that below the minimum tends to go on to extinction. The unstable equilibrium point may be rather high, especially for social animals. Extinction of populations may thus ensue even though the unfavorable environmental factor ceases to operate or does not operate to eliminate all individuals but only to reduce size of population unusually far below the optimum.

That one population may have and another may lack the *sort* of

[1] An influence of lack of subdivision is dramatically illustrated by the rapidity of extinction of the passenger pigeon and near extinction of the plains bison by man. Both had enormous but little subdivided populations.

variability required is also and clearly another factor in extinction, but a rather mysterious one in most cases. Horned rhinoceroses have survived and hornless rhinoceroses otherwise similar became extinct. Supposing that lack of horns was a factor in extinction of the latter, which is at least possible, we may say that they lost adaptation because they had no variation or mutation in the direction of horns. We certainly cannot say *why* they did not have such variation, when their relatives did have. In this and in the great number of similar cases, we can only clothe our ignorance in the words "mutation is random." (Incidentally, such cases have a bearing on the reactionary Lamarckian-Michurinist doctrine that organisms develop what they need.)

The clearest general relationship between characters of organisms and the chances of extinction and the one most often mentioned concerns their breadth of adaptation or specialization. Narrowness of adaptation means that the organisms range over a smaller variety of environmental conditions. The adaptation is therefore more likely to be affected by environmental change than in the case of organisms that tolerate or thrive in a wider environmental range. Under relatively constant conditions or with a general deterioration of the whole range of both, the more narrowly adapted animals have the advantage, but with more specific change the more widely adapted animals are less likely to become extinct. (This is, of course, survival of the more or the less specialized, discussed in Chapter VII.) In this, as in so much of evolution, there is a balance, and neither broad nor narrow adaptation has become the general rule although the more usual tendency is for narrowing of adaptation.

It does not follow automatically that because a narrower adaptive zone is more liable to change the populations in it are more liable to extinction. Extinction ensues only if the populations fail to change adaptively. Many narrowly adapted organisms are still able to make extensive changes and thus to avoid extinction in a changing environment, but on an average narrowly adapted forms are also more specialized in another sense: they are, one might say, less plastic; the sorts of change they can undergo are more restricted. It is this combination of factors, of two different things both usually called "specialization," that makes such a situation likely to lead to extinction.

Specialization in the sense of limitation of further change may also occur without specialization in the sense of narrowed adaptation, although the two do tend to be associated. It is not at all clear that any one lineage of one-toed grazing horses has a narrower adaptive range than had a lineage of three-toed grazing horses. Indeed the adaptive range of the former may be wider and they did survive after the three-toed horses became extinct. But the one-toed horses do have less chance of further adaptive change and are more specialized in that sense of the word.

The word "specialization," one of many ambiguous and abused terms too loosely used in evolutionary studies, is also often applied in a relative way to comparatively large changes from an ancestral or modal condition. Thus man is "specialized" in being much bigger than most or than early primates, but a whale the size of a man would be considered "unspecialized" in size. It may even be considered that *Nannippus* is "specialized" because it is smaller than its immediate ancestors while a *Mesohippus* of the same size is "unspecialized" because it is smaller than its immediate descendants. Specialization in this sense has no necessary relationship with either of the other two sorts of specialization or with extinction.

Extinction in which specialization (of any sort) is involved is usually said to be due to overspecialization, but this way of putting things confuses the issue. It seems to imply that overspecialization is a definable extra degree of specialization or that it is something quite different, even moralistically so, "bad" as opposed to "good" specialization. Of course it is nothing of the sort. Overspecialization is just specialization that has become disadvantageous because the environment has changed. Precisely the same characters may be highly adaptive at one time or in one group and inadaptive at another time or in a different group. To call them now "specialization" and now "overspecialization" only obscures their real relationship to survival and to extinction.

The sorts of environmental changes that may be involved in extinction are legion. The environment is extremely complex and any sort of change in it may require changed adaptation which in turn may lead to extinction if organisms fail in adaptive change. Changes in biotic environment especially likely to lead to extinction include appearance of new competitors or predators, disappearance of types of animal or vegetable food, and appearance of new pathogenic organisms. (The "appearances" in question are more often by spread or migration than by immediate evolution *in situ*.) The changes in physical environment most often discussed are those of climate, heat or cold, aridity, etc., but other physical changes may have been equally or more important, such as withdrawal of the great epicontinental seas in times of high continents (like the present) or disappearance of high alpine environments in times of low continents. Many lists of such possibilities have been compiled, but an attempt to compile a complete list, even in general terms, would be futile. Changes possibly concerned may be as varied and complex as the environment itself, which is so intricate and diverse as to defy adequate description.

Special interest attaches to competition as a factor in extinction both because it seems to have been common and because it more often can be specified in relation to the fossil record than can any other factor. It is quite common for two groups of similar adaptive type to be found together and for one to decrease to extinction while the other increases. In such cases the most likely hypothesis, at least, is that competition led to extinction of one group, although of course this may not really have been true in a given case. Some examples were mentioned in other connections in Chapters VI and VII. A particularly good example, as its compiler says "almost too perfect," is given by the replacement of multituberculates by rodents as discussed by Jepsen (1949). Table 23 (from Jepsen) indicates numbers of genera and species of the two groups known in the area (the Rocky Mountain region of North America) where the earliest known rodents and last known multituberculates occur. The two groups are markedly different in ancestry and many features of anatomy, but strikingly similar in rodentlike adaptation. The decline of the older group as rodents appeared and began to expand is evident in the table. Moreover, multituberculates are relatively abundant in all known late Paleocene local faunas but one (Bear Creek, Montana), where they are absent; precisely there is the only known occurrence of Paleocene rodents.

In such cases a complete explanation would also state why the succeeding competitor was superior. This can seldom be clearly stated, but some suggestion is often possible. In the example just given, Jepsen points out that rodent incisors are clearly more efficient mechanically than those of multituberculates. A frequent statement regarding broad competitive replacements is that the replacers are more "progressive" and the replaced more "primitive," which even if it means something, certainly explains nothing. In the case of South American–North

TABLE 23

VARIETY OF MULTITUBERCULATES AND RODENTS IN EARLY CENOZOIC DEPOSITS OF THE ROCKY MOUNTAIN REGION

(After Jepsen)

Ages			MULTITUBERCULATES		RODENTS	
			Genera	*Species*	*Genera*	*Species*
Eocene	late		0	0	13	31
	mid		0	0	9	19
	early	B	0	0	3	8
		A	3	5	1	4
Paleocene	late		7	11	1	1
	mid		6	17	0	0
	early		5	7	0	0

American faunal interchange and ensuing competition, in which the North American forms were much more successful, I (1950b) have pointed out that the latter do not seem in any clear, objective way to be more progressive or efficient but that they were the winnowed products of a long series of intercontinental competitions while the South American forms had been isolated for 70 million years or so.

Replacement by competition is one aspect of what Arambourg (1950) calls "paleontological relays," that is, the successive occupation of the same broad adaptive zone by different taxonomic groups (see also Chapter VII). In all, three cases may occur: (1) replacement by competition, (2) immediate replacement, and (3) delayed replacement. The second case involves extinction of one group immediately followed by spread of another, as if the presence of the first had inhibited and its extinction promoted the development of the second. Colbert (1949b) shows that following extinction of the phytosaurs the crocodilians, which had already arisen as a small upland group, invaded the lowlands and developed the phytosaurlike adaptations which they still have today. In such cases of immediate succession, the precision of the time scale is usually insufficient to rule out the possibility of competition or of delayed replacement. Recent examples show how rapid extinction by competition may be. In limited areas like islands, at least, it may ensue in a few years or even months. With such rapid replacement, chances would be against finding the competing forms together at the same stratigraphic level.[2]

The third possibility, which also has often occurred, is delayed replacement, such as that of ichthyosaurs by cetaceans. In such cases rise of the later group obviously had no effect on the earlier. Even when replacement or relay occurs, this does not always or necessarily involve competition.

Extinction, like origin of new groups, has been more common at some times than others and has several times reached peaks so high and has affected so many different groups that coincidence seems ruled out (see Chapter VII, episodic evolution) One such time of mass extinction, affecting some terrestrial but more marine animals, was roughly localized at the Paleozoic-Mesozoic transition and another, affecting some marine but more terrestrial animals, roughly at the Mesozoic-Cenozoic transition. (It is no coincidence that these approximate the era boundaries; the boundaries were placed there in the geological time scale because of the extinctions.)

It should be emphasized that these mass extinctions are not instantaneous or even brief events. They extend over periods of tens of millions of years (see, e.g., discussion by Watson, Prenant, Westoll, Simpson, Cuénot, and Arambourg following Arambourg, 1950). This makes the phenomenon all the more mysterious, because we have to think of environmental changes that not only affected a great many different groups in different environments but also did so very slowly and persistently. The only general and true statement that can now be made about, say, the extinction of the dinosaurs is that they all lost adaptation in the course of some long environmental change the nature of which is entirely unknown. There have, of course, been many speculations, some of them very ingenious [3] but none of them probable.

Aside from cases of apparent competition and some extinctions that seem to have been secondary effects of extinction of a food supply, the

[2] Extinction of dinosaurs and spread of Paleocene mammals cannot be explained in this way, because early Paleocene mammals seem certainly too different from any dinosaurs to have caused such rapid competitive extinction of the whole group.

[3] As an example of such ingenuity, Cowles (1945) has demonstrated that heat-sterilization can occur in recent reptiles and suggests that this happened to the dinosaurs. But further studies by Colbert (1946) make that application highly improbable, and most students think that if there was any climatic change at the end of the Cretaceous it was toward generally cooler climates.

sad fact is that explicit assignment of immediate causes to particular instances of extinction is almost always unconvincing. We may say with full assurance that all were failures of adaptation to changing environments, but in few cases can a candid paleontologist even attempt to say precisely what changed and exactly why adaptation failed. This is not surprising in view of the extreme complexity of possible changes and adaptive responses, the improbability that any one, clear-cut factor acted alone, and the fact that many of the factors can seldom or never be determined from the fossil record. We know, for instance, that viruses infecting a species biochemically unable to resist them could cause extinction, but we cannot now and probably can never say that this is what happened to any one extinct species. Our failure to be explicit about particular instances of extinction is not because such an event is esoteric and inexplicable in detail but, quite the contrary, because there are so many possible detailed explanations that we cannot choose among them.

RELICTS

Relicts are in some cases organisms on their way to extinction and groups in evolutionary traps or blind alleys are sometimes especially susceptible to extinction. These topics will therefore be briefly reviewed at this point even though they take us away from the subject of extinction, strictly speaking.

At least four sorts of groups called relicts may be distinguished:

1. Numerical relicts, rare survivors of a group once abundant
2. Geographic relicts, groups occupying a much smaller geographical area than their ancestors and earlier relatives
3. Phylogenetic relicts, groups evolving at exceptionally slow rates and so surviving from remote times with little change
4. Taxonomic relicts, groups much less varied than previously.

Many groups have become relicts in all four senses at once. Indeed, this probably is usual for a group on the verge of extinction. *Latimeria,* little-changed recent survivor of the once abundant, world-wide, and highly varied group of coelacanth fishes, is certainly a phylogenetic and taxonomic relict and almost surely a numerical and geographical relict. *Sphenodon* is a relict in all senses (least markedly so in a taxonomic sense). There has, however, been a great deal of confusion and idle theorizing because of belief that relicts must or usually do belong to all four categories. This is not a valid generalization. *Limulus* is a phylogenetic relict, but it is decidedly not rare, it still has a rather wide distribution, and its family was never particularly varied. Living proboscideans are numerical, geographic, and taxonomic but decidedly not phylogenetic relicts. Opossums are phylogenetic relicts, and their distribution has decreased although hardly to geographic relict degree, but they are highly abundant and exuberantly varied. Decrease in numbers and area of wild horses has not reached relict proportions and they are not phylogenetic relicts, but they are far less varied than formerly. A numerical relict is almost perforce a geographic relict and a geographic relict is likely to be a numerical relict, but aside from this association all possible combinations seem to occur commonly.

We may thus discard the view of Rosa (e.g., 1931) that any group of narrow distribution is a geographic relict and necessarily also a phylogenetic and taxonomic relict, along with Willis's (e.g., 1940, on which see Wright, 1941) insistence that a group of narrow distribution cannot possibly be a phylogenetic or taxonomic relict. Rosa's theory that species arise all over the world at once and later become restricted and Willis's theory that species arise in small areas and then spread to a maximum range never to be restricted are both at this date mere curiosities of scientific history that require no further discussion.

It is abundantly evident, in spite of extreme irregularity in various cases, that the usual geographic history of a group is origin in a more or less restricted area, rather rapid spread to a first maximum range which may, however, be considerably expanded later as the environment evolves, and final geographic restriction, often quite slow, up to extinction. A group of limited numerical, geographic, and taxonomic extent may be young, as Willis would invariably have it, or may be old, as Rosa would invariably have it. In the latter case it is a relict group in some sense (not necessarily a phylogenetic or taxonomic relict) and is probably verging on extinction. Aside from that point, the extremely interesting and important subject of historical biogeography is not pertinent here and cannot be more than mentioned in the limits of this book.

Phylogenetic relicts, in the sense that they are examples of arrested or slow evolution are, on the other hand, extremely pertinent to the themes of this book, so much so that much of Chapter X will be devoted to them. Numerical, geographic, and taxonomic relicts have less bearing on these themes, aside from the fact that they are likely to represent groups near extinction.

A fifth and really quite distinct use of the word "relict" is to apply it to forms that reflect former environmental conditions. Thus *Mysis relicta* and some other species in lakes of northern Europe are of marine ancestry and have adapted to fresh water as the lakes, formerly arms of the sea, were cut off and water freshened (Samter, 1905). There is no particular reason why such forms should be relicts in other senses of the word, nor evidence that they usually are so. They do have peculiar interest as groups that have succeeded in surviving radical environmental change *without* losing adaptation. They further exemplify the fact that such change, even when radical, is not in itself the cause of extinction, which occurs only to groups whose own characteristics single them out for extinction when the environmental change occurs.

Whole faunas and floras or considerable parts of these may also be considered relict if they now occupy much more restricted areas than formerly (are geographically relict) and especially if they are sorts of associations now uncommon and formerly widespread (are also environmentally or ecologically relict). The members of such associations are not necessarily all or any of them relicts in any sense of the word, although some of them are likely to be. For instance, in Moravia, Burr found a "faunal island" in which the Orthoptera are quite different from those around them but resemble those of the Volga valley. This is interpreted as a relict fauna surviving from an earlier time of more widespread steppes (discussion and references in Hesse, Allee, and Schmidt, 1951). The surviving *Metasequoia* forest in China, described by Chaney (1948), is a relict flora in the sense of representing an association that was widespread in the earlier Cenozoic and is now extremely restricted. *Metasequoia,* itself, is a numerical, geographic, and phylogenetic relict, but some plants in the relict fauna are not themselves relicts (as genera, at least) since they still occur widely but in different associations and do not belong to such ancient groups as that of *Metasequoia.*

A classic example of a supposedly relict fauna is that of the marsupials of Australia. It is still stated, even by some of the soundest authorities (e.g., Young, 1950) that this is a survival of a sort of fauna world-wide in Cretaceous (or other remote) times, and now restricted to Australia.

The statement simply is not true: (a) there is now no reason to believe that marsupials, as a group, are older than placentals or that an exclusively or mainly marsupial fauna was ever world-wide; (b) no fauna at all like that of Australia has ever occurred anywhere else, and that fauna is now as widespread as it ever was; (c) most Australian marsupials are not phylogenetic relicts but have undergone very extensive progressive change since the late Cretaceous; (d) the quite different marsupials of the rest of the world are not extinct or extremely restricted, since true opossums (Didelphidae) are still very widespread and abundant, extraordinarily successful in adapting even to radical environmental changes caused by man. The great interest of the Australian fauna and flora is that they represent the largest surviving[4] island biota. The biota is not relict in any meaningful sense of the term.

EVOLUTIONARY TRAPS AND BLIND ALLEYS

The concept of traps and blind alleys in evolution is that groups in some situations and with some characteristics become extremely restricted as to further possibilities. It becomes unlikely or even impossible that they can themselves change appreciably or that they can give rise to progressively changing branches. They thus either survive as, eventually, phylogenetic relicts or they become extinct in the face of environmental change to which they cannot adapt. "Trap" may perhaps be applied more often to environmental conditions and "blind alley" to factors in the organisms and populations, but there is no well established difference in usage and the distinction is not always clear because, again, adaptation is really involved in either case.

Islands tend to be evolutionary traps (see Mayr, 1942). Large islands (Australia, Madagascar) may themselves be important centers of evolution, but still are likely to become traps if isolation is strong and sustained. Islands with little or brief isolation (e.g., Sumatra) may not be serious traps for most organisms. Small, strongly isolated islands are almost sure to be traps for the great majority of organisms that reach them. They are colonized, directly or indirectly, from the continents or from larger islands. The colonists become very specifically adapted to a relatively small number of niches. Thereafter, since successful colonization is infrequent, a rather static, closed ecological situation

[4] Although now rapidly changing and with many members extinct or nearly so because of human intervention, both cleverly intentional and stupidly inadvertent.

persists. Populations are likely to be small, with little pooled variability available for change and with the possibility of gene drift. Successful return to continents or larger islands by organisms so situated is extremely unlikely. If successful invasion of their own islands should occur, they are particularly liable to rapid extinction. Man and the other organisms introduced by man have caused decidedly more extinction on islands than on continents, a statement that applies even to Australia, largest of islands.[5]

In the cases of some, especially large islands, escape from the trap may be improbable but not prohibitively so for all members of the biota. The armadillo *Dasypus novemcinctus*, which evolved as an island form in the great island of South America, is a highly successful invader in North America. I think it probable that some of the more distinctive Australian marsupials would have made their way into the Asiatic fauna had a migration route opened, and some Australian plants (perhaps *Casuarina* and various eucalypts) could probably have crossed such a route as well. Return invasion from smaller, strongly isolated islands or island groups (e.g., Galápagos, Hawaii) is so improbable as perhaps to amount to impossibility, although it is unlikely that any event of the sort can be permanently impossible in an absolute sense, with probability technically precisely zero.

Some other environmental situations are also frequently traps. Lakes and river systems may be so, especially if they become truly isolated and develop peculiar ecological conditions (e.g., Great Salt Lake and its drainage basin). Mountain peaks, especially if they, too, are isolated, have some of the adaptational characteristics of lakes and islands and are also likely to be evolutionary traps. The fact that oceanic islands (and often also continental ones), lakes, and mountain peaks are all temporary in the sweep of geological history adds to the probability that they will be traps and that populations specially adapted to them will become extinct without issue. Incidentally, it also makes it unlikely that they will be refuges for phylogenetic relicts.[6]

[5] Australia is, of course, a continent to geographers and to its rightly proud inhabitants, but biologically it is an island. So was South America, but it is now a continent biologically as well as geographically.

[6] The dodo and other flightless birds of oceanic islands have sometimes been called "relicts," but they were not really such. They were phylogenetically young forms evolved on and adapted to the islands where they occurred. New Zealand does harbor some phylogenetic relicts, notably *Sphenodon*, but it has had the geological

Some genetic mechanisms tend to fix organisms in blind alleys. This is likely with mechanisms producing change that is irreversible or, at least (since none is theoretically completely irreversible), very unlikely to be reversed, and especially with those that tend to fix a current, advantageous adaptive status at the expense of pooled or potential variability. Noteworthy in this respect are genetical factors reducing recombination, such as chromosomal inversions and, particularly, apomixis and self-fertilization. Inversion and the various other factors reducing crossing over seem about equally common in animals and plants, but it is difficult to judge the extent to which they have run either into blind alleys. *Drosophila,* despite the indubitable validity of its demonstration of mechanisms involved in progressive evolution, may well have lost long since its own capacity for further, marked progression and it has reduced crossing over (little or no crossing over in the male, many inversions, etc., in most races). However, we do not know that crossing over is not equally reduced in some species still adaptively plastic.

Apomixis (mostly in the form of parthenogenesis) and self-fertilization do occur in animals but are relatively infrequent in them, which may well be correlated with the fact that animals do, on the whole, seem to be less liable than plants to be hemmed into evolutionary blind alleys. Both apomixis, in a variety of forms, and self-fertilization are common in plants. When they are obligatory and not alternative to ordinary sexual reproduction, their effect is greatly to reduce variation, which improves adaptation of the population at a given point but stringently limits capacity for change. Obligate apomicts and self-fertilizing plants have not been involved in major progression in the plant kingdom and they do seem, on the whole, to be in blind alleys (Stebbins, 1950).

Polyploidy is also less common in animals than in plants. In both animals and plants it tends to be associated with apomixis (White, 1951). It may but does not necessarily lead into blind alleys. Stebbins (1950) concludes that polyploidy has played an important role in the origin of families and orders of flowering plants, but that it produces variations on old themes and not major new departures. In a study of chromosomes and evolution of the pteridophytes, an old group and one with a number of phylogenetic relicts, Manton (1950) finds that they tend

stability of a continent and from this point of view compares with Australia and not with typical oceanic islands.

(now) to show extreme polyploidy. She concludes that the group as a whole is running down, that its lines have entered or are entering blind alleys, in considerable part because of high polyploidy. (*Equisetum*, a famous phylogenetic relict, has the haploid number 108, and numbers higher than 250 occur in pteridophytes.)

It is also true that the process of specialization has limits and that an organism as specialized as possible, morphologically and physiologically, would appear to have no place else to go, one might say, and hence to be quite at the end of a blind alley. A horse cannot have less than one toe, a land animal cannot become heavier than the structural limit of carrying power of bone, and so on. Such limitations undoubtedly exist and do sometimes involve blind alleys, but in application to specific cases difficulty arises from the subjective nature of our judgments of specialization and from the incapacity of our imaginations. The first mammals were, with respect to their contemporaries, extremely specialized reptiles. Some of the very characters, such as homothermy and lactation, that now are clearly seen to have led them *out* of a blind alley and onto a broad, richly branching series of boulevards were physiological specializations pushed to an absolute limit. Obviously reaching such a limit does not *ipso facto* mean entering a blind alley.

The teeth of early mammals were also very highly specialized in comparison with ancestral and contemporaneous forms. They were not anywhere near a limit, as we know because we witness what followed, but we almost certainly would think a limit had been reached if we were not later witnesses. Evidently judgment that limits have been reached in *all* respects is not likely to be reliable. Advocates of the view that specialization inherently leads into evolutionary blind alleys sometimes demand examples of specialized groups that were not in blind alleys. When an example like the early mammals is given, they reply that those forms were not *really* specialized. When terms are redefined so as to force agreement with the postulate under consideration, discussion bcomes futile.

In other terms, an evolutionary blind alley is an adaptive zone which comes to an end sooner or later (as all do eventually) and from which it is unlikely that another will be reached. A group in a narrow zone, hence specialized in that sense, is less likely to reach another zone than is a group in a broad zone. A group that has progressed far along a zone diverging from others, hence specialized in *that* sense, is also less likely to reach another zone than one less divergent. But this is only a generalization of probabilities. Groups in narrow zones do sometimes reach others, and radical divergence is the pattern of origin of major changes in evolution on the relatively rare occasions when they do occur.

THE IRREVERSIBILITY OF EVOLUTION

A group of organisms does not return entirely to the different condition of its ancestors. In broadest statement, that is the principle of irreversibility of evolution, sometimes called "Dollo's Law." [7] Equally true but less often stated is a complementary principle of the irrevocability of evolution: influence of a different ancestral condition is not wholly lost in a descendant group. (This principle was also involved in Dollo's apothegms.)

Discussion of the principle of irreversibility is seemingly endless, and no review of the large literature will be attempted here. Much of the discussion has revolved around points that are trivial or are only verbal quibbles. There has been much solemn dispute over whether an adaptive trend in some descriptive character can be reversed. Of course it can be and often is, witness the horses that became smaller although their immediate ancestors had a definite trend toward larger size. But they did not become like their still earlier ancestors of their own size. Evolution from fish to land mammals was reversed in many functional and anatomical characters when land mammals gave rise to whales, but whales are not fishes, far from it. Some students have such a respect for law that they insist that apparent exceptions to their (or even to Dollo's) concept of "Dollo's Law" are inadmissible, for instance that a small animal cannot really be the descendant of a larger one, even if everything points conclusively that way, because that would be against the law. Such quibbles do not merit much serious con-

[7] It has already been remarked that Dollo was not really the first to state this principle. Dollo's statement was also more rigid and explicit. It early became clear that Dollo's own version of his "law" is open to very numerous exceptions, and there has grown up a sort of kindly tradition (e.g., Carter, 1951, also repeatedly voiced by Gregory) that all he meant to say was that lost organs are never regained. In fact the question came up during Dollo's lifetime and he flatly stated (in his printed work, which seems no longer to be read at first hand) that this is *not* all he meant to say. It was Dollo's student O. Abel who confined "Dollo's Law" to the trivial case of reappearance of lost organs. Of course it is what we find to be true that matters and not what Dollo said, but it is poor justice to his memory to falsify what he said.

sideration, nor does semantic argument whether, when lost structures do reappear (as in tetraptera mutants in *Drosophila* or the polydactylous guinea pigs of Wright, 1934) they are "really" the *same* structure (e.g., Gregory, 1936).

There is nothing in the analyzably separate factors of evolution that prohibits reversion. Back mutations occur. Lost combinations can be reconstituted by recombination. Variation is practically always present on *both* sides of the mode. Selection can reverse its direction. As regards particular characters and trends, therefore, if genetic variation and selection really are essential factors of evolution, evolution should be reversible—and it is. Evolution is readily reversible for particular features, especially those like size that have a very broad genetic basis not necessarily or probably homologous in different groups. It is also more or less reversible to a condition of the immediate ancestry in which the genetic system was essentially the same. This does not alter the broader fact that evolution does not reverse itself exactly or for the whole organism, that it does not double back on itself and repeat quite the same sequence a second time.

The statistical probability of a complete reversal or of essential reversal to a very remote condition is extremely small. Functional, adapted organisms noticeably different in structure have different genetic systems. They differ in tens, hundreds, or thousands of genes, and such genes as are the same have different modifiers and are fitted into differently integrated genetic backgrounds. A single structure is likely to be affected by many different genetic factors, practically certain to be if its development is at all complex. The chances that the whole system will revert to that of a distinctly different ancestor, or even that this will happen for any one structure of moderate complexity, are infinitesimally small (Muller, 1939).

The genetical factors are so complex and so constantly changing that extensive reversion is almost impossible. The same is true of the environment. It has thousands, or more likely millions, of pertinent elements, all subject to change in various directions. Similar constellations of elements may recur, and when they do more or less similar adaptations recur, but for the environment to return to an earlier, distinctly different total condition is virtually impossible.

Beyond these considerations is another, still more general and fundamental, which I have elsewhere (Simpson, 1950a) expressed as follows: "That evolution is irreversible is a special case of the fact that history does not repeat itself. The fossil record and the evolutionary sequences that it illustrates are historical in nature, and history is irreversible. . . . Historical cause embraces the *totality* of preceding events. Such a cause can never be repeated, and it changes from instant to instant. Repetition of some factors still would not be a repetition of historical causation. The mere fact that similar conditions had occurred twice and not once would make an essential difference and . . . the sorts of existing organisms . . . would be sure to be different in some respect."

We are dealing here with something quite different from a reversible chemical reaction, which may be set up in an open system as far as entropy is concerned and in which all pertinent variables can then be repeated at will.[8] Evolution is irreversible because it results from its own past and its past is irrevocable. That irrevocability has conditioned what exists and was the ultimate cause, within our realm the most causal of causes, for the extinction of what no longer exists. It also conditions the fact that evolution can run into blind alleys, from which there is no turning back, and that perhaps some day all will end in a blind alley.

[8] But chemical reactions are also historical in nature and inherently irreversible in a closed system where the law of increase in entropy applies. Blum (1951) has shown that the law also applies to life and has related this to the irreversibility of evolution.

CHAPTER X

Horotely, Bradytely, and Tachytely

It is abundantly evident that rates of evolution vary. They vary greatly from group to group, and even among closely related lineages there may be strikingly different rates. Differences in rates of evolution, and not only divergent evolution at comparable rates, are among the reasons for the great diversity of organisms on the earth. Among the living primates there are, for instance, some rather unspecialized or primitive prosimians (i.e., little changed from Eocene progenitors), a larger number of divergently specialized prosimians, many monkeys of different degrees of progression and divergence, a few apes, and the unique species of man. Important as is the purely divergent evolution, it is also clear that differential rates are involved. At the extremes, the lineages of the more primitive living prosimians have evolved less rapidly as regards the whole of their structure and adaptive position than has the lineage of man.

Any one group, large or small, seems to have a fairly characteristic mean or modal rate of evolution and a certain range of less common rates on each side of this. The rates have a distribution pattern which is open to investigation and which is an important element in evolution. It has long been noticed, further, that there are a few lines in most large groups that seem to evolve at altogether exceptional rates. Some hardly change at all over long periods of time. Others change, over short periods of time, at rates so rapid as to be almost beyond comparison with those usual among their relatives. It must be considered whether these extraordinary rates are merely the extremes of normal variation in rates or whether they reflect special circumstances different from those in the distributions of usual rates. In either case they must involve different intensities, combinations, or both as regards evolutionary determinants, and it would be of extreme interest, perhaps even of considerable practical value, to identify these.

DISTRIBUTIONS OF RATES OF EVOLUTION

In Chapters I and II the examples of how rates may be estimated and represented showed that: (a) related lines of descent commonly differ in evolutionary rates; (b) within larger taxonomic groups such as orders or classes there is generally an average or modal rate typical for the group; and (c) the average rates may differ greatly from one group to another. Since rates do vary broadly around a group mode, rate distributions of different groups tend to overlap widely even though the modes are quite distinct. Mammals seem surely to have a much higher modal rate of evolution than molluscs, but some mammals have evolved more slowly than some molluscs. It even seems clear that some mammals have evolved at rates below the mode for molluscs and some molluscs at rates above the mode for mammals.

Such comparisons are hazardous because there is no absolute way to measure equivalent amounts of evolution in two groups anatomically so different and changing in such different ways. The two cannot even both be studied authoritatively by one specialist. This sort of objection is often made, and of course it carries a great deal of force. It is nevertheless not as strong or as fatal to comparisons as sometimes claimed. For instance, it is often said (e.g., again by Hutchinson, 1945, in criticism of the predecessor of this book) that molluscs may have evolved very rapidly in the soft parts without having this reflected in the shell. But in molluscs the *whole* skeleton is usually available and studied in fossils (which is very rarely true of, say, fossil mammals) and their skeleton reveals a great deal about the soft parts, directly (muscle scars, etc.) or indirectly (growth rates, thickness and chemical nature of shell, twisting of body, etc.). Percentagewise a mollusc shell probably tells more about the soft parts than do the usually available parts of a vertebrate. It is extremely unlikely that any marked progressive change in soft parts could occur without obvious changes in the shell.

We need not, then, bend over backward in denying validity to comparisons of rates between groups, even such very disparate groups as molluscs and mammals, and of course this applies all the more strongly to comparisons between groups that are similar, such as different sorts of molluscs or of mammals. There is, too, another sort of comparison that is really very little affected by this particular difficulty. The rate distributions can be compared in any two groups without regard for the absolute values of the rates. It was seen in Chapter II that survivorship curves are quite comparable even when the absolute rates involved are altogether different. There is little reason to doubt that both the resemblances and differences in shapes of such curves can be significant facts that are independent of our ability (or inability) to equate individual rates in the groups compared.

Survivorship tends to be negatively correlated with rate of evolution. By and large, the longer a genus, say, endures the more slowly its included populations are evolving. The correlation cannot be exactly —1.00 and it cannot be well measured, but it is a reasonably safe approximation when we can average out fairly large bodies of data. This provides a very convenient way to approach the problems of the distributions of evolutionary rates. By assuming that evolutionary rates are the reciprocals of survivorship, it is possible to calculate from any survivorship curve (or tabular data) what percentage of the population has evolved at any particular rate. For purposes of comparing different distributions, it is further possible and desirable to divide the scale of rates not in absolute but in relative terms, e.g., in decile classes of range.

Such rate distributions for pelecypod and mammalian land carnivore genera have been calculated from the survivorship data of Figure 5, Chapter II, and are graphically presented in Figure 39. For reasons that will become clear, these rate distributions are based on extinct genera, only. The two curves have a characteristic difference: the carnivore mode is more strongly peaked and classes just below the mode are correspondingly lower. Nevertheless the two are strikingly similar in these respects: (1) both have a single mode from which frequencies drop in both directions; (2) both are more peaked (leptokurtic) than a normal curve; (3) both are strongly asymmetrical with the mode much nearer the upper (faster rate) end of the distribution (negatively skewed). It happens also that in both the mode is in the ninth decile class, and that the amount of skew is comparable in the two.

Rate distributions of this sort have so far been made for only a few groups, although the number has increased since 1944 when these were first published. Those that have been studied all agree in the general characteristics (1)–(3), above. It is probable that these are generaliza-

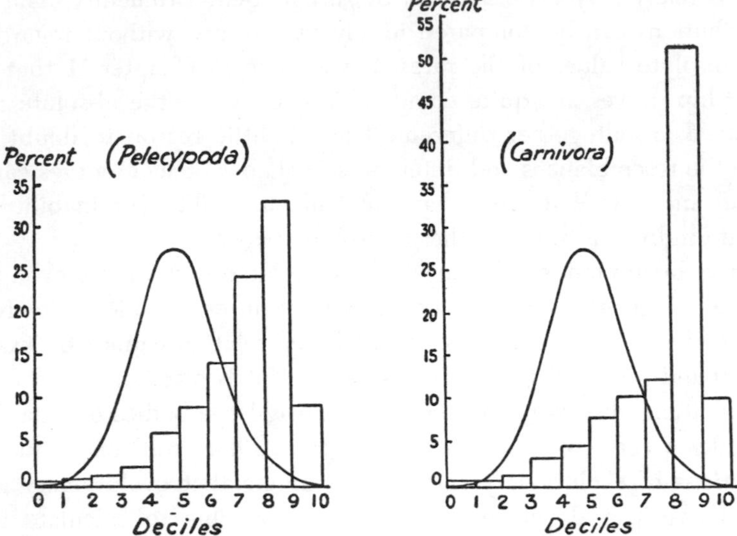

FIGURE 39. FREQUENCY DISTRIBUTIONS OF RATES OF EVOLUTION IN GENERA OF PELECYPODS AND LAND CARNIVORES. Histograms based on survivorship of extinct genera (see Figs. 5–6) and on the postulate of −1.00 correlation of survivorship and rates of evolution. Ranges (abscissal scale) divided into deciles to produce histograms strictly comparable in form although absolute rates are very different. Normal curves equal in area to the histograms drawn for comparison.

tions true of most distributions of evolutionary rates. Distributions may, however, differ markedly in degree of leptokurtosis (as do the pelecypod and carnivore generic distributions to some extent) and of negative skew (as these do not to any marked degree). Figure 40 shows a distribution that is near the extreme in both respects, rates of evolution in the 1,374 extinct species of centric diatoms, calculated from raw data published by Small (1946). Kurtosis is very great, with no less than 73 percent of the total frequency in a single decile class as against about 26 percent in each of the two middle decile classes of the normal curve. Skewness is also extreme, with the mode in the tenth decile class (it is, of course, at the fifth decile, between fifth and sixth decile classes, in the normal curve). If decile classes, only, are shown, such a curve seems to be J-shaped and an exception to (1), above, but the frequency does fall off to the right of the mode, as suggested schematically at the top of the figure. Finer subdivision would show that the mode is near the

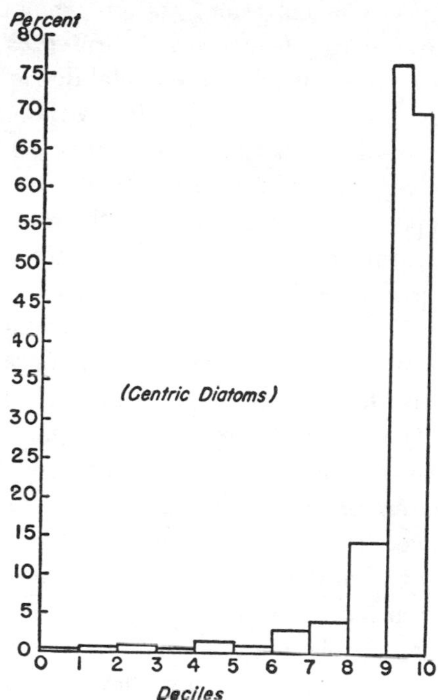

FIGURE 40. DISTRIBUTION OF EVOLUTIONARY RATES IN EXTINCT SPECIES OF CENTRIC DIATOMS. (Calculated from tabulation of raw survival data compiled by Small, 1946.)

ninth decile and that frequency does fall rapidly from it to zero at the tenth decile. There is also some slight irregularity in the dropping off of frequencies to the left (in the fourth and sixth decile classes), but this is evidently mere sampling fluctuation here where frequencies are very low. The distribution is of the same type as those of Figure 39, but more extreme.

The peakedness of these distributions shows that among related lineages, most of them tend to evolve at about the same rate. In extinct pelecypod genera about a third, in extinct carnivore genera about half, and in extinct centric diatom species about three-fourths evolved at rates included in only one-tenth of the whole range of variation in rates. These concrete data confirm the vague and subjective estimation of many paleontologists that each group of organisms does have a characteristic rate of evolution approximated by most lineages in the group.

The asymmetry of the distributions shows that the characteristic (modal) rate is decidedly nearer the maximum for the group than the minimum. Rates above the modal class do not extend far beyond it (are not much faster) and are, moreover, exhibited by relatively few lineages. In pelecypods and carnivores only 9 to 10 percent of the genera had rates faster than the modal decile class; none of the diatom species have rates above the modal decile class. Rates below the mode are more numerous than those above and also extend farther (are much slower relative to the mode than those above the mode are faster). In pelecypods more than half the rates are below the modal decile class, in carnivores more than a third are, and in centric diatoms more than a quarter. Although it was less clearly noticed before rate distributions were objectively calculated, this tendency, too, agrees with opinions occasionally expressed on the basis of subjective impression by earlier paleontologists (e.g., Matthew, 1914).

These distributions, based on taxonomic units that had run their full span and had become extinct, show that there is a definite range and pattern of such rates. The pattern is similar in different groups (so far as yet demonstrated), but it also has characteristic peculiarities in each. It gives a picture of a sort of evolutionary metabolism, one might say, for each group as a whole. (Of course I do not mean that it is an internal thing, like individual metabolism; it reflects the evolution of adaptation and the turnover of taxonomic units in the course of this.) It is this distribution that I have called "horotelic" (Simpson, 1944a). Organisms involved in it have horotelic rates, and evolution at rates so distributed is horotely.[1]

Further study of a number of groups, including the pelecypods and diatoms, shows that some of their lineages have evolved at rates much slower than any in the horotelic distribution as calculated from fossil records of extinct taxonomic units, only. Further studies show that these extremely low rates are part of a statistical excess of low rates in general in comparison with the horotelic distribution. This low-rate, nonhorotelic excess is bradytely and the rates involved in it are bradytelic. Other considerations strongly suggest that some phases of evolution have also involved rates higher than any in distributions obtained in this way. That exceptionally fast evolution is tachytely and it moves at tachytelic rates.

[1] When I proposed the terms horotely, bradytely, and tachytely, I apologized for adding to the technical terminology of evolution and explained that I had been quite unable to discuss rates clearly without this invention. As expected, at least one critic did object to the use of new terms, but apparently most others agreed that they make for clarity as they have come into rather general use.

Bradytely and tachytely are to be separately discussed in the remainder of this chapter. Before their detailed consideration, however, it may be pointed out that, although these concepts are derived mainly from analysis of the fossil record, it is sometimes possible to apply them with reasonable probability to groups inadequately or not known as fossils. Thus Ross (1951) has made a careful analysis of a tribe of caddis flies (Hydrobiosini) and its probable zoogeographic history. He adduces cogent evidence that generic splitting of the group began in the late Cretaceous and that of its 13 genera: (a) one has hardly changed at all since that time; (b) eleven have changed conspicuously and to varying degrees without becoming extremely different, and (c) one has undergone extreme change and reorientation of characters. His reasonable conclusion is that (a) is bradytelic, (b) represent together the horotelic distribution, and (c) is tachytelic (or it might be better to say, its lineage has been tachytelic at some time since the late Cretaceous—persistence of tachytely for so long a period is unlikely).

THE PHENOMENON OF BRADYTELY

Evolution occurs at a great variety of rates and the lower limit of rate is obviously zero. In view of the constant flux of the environment and constant mutation and genetic recombination in organisms, a sustained rate of precisely zero would hardly be expected. It would, however, be expected that rate distributions like those of Figures 39 and 40 would tail off to the left to a point not far from zero and would include a few, rare lineages that had not changed appreciably for long periods of time or had fluctuated without appreciable additive effects. Such lineages do occur, as has long been recognized. They are of great interest and the extreme slowness of their evolution calls for explanation.

Further analysis of rate distributions has, however, revealed that there is something more to explain than merely the slow rate of an occasional lineage in the tail of a falling distribution curve. In certain groups, although not in all, it turns out that rates near zero are actually more frequent than are slow rates in horotelic distributions calculated as above. It also appears that there may be a discontinuity in survivorship. In some groups, if the history of lineages arising at a given time is followed, their extinction rises to an early peak and then tapers off gradually, as expected; but there finally comes a time when *no* more extinctions occur over geologically long periods of time. This point

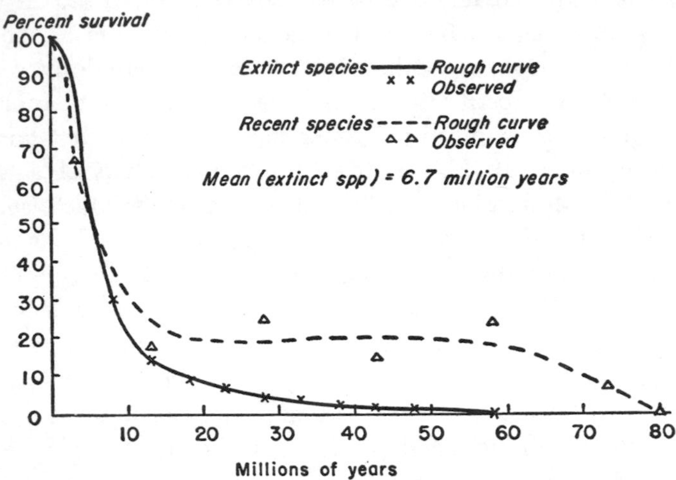

FIGURE 41. SURVIVORSHIP CURVES FOR SPECIES OF CENTRIC DIATOMS. The continuous line represents survivorship among species now extinct. The broken line approximates survivorship on the basis of species now living. The large discrepancy between these two lines from about 15 million years upward is due to bradytely. (Calculated from raw data compiled by Small, 1946.)

should be reached when the number of existing lineages drops to zero. But in some cases lineages do continue to survive; their number has not reached zero. Survivorship then has not dropped gradually to zero but reached a definite point in the history of each group where a discontinuity in survivorship occurred.

This odd phenomenon appears and can be analyzed in various ways. For instance, of the 54 broad genera of pelecypods first appearing in the Devonian, 30 became extinct in that period, 10 in the next, and so on down to 1 in the Jurassic. None became extinct thereafter, but 4 survive today, and will for some unmeasurable time longer (see Table 10 in Chapter II). Even when such a gap does not appear in our often rather crude data, an effect of what is surely the same phenomenon is seen in discrepancies between rate distributions calculated in different ways.

If the rates of evolution of living genera are distributed in the same way as rates of extinct genera, then a cumulative curve of ages of recent genera will be exactly the same as a survivorship curve for extinct genera, aside from fluctuations of sampling. The examples of such curves given in Chapter II showed that this definitely is not true for curves based on pelecypod genera. Schindewolf (1950b), although critical of those pelecypod data (and with some reason), obtained the same sort of discrepancy with better data on prosobranch gastropods. Lest it be thought that this result depends on use of genera or is limited to molluscs, it is noted also that calculation of the same curves for Small's species of centric diatoms produces the same sort of discrepancy in still greater degree (Fig. 41).

The discrepancy arises from the fact that more genera or species have survived from remote times in the past than would be expected from the horotelic rate distribution for extinct genera. In other words, there have been more lineages with rates near zero than that sort of distribution indicates. The same discrepancy is clearly shown by comparison of actual age composition of the recent fauna with expectation on the basis of survivorship among extinct forms (Table 24, here, also Table 12 in Chapter II).

TABLE 24

EXPECTED AND REALIZED SURVIVAL INTO THE RECENT OF BROAD PELECYPOD GENERA OF VARIOUS AGES

Time of Appearance (in millions of years previous to present)	*A. Percentage of Expected Survival, from Survivorship in Extinct Genera*	*B. Percentage of Realized Survival*	*Difference B—A*
25	92	78	−14
50	63	62	− 1
100	24	39	15
150	8	30	22
200	2	24	22
250	1	19	18
300	0	13	13
350	0	8	8
400	0	4	4
450	0	0	0

In carnivores actual survival is somewhat less than expectation throughout (Table 12), but this has affected genera of all ages in about the same proportion. As a result, the recent carnivore fauna is less varied than would have been expected from the Tertiary record, but its age composition, as proportions of genera surviving from various times in the past, is almost exactly as expected (Fig. 42). Rate of ex-

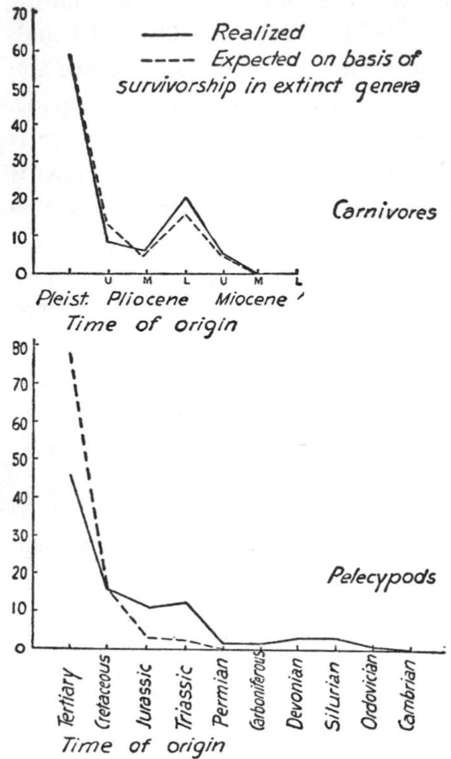

FIGURE 42. REALIZED AND EXPECTED AGE COMPOSITIONS OF RECENT PELECYPOD AND LAND CARNIVORE FAUNAS. Percentage of living genera arising (first appearing in fossil record) in stated epochs and periods as observed in the recent fauna (counting only genera that are known as fossils) and as would have been predicted on the basis of survivorship in extinct genera.

tinction was higher than expectation from late Pliocene to Recent, but it did not tend to affect later genera more than earlier, or earlier more than later. Moreover, no living genus is older than the maximum span for extinct genera. In this group the discrepancy under consideration does not occur, and the inference is that there are no bradytelic carnivores.

In pelecypods and gastropods, on the other hand, actual survival is greater than expected survival (Table 12, Chapter II) and these groups are today much richer, more varied, than expected on the basis of their horotelic distribution. Survival of late genera, those arising in the Tertiary, is near expectation: a little low for pelecypods and high for gastropods but in both cases as near as would be probable with sampling effects. Survival of pre-Tertiary genera, from the Ordovician for the very broad pelecypod genera and from the Triassic for the narrower gastropod genera, is, on the other hand, much greater than horotelic expectation. The result in terms of percentage composition of the recent fauna is that there is a much smaller proportion (not absolute number) of young genera and a much higher proportion of older genera than the expectation, as is well shown in Figure 42 for pelecypods.

Moreover, even though their spans have not yet ended, some living genera have already had much longer spans than any extinct genera of the same group. Some broadly defined living genera of pelecypods have been in existence for upward of 400 million years, but the longest-lived of extinct genera, defined equally broadly, endured no more than 275 million years. In the centric diatoms, there are living species on the order of 75 million years of age but no extinct species more than about 50 million years old. It is such facts that indicate a discontinuity in survivorship. On the other hand, in the prosobranch gastropods (Schindewolf's data as previously noted) the oldest living genera are about 200 million years old and there is at least one extinct genus that endured for 250 million years or somewhat more. This shows that there are complicating factors,[2] but does not alter the fact that this group, too, shows clear-cut excess of survivorship of old genera in the recent fauna.

The phenomenon here noted has of course long been discussed in a more general way or in other terms and techniques. T. H. Huxley (e.g., 1870) pointed out as early as 1862 that certain types of life are unusually "persistent." The phenomenon of "arrested evolution" (Ruedemann, 1918) has long engaged the attention of paleontologists. Their earlier analysis was, however, even more unsatisfactory than at present because the usual criterion of "persistence" was quite arbitrary and it was not shown whether the rates of evolution involved were really slow for the particular group in which they appeared. Thus Ruedemann considered as "persistent" and cases of "arrested evolution" any genera, from protistan to reptilian (there are no avian or mammalian examples) that

[2] A really detailed analysis of this group cannot be attempted here, and should be done as a separate study by a specialist in fossil gastropods, but the suggestion may be made that high Triassic mortality (related to the Permian-Triassic crisis for marine vertebrates) may in this case have caused the unusual extinction of bradytelic lines. The five genera surviving from Ordovician to Triassic were probably bradytelic and the three then surviving from the Silurian were possibly so. Four of the former and two of the latter became extinct in the Triassic, and the last Ordovician and Silurian genera disappeared in the Jurassic. Except for these few genera, the maximum span for extinct genera is under 200 million years, hence somewhat, at least, less than for the oldest living genera.

appeared in more than two geological periods, in spite of the fact that this might mean survival for anything from 30 million years upward and that in some groups survival of broadly defined genera for such spans is the rule rather than the exception.

Rensch (1947) has demonstrated that the mean age of living higher categories (families to orders) is generally greater than the mean span of extinct units of the same categorical rank and in the same group, often two or three times as great. Since the long persistence of a family or an order does not preclude a fairly high rate of evolution within the group, these examples do not necessarily represent essential arrest of evolution, but the latter phenomenon is probably involved in such averages. It is quite certainly involved in the elaborate series of studies by Small (e.g., 1946), begun in the 1920's and still continuing, on "short" and "long" species, mostly in diatoms. Diatoms do show an unusually high degree of bradytely, as demonstrated above by a different analysis of some of Small's data, and Small seems to be quite right in concluding that two sorts of species, different in some quantitative way as regards rate of evolution, can be distinguished among these organisms.[3]

Thus there is nothing really new in the observation of what I have called "bradytely" (in print since 1944), but it does seem that rate distribution and survivorship methods put its study on a more nearly objective basis and permit clearer analysis of it.

A next step is suggested in Figure 43. In this particular group, the pelecypods, it appears that any living genus more than about 250 million years old, i.e., dating from about the Mississippian or earlier, is almost surely bradytelic. Such genera include *Nucula, Nuculana* (or *Leda*), *Volsella* (or *Modiolus*), and *Pteria* (or *Avicula*).[4] This is sufficiently clear and these individual genera can be labeled "bradytelic" on a

[3] It is necessary to add that Small's criterion for "shorts" and "longs" is arbitrary and does not really separate the two rate distributions. His conclusions as to the bearing of these data on evolutionary theory further seem to me beside the point and not really supported by the data. These personal opinions should not be allowed to obscure the real value of Small's laborious compilations.

[4] Shimer and Shrock (1944), although reflecting fairly modern nomenclature and constriction of genera, still record *Nucula* and *Nuculana* from the Silurian, *Volsella* and *Pteria* from the Devonian. Of course there is *some* difference between the earliest species and any living species in each of these genera, and sooner or later someone is sure to call that difference "generic." This will obscure but not alter the fact that the genus, as here recognized, has not shown appreciable progressive change since the Silurian or Devonian, although it has continued to undergo continual, noncumulative speciation, as do all long-continued or widespread groups.

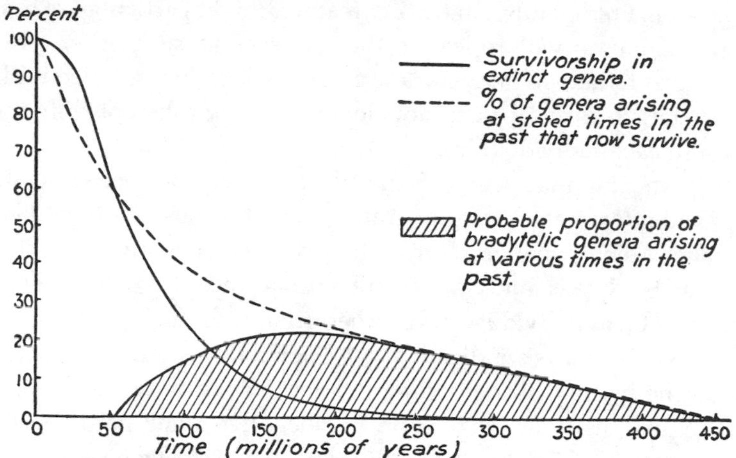

FIGURE 43. BRADYTELY IN PELECYPODS. Survivorship in extinct genera, continuous, unshaded curve, and analogous curve based on recent genera, broken curve, see Figure 5. The shaded area represents the excess of realized over expected survivorship caused by the nonextinction of bradytelic phyla and so is approximately proportional to the rise of bradytelic lines at the stated times in the past. (The apparent absence of bradytelic lines arising in the last 50,000,000 years may be real, but is probably an artifact caused by differential extinction of horotelic genera in this period, the extent of which cannot be estimated from the available data.)

reasonably objective basis. But the situation in, say, the Jurassic, around 150 million years ago, is different. The curves and tabular survivorship data give good reason to believe that perhaps three-fourths of the genera that survive from Jurassic to Recent and perhaps about one-fourth of the genera that arose in the Jurassic are bradytelic. But this is a statistical result, and like many such results it cannot be directly applied to the individual. There is no possible way from these data to tell which of the genera first appearing in the Jurassic are bradytelic and which are horotelic in the terms of this analysis. Such designation could only be made if some separate and individual criterion of bradytely could be found. If some Jurassic-Recent genera were found not to have changed at all and others merely to have changed very slowly, the former would include the bradytelic genera, but such a difference cannot be very clear-cut among lineages all of which have changed so little as not to warrant calling the change generic in broad, conservative classification. Presumably some other features might particularize bradytelic genera but, as will appear in the next section, such criteria really diagnostic as to individual genera are not available. It is, indeed, likely that the statistical result does not depend on any inherent differences uniform in each individual line.

It thus appears that some of the oldest genera or species in some groups and after analysis of their rate distributions can be identified as bradytelic. Among younger forms, however, the analysis may clearly show that bradytelic lines were still arising and yet give no way of designating them individually. In other groups, as among prosobranch gastropods, the analysis indicates that bradytelic lines are present, but none can be individually designated, even among the oldest, because the same analysis shows that none is older than some horotelic lines. We can only go so far as to say that lines that have shown no appreciable change while most allied lines did show change must include the bradytelic lines, perhaps along with some that are slow horotelic, and may be more likely bradytelic than not.

Since this concept of bradytely was first published, many of those who have discussed it either have assumed that no distinction was intended between bradytely and very slow horotely or have denied that such a distinction exists. Most of these students have overlooked, or at least have made no reference to, the fact that bradytely is defined by a difference between rate distributions, a difference that does objectively exist. (There are, of course, noteworthy exceptions, e.g., Schindewolf, Carter.) It seems pointless to discuss whether a difference exists, unless it can be shown to be an artifact of the method, which has not been demonstrated or even suggested.[5] It is also somewhat beside the point to designate some particular slowly evolving line as bradytelic, to show that its evolution had slowed down from a more usual, presumably horotelic line, and to deny any distinction between the two on that basis. The impossibility under some circumstances of designating particular lines as bradytelic has been pointed out above. It is also quite clear that all bradytelic lines are derived ultimately from horotelic or

[5] Hutchinson (1945) did suggest an artifact of materials rather than of method: that the bradytelic genera are merely those in which evolution at normal rates affected the soft parts only, and so did not appear in fossils. This seems quite impossible, for reasons already mentioned in part. It may be added that this explanation cannot possibly apply to the diatoms, which include much bradytely in spite of the fact that living and extinct species are commonly based on identical criteria.

often from tachytelic ancestors. It is not to be assumed that they became abruptly bradytelic, although that may sometimes occur. Discontinuity in that sense, in application to change of rate within one line, is not involved in the evidence for or concept of bradytely.

It is very much to the point, however, to inquire what the difference between bradytely and horotely is, whether, for instance, it is quantitative and statistical, with intergradation and yet with mean differences and possible discontinuity of average outcome, or whether it is qualitative, with inherent discontinuity of sort and of underlying factors. Of these two alternatives, the evidence strongly suggests that the former is correct. In *this* sense, I agree with, for instance, Westoll (1949) that bradytely and very slow horotely are indistinguishable as seen in individual instances, but it remains true that in certain groups the prevalence of such exceedingly slow rates has produced a statistical discontinuity in rate distributions which is a collective phenomenon distinguishable from collective horotely. An opposing view is taken by Small (1946 and numerous other papers in the same series), who thinks his "shorts" and "longs" are inherently different as applied to individual cases, the short or long duration being "very largely determined by what happened to one nucleus" when the species arose by saltation. Extensive and conclusive evidence against such a view is given throughout this book, and the phenomenon of bradytely neither demands nor warrants the postulate of predetermined life span or any other factors than those normally involved in the complex evolution of adaptation.

FACTORS OF SLOW OR ARRESTED EVOLUTION

Bradytely is essentially a statistical effect produced by the prevalence in some groups of organisms of lines with extremely low rates of evolution, or with changes fluctuating on a small scale and not appreciably cumulative. The first point, then, is to consider the circumstances of very slow or arrested evolution in general.

Certain genetic and other factors in organisms set limits to rates of evolution. It is therefore natural to suppose that slow evolution is primarily conditioned by these factors, and this has often been suggested. Nevertheless there is much evidence: (a) that the limits set by these factors in real populations are generally permissive of average or even high rates, and (b) that very slowly evolving lines are not particularly likely to have these factors in unusual degree.

Low mutation rate is one such factor often suggested (e.g., Dobzhansky, 1941, but the suggestion was practically withdrawn in 1951). But what appear to be low rates are sufficient to account for sustained and even rapid evolution (see Chapter IV). Mutation rates have not been directly determined in many slowly evolving lines, but the evidence tends to oppose the postulate that they have low rates. *Drosophila,* itself, may now be a very slowly evolving group, but seems to have rates of mutation comparable with those of any other group studied in this respect. The facts that very slowly evolving groups often arise by rapid evolution, that they also often split to give rise to rapidly changing branches, and that they do not have particularly low variation (below) all suggest indirectly that their mutation rates are more or less normal. Stebbins (1949, 1950) points out that slowly evolving plants may be rich in genetic variability, speciating luxuriantly and (e.g., *Sequoiadendron*) readily yielding horticultural varieties.

Factors that reduce genetic variability, notably apomixis, self-fertilization, and reduction of crossing over, certainly limit the possibilities for change and therefore may characterize groups that change but little ("blind alley" groups, Chapter IX). In the longer run, however, such groups generally do finally change or become unusually liable to extinction. Limitation of variation may, indeed, be a factor in bradytely of such groups as diatoms, but even in such cases it is an insufficient explanation and it does not apply at all to the majority of slowly evolving animals and plants. Most of these, even in lines surely bradytelic and with essentially no progressive evolution over very long spans, are sexually reproducing and show usual or even high degrees of variability. Opossums, which have evolved more slowly since the Cretaceous than any other living mammals, have quite as variable populations as the average of rapidly evolving mammals. Within the established adaptive type they also have undergone repeated, wide speciation, which implies a considerable store of variability. Stebbins (1949) also notes that slowly evolving plants may speciate freely. *Selaginella,* one of the most slowly evolving of all plants (little changed from *Selaginellites* in the early Mississippian) has over two hundred living species (Manton, 1950).

Long life spans or long generations, also frequently suggested as correlated with slow evolution, may be ruled out even more conclusively. As noted in Chapter V, there is surprisingly little evident correlation between lengths of generations and rate of evolution. Here it may be added that most of the outstanding examples of slow or arrested evolution occur among organisms with relatively short generations. Bradytelic lines do not seem to have generations consistently different in length from their horotelic and tachytelic allies, and again it is pertinent that bradytelic lines may arise from and give rise to tachytelic lines. Surely change of length of generations is not a significant factor.

Another common approach, exemplified by Ruedemann (1918, 1922a, 1922b), who devoted much attention to this subject a generation ago, is to seek particular morphological and physiological characteristics associated with low-rate lines. Small size, nocturnal habits, and carrion-feeding are among the many suggestions of this sort that have been made. Some such characters may have a bearing (although it is not likely that those just mentioned have much if any), but obviously they are not really explanatory. Among related lines fully comparable in such respects, some may evolve very slowly and others rapidly. Groups almost wholly different in characters of this sort, such as some plants and some animals, may have very similar slow—or fast—rates of evolution. It is perhaps true that sessile animals have evolved more slowly on an average than mobile animals, but this is a slender and probably misleading clue in view of the facts that similarly sessile animals have evolved at very different rates and that most plants are sessile but that they have run as broad a gamut of rates as have animals.

In such comparisons as that of sessile and mobile animals a certain confusion may arise. It is one problem that whole groups (say, corals) may have an average rate that is lower than that of another group (say, fishes) and quite a different problem that one line (among corals or among fishes) may have a very much lower rate than another allied and similar line. Both problems call for study, and they are not wholly unrelated, but it is stultifying to confuse the two. The frequently repeated generalization that primitive organisms have evolved more slowly than advanced ones lends itself to that confusion and also involves a howlingly circular argument as applied to groups now living: we call some of them "more primitive" precisely because they have evolved more slowly. If the statement is applied more correctly to older, hence literally more primitive, and younger members of the same group, it simply is not true. There is no definite tendency for average rates within given groups to accelerate. Changes in such averages are highly variable and if there is an over-all balance it may be on the side of deceleration of rates. For instance, the later lungfishes evolved much more slowly than the primitive members of the group (Chapter II, and Westoll, 1949), and this seems to be a rather common pattern in long-lived groups.

As applied to individual lines, the idea that more primitive forms evolve more slowly is also a misconception. It is an interesting fact that bradytelic lines, as far as identifiable as such, or any extremely slowly evolving lines are usually seen to have been progressive, relatively advanced types up to the time when their evolution was arrested. It is their long persistence without much change after their allies have gone on to explore other avenues of modification that gives them an archaic aspect or makes them primitive in comparison with contemporaries of later phases of their history. Lingulids had a fairly high type of invertebrate structure for the Ordovician and limulids were advanced for the Triassic. Sphenodonts were about as advanced as any reptiles in the Triassic and so were crocodiles in the Cretaceous. *Lingula, Limulus, Sphenodon,* and *Crocodylus* are primitive as of now just because they are (with little question in these cases) bradytelic and have been for long periods. But such groups have been passed in the race; they did not have an initial handicap. Stebbins (1949) has pointed out that the same generalization is true of plants: the lines with arrested evolution were specialized and advanced when their evolution was arrested.

Another approach has been to designate certain places or broad environmental situations as conducive to slow evolution: oceans, islands, tropical forests, and so on. It is probably correct that sustained, slow, or arrested evolution is more likely to be found in some such environments than others. (Islands happen to be one of the places where it is less likely, contrary to some statements, see Chapter IX.) But the approach to the problem is still obviously inadequate, for very rapid evolution may occur in exactly the same situations (many marine groups, epiphytes in tropical forests, etc.).

It is, in fact, clear once more that the key to this evolutionary problem is not in the organisms or in their environments but in the relationship between the two. When lines persist with arrested evolution, they have maintained adaptation without appreciable change. For this to occur, there must be a balance between breadth of adaptive zone and its variation or deviation in time. If organisms are very broadly adapted, if they can, for instance, live on a variety of foods or grow on a variety of soils and in a variety of climates, environmental factors may fluctuate greatly without requiring any adaptive change for survival, as long as *some* part of the broad zone persists. If adaptation is narrow, the zone must be a very uniform and stable one in space and time for evolution in it to be arrested and its occupying populations to persist.

Arrested evolution may thus occur in a great variety of environments and organisms and with quite different sorts of adaptation and specialization. It is, however, more likely to occur with broad adaptation than with narrow, because adaptation in a broad zone is more likely to persist than in a narrow one. Both broad and narrow may persist, however. The lungfish zone would seem to be quite narrow, and yet it has persisted without marked change since the Devonian and all lungfishes have apparently been bradytelic in it for perhaps 200 million years. It may well be that if a major zone ("major" in being quite distinctive from all others) is rather narrow, *only* bradytelic groups will eventually be left in it. On the other hand, the zone of the opossums is quite wide, and the opossums have survived repeated and radical environmental change without themselves changing much.

Evolutionary change is so nearly the universal rule that a state of motion is, figuratively, normal in evolving populations. The state of rest, as in bradytely, is the exception and it seems that some restraint or force must be required to maintain it. This force undoubtedly exists and is the same as the force that usually orients evolutionary change: selection. Selection on groups with arrested evolution must be centripetal and must be more effective than mutation, cyclic or fluctuating environmental change, or any other factor making for change. Whether adaptation is broad, as it usually is, or narrow, such a group must be perfectly adapted to its zone and almost any change must be quickly detrimental and effectively checked by selection.

The most slowly evolving groups do seem all to be very highly and specifically adapted to a particular zone. Their typical history is one of rather rapid shift into a new, stable and persistent zone, still rapid adjustment (postadaptation) in the zone, with weeding out of less adaptive characters and lineages, and then a long, eventless course of relatively unchanging continuation. That they are actually held by force, so to speak, is confirmed by the fact that many of them (not all) have given off branches that rapidly evolved into different zones when the opportunity, or perhaps it is better to say the occasion, arose. Thus bradytelic *Ostrea* (*sensu lato*) repeatedly gave rise to relatively very rapidly evolving coiled (*Gryphaea*) and spiral (*Exogyra*) lineages, among others. Opossums gave rise rapidly and early to a great variety of divergent groups in South America: Caroloameghiniidae, Borhyaenidae, Polydolopidae, Caenolestidae, each family except the first (as far as known) highly varied. Typically these rapid side-branches, or most of them, become extinct while the parent group continues unchanged.

With this background, let us turn back briefly to the characteristics of the populations and environments involved. As was found true of extinction, the balance maintained here, lost in extinction, is compounded of such extremely numerous, varied, and complex factors that it may be quite impossible to designate them explicitly in a specific instance. For the populations, the prime essentials are that these be extremely well integrated genetically and that any deviation be subject to effective counter-selection. It might be expected that such groups would have especially elaborate systems of modifiers or polygenes, but as far as I know no concrete comparisons pertinent to this relationship have been made or are yet possible.

If selection, and apparently quite feeble selection in some cases, is to be effective, gene drift must be excluded. This requires population above a minimal size, but that size need not be very large in all cases. Westoll (1949) has given some reason to believe that early lungfishes, evolving rapidly, had larger populations than the bradytelic living forms, but the population size in the latter is apparently quite large enough to enable selection to hold them in a narrow zone—the fact that the zone is narrow probably means that centripetal selection is relatively strong and so is still effective in relatively small populations. Most groups with arrested evolution have large, even extremely large, breeding populations. Outstanding exceptions seem to be *Sphenodon* and *Latimeria,* but *Sphenodon* apparently had large populations until human disturbance banished it from the mainland of New Zealand and no one has any idea how large a breeding population *Latimeria* may have.[6]

[6] It has also been remarked (e.g., Carter, 1951) that the monotremes and *Peripatus* have small populations and so do not agree with requirements for brady-

As Stebbins (1949, 1950) has emphasized for plants, there is no reason to think that large population size in itself promotes slow evolution. It makes for more effective selection, and also within limits for longer duration, and hence is propitious for arrested evolution when selection is centripetal. That selection may tend not toward change but toward stabilization is a thesis developed with skill and at great length by Schmalhausen,[7] whose work cannot be summarized here (but see also brief comment in Chapter VI). Others, e.g., Heuts (1949), have also given specific examples of mechanisms developed by selection that inhibit rather than promote change. One additional point that might be mentioned is that intraspecific selection must also be centripetal if evolution is to be arrested while a lineage continues. The group must be so organized as to minimize genetically effective competition within it, or always to favor the modal type, or both. Strong dimorphism in secondary sexual characters would not be expected in such a group among animals and in fact does not occur in any example known to me.

As to the environment, the principal requirement is that it persist and do so with less fluctuation than the maximum that can be met by existing and expressed, not potential, variation in the populations concerned. How much fluctuation may be so met depends of course on the amount and sort of expressed variation and the details of the adaptation involved but, in general and in terms of the environment itself, the more stringent it is in requirements for life, the less variation can occur without requiring evolutionary change or causing extinction. Arrested evolution is therefore less likely to occur in very difficult environments, such as deserts, impermanent environments, such as salt lakes, or highly variable environments, such as the alpine zone. It is more likely to occur (although of course rapid evolution may also occur) in the easier, more permanent, and less variable environments

---

tely. It is now clear that a large population is not a requirement but only a probability. What some students would consider a small population in subjective and relative terms may yet be such as to make centripetal selection effective, which is the essential point. As regards the monotremes, they are extremely specialized in many respects. It is improbable that they are bradytelic and their evolution may not even have been unusually slow—they merely happen to have some reptilian characters lost in true mammals. Their breeding populations were also moderate if not large before they felt the effects of man and his works. *Peripatus* is quite likely to be bradytelic, but I know of no estimates as to its breeding populations; forms so obscure in habitat are likely to appear rare when not so in fact.

[7] This is the same man occasionally cited as "Shmalgausen," which is an English transliteration of the Russian transliteration of his originally German name.

such as the ocean, its strand, the shifting but long enduring major lowland rivers, the more slowly shifting and also long enduring great forest belts, and particularly in the most nearly permanent of climatic zones, subtropical to tropical.

It now seems to me possible to explain the phenomenon of bradytely in general terms and in the light of these factors of slow or arrested evolution. Any series of related lineages or even one lineage in the course of its history is subjected to a great many environmental changes and to many changes in its own populations. Some of these changes are concomitant with or in a sense produce progressive adaptive changes in the populations. Others shunt development off into another direction, or move it into a different adaptive zone. Still others lead to loss of adaptation, hence extinction. In the case of single lineages, in most instances one of these three results will ensue. Sometimes, however, early change may, in a manner of speaking, have oriented and aimed the group down a line where adaptation is retained indefinitely without further change.

In terms of multiple lineages the process might be visualized as movement of the various sorts of organisms down a series of corridors with many baffles. Most corridors turn, forcing the organisms to turn, too (evolve new adaptation). In a straight corridor the baffles (changes in environment and population) deflect some into side corridors (new adaptive zones) and stop some literally dead (extinct). After a certain time (a long time) all those seriously liable to deflection or stopping have been weeded out. There remain only those sorts of organisms that have run through the whole repertory of baffles and are no longer subject to deflection or stopping. Suddenly at the last baffle that does stop or deflect, there is a discontinuity of outcome. The remaining sorts of organisms now persist indefinitely.

The statistical result actually observed would arise from such a situation. It is seen, too, that one of the survivors, the bradytelic lines, need not be evolving more slowly than some of the others, the horotelic lines, and might have no characters that would reveal its bradytelic nature before the final event, the disappearance of the last horotelic line of the same age as the bradytelic line. The distinction of the bradytelic line is an adaptive relationship that makes it miss the baffles, that permits indefinite survival without change in the face of all vicissitudes that do eventuate in its adaptive zone. Bradytelic lines are merely the residuum of a process that regularly reduces the percentage of unchanged groups but that stops short of reduction to zero, if bradytely does occur.

The baffles are so numerous in nature that one would expect bradytely to be in general less common than horotely and to be absent altogether in some groups. That expectation is confirmed by observations, such as the examples discussed earlier in this chapter.

TACHYTELY

It is my opinion that tachytely is a usual element in the origin of higher categories and that it helps to explain systematic deficiencies of the paleontological record. It will therefore be discussed in some detail in the following chapter, which is devoted to those topics. At this point, the need is only to try to clarify the relationships of tachytely to horotely and bradytely (a subject that I personally find both abstruse and difficult but highly significant and enlightening as regards major evolutionary processes).

It is in the nature of the method that the study of survivorship cannot reveal whether there is also a nonstandard distribution of rates generally faster than horotelic, analogous to the nonstandard distribution of low rates that it does reveal. Evolution at exceptionally high rates cannot long endure. A tachytelic line must soon become horotelic, bradytelic, or extinct. For this and other reasons, which will later be suggested, tachytelic lines, or tachytelic phases in the evolution of lines horotelic or bradytelic at other times, are poorly recorded by fossils and often not recorded at all. There are occasional glimpses in the form of fossils that happen to be found at some point in the tachytelic phase (e.g., the Devonian amphibians from Greenland) and some complete sequences of a rather marginal sort, but for the most part a given tachytelic phase is an inference from ancestral and descendant forms in a record with important gaps. Often the inference is perfectly clear and its correctness is confirmed by the scattered records that do show it directly, but the incidence of tachytely cannot be counted or calculated, even on a sampling basis, as can that of bradytely.

What distinguishes tachytely from horotely is not only that it occurs at exceptionally fast rates but also that it occurs while populations are shifting from one major adaptive zone to another, and especially when a threshold is crossed (see Chapter VI). Such a phenomenon differs from horotely only quantitatively and not sharply. It grades into horotely both in the sense that a tachytelic line gradually becomes horotelic and in the sense that the differences in degree of adaptive change involved, in rate of evolution reached, and in clarity of threshold are all on continuous scales. There is no sharp point where one can say, "Below this is horotely and above it is tachytely." Nevertheless there is a difference that may be very appreciable and significant in less marginal cases.

Well-recorded but marginal instances occur in such sequences as that of the Equidae (Chapter VIII). The changeover from browsing to grazing has a threshold effect and involves a definite, temporary increase in evolutionary rate. This verges, at least, on tachytely. The even more rapid changes from one stable type of foot mechanism to another are still more clear-cut examples of this sort of phenomenon and might well be designated as rather small-scale tachytely. Another well-recorded example of small-scale tachytely, leading to extinction in this case as tachytely often does, is that of *Gryphaea* (Chapter IX). One of the most remarkable known examples is that of the snail *Valenciennesia*, formerly classified in another suborder than the notably different *Limnaea*, but shown by Gorjanovic-Kramberger (1901, 1923) to have evolved from the latter so rapidly that the whole process occurred while a horse, *Hipparion gracile*, on adjacent lands showed no appreciable change. Here, too, change to a distinctly different adaptive zone is involved: from clear and fresh to muddy and brackish water.[8]

As Stebbins (1950) has suggested, tachytelic lines are likely to be, although they are not necessarily, narrowly adapted. They include the forms that are forced into rapid change by changes in their adaptive zones. Their adaptive evolution tends to be the opposite of that seen in bradytely, and the factors involved tend to be inverse. In the figure of speech utilized above, tachytelic lines hit the baffles and are deflected into sharply divergent corridors.

Tachytely is defined as a phylogenetic phenomenon and is to be distinguished from rapid speciation or splitting. Rapid evolution in the sense of high origination rates, such as occur on invasion of new and open habitats (e.g., in the Galápagos or Hawaiian Islands, as discussed before, or rapid development of endemics in African lakes, see Worthington, 1937, 1940) is a phenomenon quite distinct from tachytely. Nevertheless, when adaptive radiation occurs, and especially if this

[8] Basse (1938) has cast doubt on Gorjanovic-Kramberger's interpretation, but Basse's opposing evidence is weak.

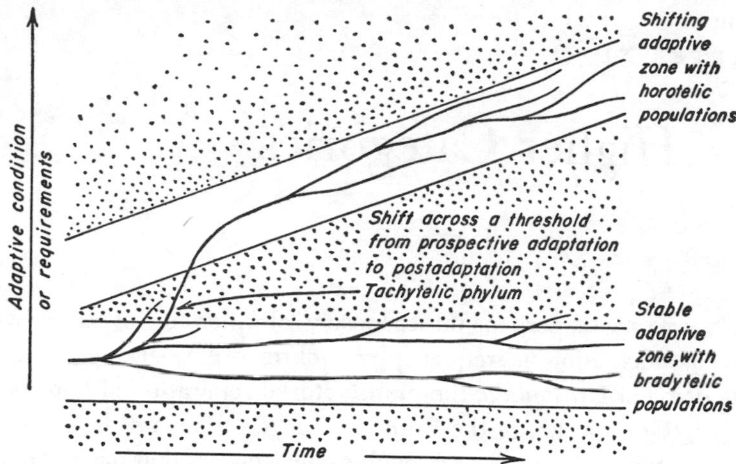

FIGURE 44. DIAGRAMMATIC REPRESENTATION OF CASES OF HOROTELY, BRADYTELY, AND TACHYTELY. This particular sequence, one of many that occur, shows a bradytelic group from which is split off a tachytelic branch which in turn gives rise to a horotelic group.

is of a basic or major sort, the shift of each divergent line into the zone it comes to occupy is commonly tachytelic.

Horotely represents a sort of normal or average turnover or metabolism in the evolution of a group of organisms. To the extent that evolution is defined as change, bradytely is a cessation of evolution without extinction. Tachytely is an episodic acceleration, figuratively a brief fever imposed on the normal metabolism. Tachytelic lines arise from horotelic or bradytelic lines, in the latter case usually (perhaps always) by splitting. A tachytelic line must soon become horotelic, bradytelic, or extinct. Horotelic lines usually arise from other horotelic lines, occasionally from tachytelic lines (but these occasional events are unusually important) or from bradytelic lines, in the latter case by splitting and usually through a tachytelic phase. Horotelic lines may remain horotelic, or become bradytelic or tachytelic. Bradytelic lines develop from horotelic or often rather directly from tachytelic lines. The bradytelic lines themselves remain bradytelic and rarely if ever become horotelic or tachytelic again, but it is rather common for them to give rise to tachytelic branches, which in turn become horotelic or extinct. A simplified pattern of one of several possible relationships between horotely, bradytely, and tachytely is shown in Figure 44.

CHAPTER XI

Higher Categories

WE TEND TO SPEAK, perhaps rather loosely, of different "levels" of evolution. The fluctuation in frequency of a chromosome arrangement in a single deme of *Drosophila*, the multifactorial separation of two species of frogs, the evolution of the Equidae, and the rise and diversification of the Reptilia, as examples, are evolutionary events that have strikingly different scope, at least. It is convenient and usual to recognize such differences in scope by putting them in taxonomic terms and relating them to the levels of the Linnaean hierarchy.

It is generally agreed, although with a great and confusing variety of ways of expression, that a crucial point in the hierarchy occurs at the specific level. The species, with its included subdivisions, is a different sort of group from those above it in the hierarchy, superspecies to kingdoms. It also happens that this point corresponds with a usual difference in techniques and approaches by the students of evolution. The experimentalists are mainly concerned with evolutionary processes and events in subdivisions of species. The neontological systematists work for the most part around the level of species, on larger subdivisions of species, on species as a whole, or on closely related groups of species. The biological paleontologists and others who might be called historical biologists (representing a subscience that has no accepted name) are less concerned with species and more with groups from genera upward.

This distinction of method has long existed and for even longer some have thought that species and larger groups have different sorts of evolution or, at least, different historical roles. (Even in antiquity it was suggested that infraspecific groups evolve, as we would now say, and that species do not.) Goldschmidt especially (1940) has emphasized the distinction and suggested that historical changes within species be called microevolution while those from species upward are to be called macroevolution. Goldschmidt uses his terms to emphasize the belief that microevolution, so defined, is something definitely and qualitatively different from macroevolution and not involved in or leading to the latter. The terms have, however, been adopted and are in current use, with some imprecision and difference of definition, by a number of students who heartily disagree that microevolution and macroevolution are qualitatively different. In the predecessor to this book, I expressed the opinions that microevolution and macroevolution are not qualitatively distinct, that there is no more reason from this point of view to draw a line at the species level than at the genus, family, etc., levels, and that macroevolution as applied to species and genera does not look very "macro" to a student of the major features of evolution. It was to emphasize these views that I suggested that the evolution of really high categories be called by another bastard term, "megaevolution."

At present I am inclined to think that all three of these somewhat monstrous terminological innovations have served whatever purpose they may have had and that clarity might now be improved by abandoning them. Although most readers understood the intended implications of "megaevolution," a few seemed to think it heralded the wonderful "discovery" of a new *sort* of evolution. The terms do lend themselves to such confusion, and indeed "microevolution" and "macroevolution" were coined to designate supposedly different sorts of evolution. It is better to recognize that there are not two, three, or more sorts of evolution but that there are innumerable levels of evolution, which reflect merely how much of the whole complex tangle of the history of life is to be taken into consideration at once. When it is necessary or useful to designate these levels—and it often is—greater flexibility and accuracy without essential loss of succinctness can be obtained by the device of the Linnaean hierarchy. One may thus speak with sufficient exactness of evolution at subspecific, familial, ordinal, etc., levels, or if broader, relative reference to more or to less detailed phases of evolution is called for, one can speak of the evolution of lower or of higher categories. The difference from speaking of micro- and macroevolution is not merely terminological, in which case it might not merit this much discussion here; it also may reflect and assist a different approach to the important problems in the study of which the terms are used.

The study of the evolution of higher categories, say for present purposes from families upward, does involve some of the most important problems of evolution. For one thing, evolution on this scale cannot be directly studied by experimental methods. When a genus in a steadily moving lineage like that of *Equus* has an average duration of 7½ million years, nothing we can do to speed up experimental evolution is going to bring such events, let alone the longer spans of families and still higher categories, down to a period men can hope to follow in experimentation. Cross-breeding, essential for most genetical analyses, is almost never satisfactorily possible at the level of genera, and absolutely never above that level. Here, then, is a domain in which the observational approach and, when available, paleontological materials are the only ones possible.

The fact that approach and materials must be different also arouses proper doubts as to whether the factors found experimentally in low-level evolution are those effective at high levels. Further, the absence of experimental control and the incompleteness of the fossil record make it difficult, although eventually not impossible, to draw final and decisive conclusions regarding widely variant, basic interpretations.

The main themes of this book relate to evolution on a broad scale, hence largely to categories above the species. The time has now come to undertake explicit consideration of what higher categories are and how they arise.

THE NATURE AND DEFINITION OF HIGHER CATEGORIES

The science of systematics has long been affected by profound philosophical preconceptions, which have been all the more influential for being usually covert, even subconscious. Most of the older systematists and some more recent ones, especially among those whose work is mainly identification of specimens for purposes not primarily biological (stratigraphy, economic botany and zoology, etc.) are covert (or occasionally overt) idealists in the technical philosophical definition of "idealist." To them a taxonomic unit is a pattern, a typological concept or morphotype, and their activity consists of forming concepts of patterns, comparing individual specimens with the idealized pattern, and deciding whether agreement is sufficiently close to place the specimens in the corresponding taxonomic category. Some systematists who scorn philosophical discussion and are quite sure that philosophy never colors *their* work are most deeply affected by this philosophical attitude.

Modern systematists who are primarily interested in organisms for their own sake, forms of life to be studied as such, are usually realists or materialists. (In the technical philosophical sense they are naturalists, and it is interesting that many naturalists by profession or avocation have not been philosophically naturalists.) To them organisms are real objects, no two of which are alike but which have certain material relationships to each other. The purpose of their activity is to find out what these relationships are, then to express the sort and degree of relationship in the formal symbolism of taxonomy. What they classify is not a specimen but a population or series of populations existing in nature, of which the specimens are samples. The characteristics of a population are those of *all* its members collectively with all their resemblances and differences. Such a population has an average condition, but it has no single, crystallized, idealized pattern or morphotype.[1] These are the basic viewpoints of what is sometimes called the "new systematics" (a term perhaps first used by Hubbs, 1934; see also, among many others, Huxley, 1940; Mayr, 1942; George, 1948; Simpson, 1951b; and references in those works).

It is a convenience to a systematist engaged in identification and cataloguing to recognize taxonomic units by the characters common to all their members. Such a procedure is, indeed, virtually necessary in practice, but it is inherited from the old systematics and it lends itself to serious philosophical confusion. The characters-in-common may become a morphotype in the mind of the classifier. He tends to think that a category is *defined* by these characters, that a low category is one with more and a high category one with fewer characters in common, and that quite different sorts of characters are involved in the two cases.[2]

Diagnostic characters-in-common do of course exist in many cases (not really in all, as will be shown) and they are significant for evolutionary study. The explanation of many of them is that they have been inherited from the common ancestor of all the groups possessing them. This interpretation may lead to two other idealistic concepts that are

[1] To a realistic systematist the technical type of a species is not a standard of comparison but only a needed device for legalistic nomenclatural purposes.

[2] This sort of subconscious hangover from the old systematics is illustrated by the fact that Mayr, one of the clearest and most forceful exponents of the new systematics and of realistic taxonomy, still had some tendency in 1942 to discuss lower and higher categories in terms of characters-in-common, a pseudo-idealistic approach into which we all may fall on occasion.

not really applicable in this materialistic world: (1) that the characters-in-common are somehow a picture of the common ancestor, which was a "generalized" organism without the sorts of "specialized" characters that distinguish its various descendants from each other; and (2) that the higher category arose as such when an organism acquired the given characters-in-common. These views are certainly incorrect, but it is surprising how often they creep into the background for students who would firmly repudiate them if they were expressed. They are expressed, accepted, and made the basis of theories of evolution by a few philosophically idealistic biologists. It is a further frequent concomitant of the idealistic philosophy and the typological approach to conclude that the "types" arose by saltation. Supposed evidence for this view is considered later in this chapter, and generally throughout this book. At this point it seems unnecessary to specify proponents of the view; that is done and the whole question summarized with great skill in Heberer (1943b).

Our recognition of a higher category is *ex post facto,* as is our designation and placing of it in the hierarchy. The Cricetidae (a family of mouselike rodents) are a family because they have become so extremely varied. If there were only a few genera or species of Cricetidae they would be members of the Muridae. As Mayr (1942) has pointed out, if pterodactyls had persisted to become the dominant flying vertebrates and as richly various as birds, they would not be reptiles, but a separate class; and if birds had stopped short with *Archaeopteryx,* they would be reptiles and not a separate class. A higher category is higher because it *became* distinctive, varied, or both to a higher degree and not directly because of characteristics it had when it was arising.

In the early and middle Paleocene, differentiation of placental mammals was just getting under way. All together, they differed from each other decidedly less than do the marsupials, extinct and recent, all of which are classified as forming a single order. If we knew no placentals after the middle Paleocene we would certainly place them in one order. *As of then* their proper comparative categorical rank was in fact that of an order. They are placed in six different orders because we recognize in them ancestors and allies of what *later* became six orders.

Among recent placental mammals, one would say that two groups could hardly be more distinctive than the carnivores and the ungulates. In the top half of Table 25 are listed characters that might be taken as diagnostic of the two groups in the Recent fauna. In order to make diagnosis possible, only land forms are included in both cases and the doubtfully ungulate tubulidentates are omitted. The danger of selecting characters that tend to prove my point has been avoided by basing the diagnosis on Rode (1946–1947) and including not only the points in his formal definitions but also everything in his text suggested as possibly distinctive in the two groups. In the first place it is evident that these higher categories are so varied in character that it is practically impossible to find characters-in-common that do not also occur in other groups. As between these two rather strictly and somewhat arbitrarily limited groups, only 3, 8, 9, and 10 of the table are diagnostic characters-in-common, and these not too clearly. If living aquatic forms or some rather lately extinct forms are taken into consideration, there are no diagnostic characters-in-common in either group. Seals do not have well-developed claws, sirenians do not have hoofs; some ungulates as late as the Pleistocene did have claws. Dental distinctions other than those characterizing lower categories also tend to disappear if aquatic and extinct forms are included: seals have no carnassials and some sirenians have no teeth at all.

Even with sirenians omitted, the recent ungulates may not be a strictly monophyletic group and their lack of fully diagnostic characters-in-common might be imputed to heterogeneity of ancestry. The fissiped carnivores are surely monophyletic, however, and I cannot find a single character that occurs in all fissipeds and in no other animals. Even for such a well-knit group as, for instance, the Equidae it is difficult or impossible to find any such character if ancestral and extinct lines referred to the family are taken into account.

Other important points are shown by the early Paleocene situation; see the lower half of Table 25. The same characters are listed as for the Recent descendants. Each group is seen then to have been very uniform in these characters, but there is not in either group any distinctive ancestral character passed on to all or even to most of its descendants. The few characters that are common to the ancestral and most descendant forms, such as large canines and brachydonty in carnivores, are merely primitive characters for most or all placental mammals. In the early Paleocene carnivores these are not new characters the rise of which produced an order Carnivora, and there are no such characters.

TABLE 25

CHARACTERS OF RECENT AND EARLY PALEOCENE CARNIVORES AND UNGULATES

Recent Carnivora (Fissipeda)	*Recent ungulates*
1. Generally carnivorous, sometimes omnivorous, rarely herbivorous or insectivorous	1. Generally herbivorous, sometimes omnivorous
2. 4–5 digits	2. 1–4, rarely 5 digits
3. Well-developed claws, usually hooded.	3. Well-developed hoofs (except hyraxes)
4. Digitigrade or plantigrade	4. Usually unguligrade, sometimes digitigrade
5. Dental formula $\frac{3.1.4\text{-}1.3\text{-}1}{3.1.4\text{-}1.3\text{-}1}$ (molars exceptionally $\frac{4}{5}$)	5. Dental formula $\frac{3\text{-}0.1\text{-}0.4\text{-}2.3}{3\text{-}0.1\text{-}0.4\text{-}2.3}$
6. Brachydont	6. Brachydont to hypsodont
7. Canines large	7. Canines usually small or absent, sometimes large
8. Premolars usually simple, not molariform	8. Premolars generally complex, often molariform
9. P^4 and M_1 almost always enlarged, carnassial	9. No carnassials
10. Molars usually simple, without grinding pattern	10. Molars complex, with grinding pattern
11. Terrestrial, arboreal, or semi-aquatic	11. Terrestrial

Early Paleocene Carnivora	*Early Paleocene ungulates*
1. Inferred to be omnivorous	1. Inferred to be omnivorous
2. 5 digits	2. 5 digits
3. Narrow, fissured unguals, unhooded, not specialized claws or hoofs	3. Narrow, fissured unguals, unhooded, not specialized claws or hoofs
4. Digitigrade or sub-plantigrade	4. Digitigrade or sub-plantigrade
5. Dental formula $\frac{3.1.4.3}{3.1.4.3}$	5. Dental formula $\frac{3.1.4.3}{3.1.4.3}$
6. Brachydont	6. Brachydont
7. Canines large	7. Canines moderate to large
8. Premolars not molariform, simple	8. Premolars not molariform, generally simple
9. No carnassials	9. No carnassials
10. Molars simple, upper molars triangular to subquadrate, no grinding pattern	10. Molars simple to moderately complex, upper molars triangular to subquadrate, no or incipient grinding pattern
11. Terrestrial	11. Terrestrial

Moreover, the characters-in-common of ancestral carnivores (creodonts) and ungulates (condylarths) in the early Paleocene are exactly the same in both groups. There is one genus, *Protogonodon,* in which on balance of resemblance in small details some species would be classified as carnivores and others as ungulates. Aside from that case, the known genera are distinct enough, but it is entirely impossible to produce a differential diagnosis of the orders.[3] Indeed, they were not *then* different orders or even families by any reasonable criteria but only became so as their descendants subsequently diverged and diversified. All the early Paleocene carnivores and ungulates put together are much less varied and have much less difference in structural plan or "type" than occurs within some single living families, say the Mustelidae or Viverridae among recent carnivores.

It is most unusual to get good samples of the ancestors of higher categories so near their point of divergence. The example proves beyond possible question that in this case, at least, the higher categories did not arise as such, that there was no "archetype," no "generalized" ancestor with the characters-in-common of the order, no "systemic mutation," but that there was simply normal speciation among forms at first closely related and closely similar in all respects. Other examples can be adduced at various levels. For instance, Matthew (1926) pointed out, but later students have mostly ignored, the fact that eohippus was *not* a horse, that it is about as good an ancestor for *Rhinoceros* as for *Equus.* In effect, there was no family Equidae when eohippus lived. The family and all its distinctive characters developed gradually as time went on. Eohippus is referred to the Equidae because we happen to have more nearly complete lines back to it from later members of this family than from other families. There is no particular time at which the Equidae became a family rather than a genus or a species; the whole process is gradual and we assign the categorical rank after the result is before us.

Although the carnivores and ungulates have, within each, no characters-in-common as a legacy from an archetypic or generalized an-

[3] I have closely studied a large proportion of known early Paleocene specimens and make this statement on that basis. Matthew (1937), who probably knew this fauna better than anyone before or since, wrote formal definitions of Carnivora and Creodonta, but these are not really diagnostic, and when he came to the Condylarthra he found himself unable to produce a definition and distinguished them from carnivores only by their lack of characters that *later* appeared in carnivores.

cestors, they do have more or less distinctive adaptive characteristics involved in the evolution of most of their members, even though worked out in a variety or ways and secondarily lost in some lines. Carnivores did, as a group, develop adaptations for meat-eating and predation and ungulates did, also as a group, develop adaptations for eating flowering plants and for running away from carnivores. It was these gradually developed adaptations, not present in the ancestry of either group when the two actually split phylogenetically, that may in a sense be said to have made these groups higher categories.

As a generalization, the development of a higher category seems always to involve the rise of some distinctive sort of adaptation related to spread into a major adaptive zone. It is rather common for the adaptation characteristic of the category to be lost in some of its lines, or for quite a different adaptive type to be developed from it, but higher categories do seem always to develop in relation to adaptations characteristic for them. The adaptive characters involved may be quite broad and varied, as in carnivores or ungulates, or may be quite specific, virtually "single characters" in a taxonomic sense, as in rodents or bats. In the former case, the characters may arise *after* the group became phylogenetically distinct; in the latter case they may arise *before* that point.

The success of the rodents, so to speak, the fact that they became a higher category of ordinal rank (at least) is based primarily on one feature: persistently growing, chisellike incisors. (Of course some functionally correlated characters, as of jaw musculature and digestive system developed along with the incisors.) This is clearly true even though similar teeth occur in other mammals and have not given rise to orders in those cases, e.g., in *Daubentonia,* a genus of primates. (This genus, all by itself, is set aside in a family, sometimes even in a suborder, of its own just because it does have this un-primate-like character, but that is another point; it is simply a peculiar lemur.) Similarly the bats have become an order on the basis of their wings.

A higher category may thus involve a key character which, as things turned out, permitted expansion in a broad and distinctive adaptive zone. In retrospect the category then may be said to have originated when the character was developed and had become normal in an isolated population. Yet the same character or one equally distinctive in other groups may not lead to a category of the same scope. Because gnawing incisors are the key to rise of the order Rodentia does not make such incisors an "ordinal character." Still less does it follow that the key character arose by one mutation, although that is a possibility in some cases, or that if it did such an event is correctly described as the origin of an order by a single mutation: the original mutant still belonged to the parent species.

The characters that distinguish higher categories are adaptive and they are the same sorts of characters, although often cumulatively greater in degree, as adaptive characters involved in speciation. Examples have earlier been given (Chapter IV). Another very striking example is provided by frogs in which the degree of suppression of a larval stage and other characters associated with elimination of aquatic life may be distinctive between genera and species and are developed by selection acting on intraspecific variation. This is the same sort of adaptation that eventuated in the Class Reptilia (see Lutz, 1947, 1948).

Significant on this and on other points is Stebbins's (1949) survey of the combinations of eight reproductive characters in the 283 families of angiosperms. These characters and combinations of them are commonly used to define families and orders, but differences in them also occur in genera of one family and in species of one genus. Again, the same sorts of characters, indeed here identically the same characters, are involved in speciation and in the origin of high categories. Also, although the higher categories have characteristic adaptive types, they do not have uniform characters-in-common or archetypes. As a third point, Stebbins establishes that although these characters separately do not have clear adaptive significance, their combinations definitely are adaptive and are best interpreted as formed and preserved by selection.

It is even more common for "blocks of characters" (Mayr's phrase) than for single characters to be involved in the rise of higher categories. When they are and when the blocks tend to be retained as such, the simple explanation is that the block as a whole, its combination of characters, has adaptive value different from or additional to that of the various separate characters. This is the reason for such diagnostic combinations of characters-in-common as do occur in higher categories. It also explains why such combinations may arise after the group has split off from others (e.g., claws and carnassials in carnivores) rather than being an inheritance from the ancestry involved in the initial branching.

In evolutionary classification it is an expressed ideal that all recognized and named groups should be monophyletic, presumably that each should be theoretically traceable to a single species as its beginning.[4] The facts that the characteristics of higher categories are adaptive and that higher categories do not arise as such entail the consequence that this theoretical ideal is rarely realized. Since selection is likely to act similarly on similar populations, adaptive characters are likely to arise in parallel in several or many related lineages. This is particularly true of characters very broadly adaptive in nature such as are most commonly but not exclusively involved in higher categories. If diagnosis is couched largely in morphological terms, a virtual necessity in the actual practice of classification, it frequently happens that several different lineages separately cross the arbitrary morphological line drawn across the evolutionary continuum. For example, numerous lineages of therapsid reptiles were all changing adaptively in a mammal-like direction. Paleontologists use the arbitrary criterion that a reptile became a mammal when a dentary-squamosal joint developed and the functional jaw movement ceased to be on the articular-quadrate joint. This line was probably crossed separately by at least five different lineages (leading to monotremes, multituberculates, triconodonts, symmetrodonts, and pantotheres, although it is just possible that two or three of these early differentiated from a single crossing of the line; there may have been some other late Triassic–early Jurassic crossings with early extinction).

Since such parallel lineages are related, they still do trace back eventually to one species. Inclusion in a higher category of the single species from which the category developed is, however, almost always impractical on other grounds. Such a species is much more like the ancestral group than it is like any typical member of the higher category later developed from it. It usually has the definitive characteristics of the descendant category barely incipient and may actually have no trace of them. It seems rather common for descendant lines already to have split (speciated) more or less extensively before definitely diagnostic characters of the higher category appear.

It is certain in some cases and probable in the majority that higher

[4] Occasional suggestion that a sexually reproducing group should be traceable to a single pair of ancestors is meaningless because it is populations that evolve and it is extremely exceptional, although of course not impossible, for a population to arise from one pair of individuals.

categories as they are actually defined and used in practice are polyphyletic in detail, that more than one single lineage or specific line crossed the arbitrary boundary as drawn by systematists. This does not alter the fact that such categories do always develop from what was originally one, single species, if the category is properly defined. The point is sometimes argued, but it is hard to see why, since such one-species origin is a biological necessity; even groups arising by hybridization are not exceptions, because hybridization is possible only between two species that ultimately, somewhere, arose from one species. "Proper definition" in this connection means mainly the exclusion of effects of convergence. In a practical way, the ideal of monophyletic classification is adequately approached if all lineages leading into a given higher category arose from one ancestral group of lower categorical rank (see also Simpson, 1944b).

THE EVOLUTION OF HIGHER CATEGORIES

The event that leads, forthwith or later, to development of a higher category is the occupation of a new adaptive zone. As a general rule, the broader the zone the higher the category when fully developed. After initial occupation of the zone, adaptive radiation into its subzones, increased specialization by adaptation to narrower and narrower subzones, weeding out of lineages, and in general the whole sequence described in Chapter VII usually follows. Occasionally it happens that the zone occupied is so narrow that significant radiation does not occur, and yet is so distinctive in adaptive type and correlated morphology of the organisms occupying it that systematists do give it high categorical rank. Thus arise the small (i.e., relatively little diversified) higher categories such as the mammalian Orders Dermoptera, Pholidota, and Tubulidentata. These are poorly known paleontologically and it is possible that they were once quite diversified, but this is not likely for all three. A few extinct groups with rather good records, notably the Orders Tillodontia and Taeniodonta, are also small, with progressive development of only one or two adaptive types in each.

When radiation does occur in a zone, it usually happens that this entails not merely an occupation but also a considerable expansion of the zone and changes in many of its features. Although the land carnivore zone of today has developed from that of the Paleocene, it is tremendously broader and so different that it does not even include any animals adaptively very similar to the earliest carnivores—the smaller bears are perhaps as nearly similar as any, and they are a late group, not a survival. Another rather frequent development, not sharply distinguished from the last, is spread from the occupied zone into others more or less different, as exemplified by the pinnipeds among carnivores.

Expansion and diversification of a group do not invariably follow reaching a new adaptive zone and they may be considerably delayed when they do occur, but they do usually happen soon, geologically speaking. In these usual cases it is true that occupation of the zone, which in retrospect is the origin of the higher category, precedes the origin of numerous genera, species, and other units that come to comprise the higher category. In *this sense,* and this only, we can agree with Wright (1949a) that "there seems to be a large measure of truth in the contention of Willis and Goldschmidt [also Schindewolf, G. G. S.] that evolution works down from the higher categories to the lower rather than the reverse." Enough has already been said to make it abundantly clear that this does *not* mean that the higher category arises as such, or that suddenly one has, say, an order without any families, which then splits into families without genera, then these into genera without species, and then finally the species arise.

It is, normally, phylogenetic progression rather than speciation that leads into the new zone, but the progressing phylum is a species, and it originally arose by normal speciation. Its progression, although unusual in outcome in that few specific lineages lead to higher categories, is of the same qualitative sort as any divergence of species after these have separated. The condition for the unusual outcome in such cases is that a "break-through" (Wright, 1949b; see also Wright, 1950) into a new major zone has occurred. The circumstances involved in access to such a zone were sufficiently discussed in Chapter VII.

The break-through usually occurs at higher rates than those of ordinary phylogenetic progression within the zone and its subzones—the latter rates are horotelic, the former tachytelic. As will be discussed later in this chapter, there are usually gaps in the fossil record of lines involved in a break-through, so that direct observation of evolutionary rates is unsatisfactory at this point, but indirect evidence shows that the rates are unusually high in most cases. Postulation that the break-through was at rates comparable to those well-recorded for lines after the break-through leads to absurdities when extrapolation backward is attempted.

The morphological difference between modern opossums and some Cretaceous opossums is slight, but some 60 million years of evolution occurred between them. If the missing pre-Cretaceous sequence changed at a comparable rate, transition from a reptile to an opossum can hardly have taken less than 600 million years; it probably took several times that long—in short it must have occurred in the pre-Cambrian, which is certainly absurd. Or if a structural unit, such as a bat's wing, be studied, it may be found that its recorded rate of evolution is effectively zero. The bat's wing has not essentially progressed since the middle Eocene, although a few of its nonfunctional elements have degenerated, and it has become more diversified. Extrapolation of this rate in an endeavor to estimate the time of origin from a normal mammalian manus might set that date before the origin of the earth.

Attempts to triangulate a date, so to speak, by extrapolation from two lines of common descent frequently also produce contradictions. As a simple example within a known record, extrapolation for hypsodonty from Pliocene equines and anchitheriines would suggest intersection, common ancestry, in about the late Eocene (*Epihippus*) stage. The same extrapolation from the later Miocene would suggest about a middle Miocene (late *Parahippus*) stage. The real intersection is between these, about late Oligocene (*Miohippus*). Use of the most rapidly evolving known line among a number of divergent groups usually produces more consistent results, showing that evolution was usually rapid in the major changes of adaptation, but still often sets incredibly remote dates and suggests that the rate of evolution is still underestimated. For instance, a relatively slow line of rodents, like the Sciurinae, would hardly warrant the assumption that there was intersection with the insectivores later than early Jurassic, and a relatively fast line, like the Hydrochoerinae, might permit placing the date as late as early Cretaceous, but even the latter date is much too early.

Some idea of probable rates of evolution in gaps in the record that correspond with major changes in adaptation can be gained by comparing the possible lengths of the gaps with the lengths of the relatively continuous record in the sixteen orders of mammals for which there is fair knowledge. These estimates are highly unreliable, because a large element of personal opinion is involved and no fully objective

TABLE 26

ESTIMATED DURATIONS OF ORDERS OF MAMMALS
(In millions of years)

Order	*Estimated Maximum Length of Unknown Origin-Sequence*	*Estimated Length of Recorded Sequence*	*Estimated Maximum Total Duration*
Marsupialia	50	80	130
Insectivora	50	80	130
Taeniodonta *	5	35	40
Edentata	15	60	75
Lagomorpha	15	60	75
Rodentia	15	60	75
Cetacea	30	45	75
Carnivora	5	75	80
Condylarthra	5	35 (50?)	40 (55?)
Litopterna *	10	60	70
Notoungulata *	15	60	75
Pantodonta *	10	35	45
Proboscidea	25	40	65
Sirenia	20	50	70
Perissodactyla	5	60	65
Artiodactyla	5	60	65
Means	17.5	55.9	73.4
Mean percent of total	24%	76%	

* Extinct

method of measurement exists as yet. The averages and consistent tendencies, shown in Table 26, should, however, give as good an indication of probabilities as can now be achieved.

The estimates of lengths of unknown origin-sequences have been made somewhere near the maximum that seems to me allowable from the record, and the actual break-through was probably much quicker in some cases. On an average, then, the origin of these orders has taken place in less, and probably much less, than a third of the time involved in later spread and diversification. The relative amounts of structural and adaptive change differ greatly. The change from a primitive land mammal, whether insectivore or ferungulate, to the earliest cetacean is much greater than any later change among cetaceans. On the other hand, eohippus is more like some condylarths than it is like *Equus*. In spite of these uncertainties and unavoidable inaccuracies, the data do force the conclusion that the break-through in all these orders involved faster rates of evolution than later rates in the same orders, probably at least twice as fast in most or all cases and quite possibly ten or fifteen times as fast in some.

The rodents are of special interest in this connection. They are a very clear-cut order to the extent that they do have a well-defined, distinctive adaptive facies or a block of characters-in-common, the essential features of which were already present in the earliest known forms. Students of the group (e.g., Miller and Gidley, 1918) have therefore had a tendency to assume that the order is extremely ancient, that rise of these basic characters and even the development of all the main subdivisions of the order must have begun a very long time before they appear in the record. Wilson (1951), on the basis of extensive special knowledge of early rodents, has recently reviewed all the evidence and has concluded that the basic rodent characters were developed during the Paleocene. No inference as to more exact dating is warranted, but origin in middle or even the beginning of late Paleocene is possible. Wilson further concludes that the progressive diversification of known fossil rodents in the Eocene (see Table 23, Chapter IX) really is what it seems to be: the record of the primary expansion of a young group of animals soon after their break-through into a major adaptive zone.

The following are some special points that bear on the evolutionary processes involved in the origin of higher categories:

1. A higher category may arise by gradual divergence and develop its distinctive adaptation rather slowly as it evolves (ungulates, condylarths, as discussed above in this chapter, but other ungulate orders divergently arising from condylarths probably did so as in 2, below).

2. More often, the rise of a higher category involves more immediate major changes in adaptive status.

3. The new adaptive status is generally unusual for its time and may begin in an environment then restricted.

4. The initial change of adaptive status often involves a threshold and in any case usually occurs at an exceptionally high rate in comparison with later rates in the group.

5. Occupation of a new zone sometimes, at least, represents a break-through by one or a few lineages of an earlier group that were divergent and exploratory in some respects but all with some tendency toward prospective adaptation for the new zone.

6. The important evolution in movement into a new zone is phylogenetic; dichotomy or multiple speciation may also occur but it is not the essential feature of the pattern.

7. The populations involved in a break-through may be relatively small.

8. The actual break-through is usually more or less narrowly regional; the geographic locus may also change in the course of the transition.

9. Effective occupation of the new zone is commonly followed by numerical increase in populations and eventually by increase in diversification.

10. Such lineages of the ancestral group as are near enough to the new adaptive type to be competitive usually become extinct during or after occupation of the new zone.

11. Animals involved in a break-through may be individually smaller than most of their earlier or, especially, later relatives.

12. It is possible that the rise of some higher categories has been significantly related to tectonic movements of the earth's crust, but this is not true of all higher categories and is not really demonstrated for any; somewhat more probable is an indirect relationship, correlation of some major tectonic phases with mass extinction, later followed by reoccupation of adaptive zones by new higher categories.

It is not likely that any one of these points *always* is involved in the origin of higher categories, and all of them were not simultaneously involved in the origin of any one higher category. It is, however probable that several of these points do apply to every instance of such origin and that any of them may do so.

Points 1 to 5 and 9 have been sufficiently discussed in this chapter and less directly in some previous chapters. Point 12 was covered adequately, for present purposes, in Chapter VII. Points 6 to 8 and 10 to 11 require some additional comments, as follows:

6. Phylogenetic diagrams are often made with a prevailing pattern of dichotomy: the Reptilia split to give rise to Mammalia and continuing Reptilia, again to give rise to Aves and, still, Reptilia, and so on. Säve-Söderbergh (1934, 1935) concluded that dichotomy is the only phylogenetic pattern, among vertebrates at least, and proposed an entirely new classification of the Vertebrata on this basis. Even if his premise were correct, classification on this basis is impractical (see Romer, 1936). Rensch (1947) has treated the origin of new organs and new

"structural plans," an essential aspect of the origin of high categories, under "cladogenesis" or phyletic branching. (His discussion of the subject is perhaps the best in print, but seems to me weakened by this approach.) It is, of course, a fact that, for instance, early Reptilia gave rise to Mammalia and later Reptilia, but this way of seeing the matter is from a viewpoint so distant that what really happened is not visible. What happened is that certain individual lines of reptiles gave rise to mammals by progressive change. Shortly thereafter all the mammallike reptiles became extinct. The lines leading to later reptiles did not branch off when mammals arose. Of course all the various lines were separated from each other at some time or other by dichotomy or multiple splitting, and doubtless those in the transition also had some speciation, but this simply does not bear on the point under discussion: progression to a new adaptive zone. It is an important point, but quite a different one, that shift from one zone to another may not affect the continuing populations in the ancestral zone.

7. The designation of populations as "small" is vague and relative. Formerly I (1944a) stressed the small size of transitional populations, and also the possibility of genetic drift in them, although commenting that drift is not always or even usually involved. Subsequent study and discussion have led to some modification of these views, although I think the essential points still stand. Genetic drift is certainly not involved in all or in most origins of higher categories, even of very high categories such as classes or phyla. It is not positively known to have been involved in any instance; in the nature of the record, it is virtually or quite impossible to obtain really positive evidence for genetic drift in a specific case. Nevertheless, it seems both possible and probable that drift has *sometimes* been involved. When it has, it is practically certain that the effect has not been the origin by genetic drift of a "hopeful monster," of the new adaptive type and new category full-blown and ready to take over. The only likely effect, and it does seem likely in a considerable number of cases, is the chance fixation of what is seen, in retrospect, to have been a key mutation in the ancestral population prospectively adaptive for the new zone. The *possibility* of occupation of the new zone is thus created. Its actual occupation is a later and slower process (see above, and Chapter VII). It is also to be recalled that random fixation of a *favorable* mutation in a small population may be both more rapid and more probable than fixation of the same muta-

tion by selection in a large population. In addition to fixation of key mutations, drift or sampling effects may be, and I think it probable that they sometimes have been, involved in the origin of key combinations of characters, as in the families and orders of flowering plants.

It is extremely improbable that the populations in a transition from one major zone to another were continuously completely isolated populations so small that drift in them was likely to be significant throughout the transition. Such transitions take time on the order of 10^6–10^7 years, and it seems to be near impossibility for an isolated population continuously very small, say on the order of 10^2 to 10^3 breeding individuals, to survive for such lengths of time. This does not exclude the possibility of drift sometime in the process because: (a) drift of characters with low selection pressure may occur in populations larger than this; (b) occasional or periodic reduction of populations to this size is consistent with indefinite survival among some organisms, at least; and (c) drift may also occur in nearly but not completely isolated demes within a species of large total population.[5]

Wright has repeatedly insisted, and I think most other students now agree, that the ideal situation for sustained, relatively rapid, progressive evolution would occur in a rather large population perhaps on the order of 10^5–10^6 breeding individuals, subdivided into many, say 10^3–10^4, incompletely isolated local populations or demes each small in size, say with about 10^2 individuals. On the other hand, absolutely maximal rates would occur in very small panmictic populations, but could very rarely be long sustained. The tachytelic rates in shift to a new major zone are fast, commonly two to several times as fast as the subsequent horotelic mode, but "fast" in this application is relative. Usually it is far below truly maximal rates. In some cases it need not have been much if any above the fastest horotelic rates. Origin of perissodactyls from condylarths, for instance, need not, as far as available evidence necessitates, have occurred at a rate significantly higher than the fastest observed rates in the Equidae, although probably well above modal equid rates. Thus it is not required that the populations concerned were continuously near the ideal size and structure for sustained rapid evolution. They may well in different cases and at different times

[5] The basis for this and other statements in the present discussion is summarized, with citations, in Chapter V.

have considerably larger or smaller and more or less subdivided than the theoretical ideal.

From another point of view, it is important to note that population in a transitional lineage may be continuously medium to large by the definition of population genetics, and yet very small in proportion to total population in the ancestral or descendant groups. The transition occurs through one or at most a few separate lineages, each of which represents one species at any one time. Such speciation as may occur during the transition (and it is probably nearly minimal at such times in most groups) usually gives rise to lines that do not lead into the new zone. On the other hand, the ancestral group is often large and widespread (5, above). Still more frequently, expansion and diversification occur within the new zone (9). Thus the population in the one or few lineages of the transition is usually relatively very small in comparison with the total population of the group arising therefrom, and also rather likely to be small relative to the total population of the ancestral group.

8. A single lineage must at all times be essentially continuous in geographic distribution. Even a lineage with a very large population is extremely unlikely to be pandemic and all lineages are likely to have quite restricted distribution at any one time. Land organisms, even in full occupation of an adaptive zone and not during the relatively restricting transition to it, rarely have single species over areas of continental dimensions, usually very much less. A single lineage of fresh-water organisms usually occurs in only one lake or one river system at any one time. Lineages of marine organisms are sometimes remarkably widespread, but even they are more likely to have quite restricted distribution: one reef or reefery, one climatic or edaphic zone of one coast, etc. Even pelagic forms of the open sea generally have restricted ranges, or move in schools or shoals covering only a very small proportion of the ocean at one time. In the case of transitional lineages, geographic restriction is likely to be all the greater because such transitions are likely to be highly habitat-specific and to involve adaptive, including environmental, relationships that are unusual at the time. It is in the nature of biogeographic distribution that such transitional lineages must always be regionally restricted and often very narrowly restricted. Densities and ranges vary so enormously for different organisms that no general statement can be made as to the actual area

required for a population of, say, 10^5, but for some groups this may be only a few square feet and for a great majority it would appear to be under, say, 5,000 square miles—which is an extremely small spot on the earth from the point of view of chances for later recovery of some of the population as fossils. Yet an average population of this size would be fully adequate in most cases for continuity of reproduction and evolution into a new zone.

Moreover, the transitional populations do not necessarily, or even probably, occur in the same spot throughout the millions of years involved in a major transition. The environment is in constant flux over such periods, even as regards its physical features. Changes in distribution of species during lengths of time comparable to this are decidedly the rule, and the rule must apply also to the species-lineages of the major transitions. (Newell, 1948, has emphasized and illustrated this feature in the evolution of invertebrates.)

10. In the fairly common situation of multiple lineages all progressing more or less toward a new adaptive type (as in 5 of the preceding list) some competition and weeding out among lines is inevitable. Firm occupation of the new zone must also inhibit further progress toward it by other lines. When multiple lines do enter the new zone, they are of quite different adaptive types in other respects. Thus the great majority of therapsid reptiles [6] became extinct in the course of progression toward the mammalian zone or while occupation of the zone was under way. The last fossil record of a therapsid (*Stereognathus*) is in the same local fauna (Stonesfield, middle Jurassic of England) as the first records of what are surely mammals (although a few early Jurassic forms may prove to be mammalian by arbitrary definition). At least three of the separate lines that did cross into the zone were sharply different otherwise and evidently noncompeting: multituberculates, rodentlike herbivores; triconodonts, small predaceous carnivores; pantotheres, insectivores. (The status at that time of two other groups, symmetrodonts and monotremes, is highly dubious for different reasons but need not alter the point here made.)

11. Phyletic increase in size is a common trend, with many exceptions. It follows that early members of what became higher categories are likely to be relatively small, also with many exceptions. This

[6] For convenience of reference, at least, the evidently heterogeneous "Order Ictidosauria" can well be retained in the Order Therapsida.

generalization is frequently found to be true in particular instances. The Mesozoic mammals are all much smaller than the average for therapsid reptiles or for Cenozoic mammals, and no doubt so were all the lines that were becoming truly mammalian. On the other hand, *Archaeopteryx* is not small for a bird (but is rather small for a pseudosuchian reptile), and the earliest amphibians were much larger than almost all recent amphibians (but did have some early descendants considerably larger than they). Among slightly lower categories, the earliest members of nearly all the reptilian and mammalian orders are rather small, at least they are in no case as large as some later members of the same orders. Early members of invertebrate higher categories also tend to be small (Newell, 1949). The rule apparently does not apply to plants which seem, indeed, to have some opposite tendency for early forms to be rather larger than later forms of the same higher category, also with many exceptions.

Space here does not permit elaboration of additional examples of the actual rise of higher categories as seen in the fossil record, but reference may be made to the cogent analysis of this sort of phenomenon in three groups of Mesozoic reptiles (Procolophonidae, Phytosauria, early Crocodilia) by Colbert (1949b).

DISCONTINUITIES IN THE PALEONTOLOGICAL RECORD

The chances that remains of an organism will be buried, fossilized, preserved in the rock to our day, then exposed on the surface of dry land and found by a paleontologist before they distintegrate are extremely small, practically infinitesimal. The discovery of a fossil of a particular species, out of the thousands of millions that have inhabited the earth, seems almost like a miracle even to a paleontologist who has spent a good part of his life performing the miracle. Certainly paleontologists have found samples of an extremely small fraction, only, of the earth's extinct species, and even for groups that are most readily preserved and found as fossils they can never expect to find more than a fraction.[7]

In view of these facts, the record already acquired is amazingly

[7] The nature of the fossil record and its deficiencies are naturally fascinating subjects to paleontologists and they have written about it extensively, although I do not know of an adequate recent summary in English. Recent summaries in German, not in very close agreement, include those of Schindewolf (1950a), Stromer (1940–1941), and Weigelt (1943).

good. It provides us with many detailed examples of a great variety of evolutionary phenomena on lower and intermediate levels and with rather abundant data that can be used either by controlled extrapolation or on a statistical sampling basis for inferences as to phenomena on all levels up to the highest. Among the examples are many in which, beyond the slightest doubt, a species or a genus has been gradually transformed into another. Such gradual transformation is also fairly well exemplified for subfamilies and occasionally for families, as the groups are commonly ranked. Splitting and subsequent gradual divergence of species is also exemplified, although not as richly as phyletic transformation of species (no doubt because splitting of species usually involves spatial separation and paleontological samples are rarely really adequate in spatial distribution). Splitting and gradual divergence of genera is exemplified very well and in a large variety of organisms. Complete examples for subfamilies and families also are known, but are less common.

In spite of these examples, it remains true, as every paleontologist knows, that *most* new species, genera, and families and that nearly all new categories above the level of families appear in the record suddenly and are not led up to by known, gradual, completely continuous transitional sequences. When paleontological collecting was still in its infancy and no clear examples of transitional origin had been found, most paleontologists were anti-evolutionists. Darwin (1859) recognized the fact that paleontology then seemed to provide evidence against rather than for evolution in general or the gradual origin of taxonomic categories in particular. Now we do have many examples of transitional sequences. Almost all paleontologists recognize that the discovery of a complete transition is in any case unlikely. Most of them find it logical, if not scientifically required, to assume that the sudden appearance of a new systematic group is not evidence for special creation or for saltation, but simply means that a full transitional sequence more or less like those that are known did occur and simply has not been found in this instance.

Nevertheless, there are still a few paleontologists, and good ones (e.g., Spath, 1933; Schindewolf, 1950a), who are so impressed by how much has been found that they conclude that most, at any rate, of what has not been found never existed, and there are some neontologists, also some good ones (e.g., Clark, 1930; Goldschmidt, 1940), who accept this interpretation. It is thus still too soon for the rest of us to take the discontinuities of the paleontological record for granted. Even apart from that, the recognition and interpretation of such discontinuities is interesting and is a necessary, frequently also a practical and useful, part of the paleontological profession. Moreover, it is a fact that discontinuities are almost always and systematically present at the origin of really high categories, and, like any other systematic feature of the record, this requires explanation.

Within a single stratigraphic sequence, there are usually many geologically brief interruptions of deposition, with or without erosion of some of the strata already deposited. These local hiatuses are usually so brief that significant evolution did not occur during the time they represent. When appreciable change did occur, the explanation for the minor discontinuity is apparent. In especially favorable cases, the length of the interruption can be closely determined in terms of evolutionary change or of thicknesses of strata. An unusually clear but otherwise typical example is included in Brinkmann's data on ammonites, previously mentioned in Chapter I. In some lineages of *Kosmoceras* certain characters evolved rather steadily, with nearly rectilinear regression against thickness of strata. Thus the number of inner ribs on the last convolution in *Kosmoceras* (*Zugokosmoceras*) increased steadily from about 25 to about 40 in the 30 centimeters of strata from 850 to 880 of Brinkmann's record (Fig. 45). But at 864.5 the regression line is abruptly offset. The slope remains the same within limits of sampling error (the character was of course variable and the regression line is a smoothing of mean values), but the line jumps from about 32 to about 33.

At the same point other characters of these ammonites, evolving at other rates and in other directions, also have their regressions offset to about the same relative degree. It is out of the question that the offset is due to a mutation rapidly spread through the population. It certainly represents an interval of nondeposition. By separating the lines at the offset until one is a continuation of the other, the actual evolutionary trend is restored and it can be seen that the hiatus corresponds with about 2.5 cm. of strata in this deposit (Fig. 45, bottom).

A large scale example of discontinuity mainly at the generic level is provided by the successive Puerco and Torrejon mammalian local faunas, early and middle Paleocene, respectively, in the Nacimiento

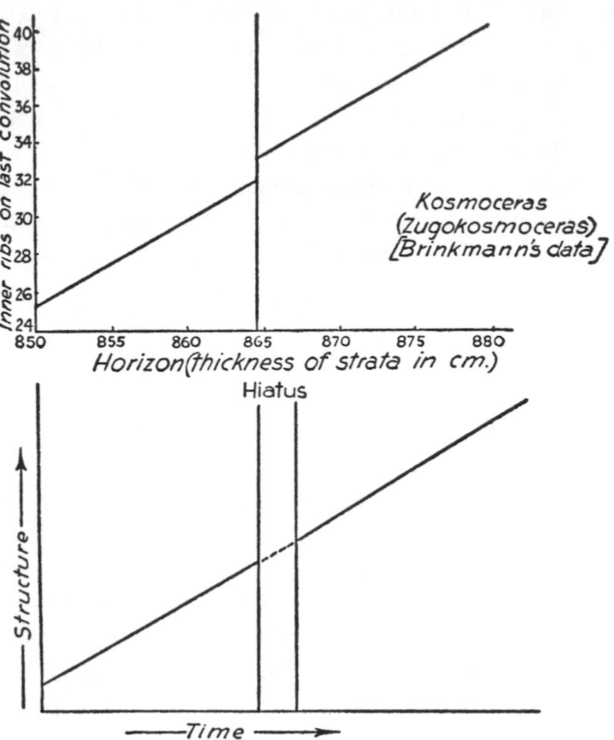

FIGURE 45. AN APPARENT SALTATION IN AN AMMONITE PHYLUM AND ITS INTERPRETATION AS CAUSED BY A DEPOSITIONAL HIATUS IN THE BEDS FROM WHICH SPECIMENS WERE COLLECTED.

formation of New Mexico. As listed in Table 27, there are fifteen single or closely related lineages that occur in both faunas (Simpson, 1936b; Matthew, 1937; and the collections). Most of these have changed so much from Puerco to Torrejon that the later forms are conservatively placed in different genera. In three lines earlier and later forms are at present placed in the same genus, but the species are sharply distinct and may well be placed in separate genera on further revision. There is no consistent lithological difference in the beds in which the two faunas occur and there is no recognized hiatus or unconformity between them, although one or more hiatuses probably are present.

On the face of the facts, this is remarkably good evidence for the wholesale discontinuous or saltatory origin of new genera. There are, however, some 170 feet of nonfossiliferous beds between the highest known Puerco and lowest known Torrejon fossils (Sinclair and Granger, 1914; later collecting may have reduced this figure, but not significantly as far as I know). Matthew and others long maintained that the time represented by these barren beds and by probable hiatuses in them was considerable and that during this time transitional evolution linked the Puercan and Torrejonian genera and species. That inference was brilliantly confirmed by discovery of another local fauna, the Dragon, in Utah, intermediate between Puerco and Torrejon in age although somewhat nearer the latter (Gazin, 1941).

TABLE 27

KNOWN EARLY TO MIDDLE PALEOCENE MAMMALIAN LINEAGES FROM NEW MEXICO AND INTERMEDIATE FORMS FROM UTAH

Orders	*Puerco Fauna*	*Dragon Fauna*	*Torrejon Fauna*
Multituberculata	*Kimbetohia*	– *Ptilodus*	– *Ptilodus*
	Eucosmodon	–	– *Eucosmodon*
Taeniodonta	*Wortmania*	– Cf. *Psittacotherium*	– *Psittacotherium*
	Onychodectes	– *Conoryctella*	– *Conoryctes*
Carnivora	*Loxolophus*	– *Tricentes*	– *Tricentes*
	Chriacus	–	– *Chriacus*
	Protogonodon [a]	– *Protogonodon*	– *Claenodon*
	Eoconodon	–	– *Triisodon*
Condylarthra	*Choeroclaenus*	–	– *Mioclaenus*
	Tiznatzinia	– *Ellipsodon*	– *Ellipsodon*
	Oxyacodon	– *Dracoclaenus*	– *Protoselene*
	Anisonchus	– *Anisonchus*	– *Anisonchus*
	Conacodon	– *Haploconus*	– *Haploconus*
	Carsioptychus	– *Periptychus*	– *Periptychus*
	Protogonodon [a]	– *Desmatoclaenus*	– *Tetraclaenodon*

[a] This is the remarkable case, previously mentioned, in which some species of a genus are nominal carnivores and others of the same genus nominal condylarths.

The Dragon collections are much smaller than those from the Puerco and Torrejon, but do contain eleven of the fifteen Puerco-Torrejon lineages. In every case the Dragon species are distinct from those of the Puerco or Torrejon. In every case where the material is adequate for significant comparison they are more primitive than Torrejon species, and when the lineage is known from both Puerco and Torrejon (i.e., in the lineages shown in the table) the Dragon species are intermediate between those of the Puerco and Torrejon. In four lines the Dragon species are so precisely intermediate that reference either to a Puerco or to a Torrejon genus would be about equally justified. In three of these lines (the other is *Wortmania-Psittacotherium* and no generic assignment was made for the Dragon form), Gazin has solved the dilemma by placing the Dragon species in separate, intermediate genera. In four lines the Dragon forms are more primitive than those from the Torrejon but resemble the latter slightly more than they do the Puerco ancestors and are therefore placed nominally in Torrejon genera. In one line slightly greater resemblance is to the Puerco species and nominal reference is to a Puerco genus. In another, Puerco, Dragon, and Torrejon species have not been distinguished generically. (In the case of *Ptilodus,* reference to the Torrejon genus has no bearing, since the parts diagnostic in the Puerco ancestor are unknown in the Dragon specimens and differential comparison has not been possible.)

Although complete faunal succession is still lacking, discovery of the Dragon fauna can surely leave no doubt that transition and not saltation occurred between the Puerco and Torrejon. The example is parallel to many other cases of apparent saltation between successive faunas, and it is reasonable to conclude that the explanation applies widely.

Another, probably even more frequent cause of discontinuity in evolution as seen in the fossil record is successive migration. The Equidae provide a remarkably clear example in which the explanation of the discontinuity is known with full certainty.

One of the first and greatest triumphs of evolutionary paleontology was Kowalevsky's study (1874) of the evolutionary history of the horse and other ungulates.[8] The Old World sequence is [*Hyracotherium*]–*Paleotherium*–*Anchitherium*–[*Hypohippus*]–*Hipparion*–*Equus.* The genera in brackets were not in Kowalevsky's original supposedly direct phylogeny; *Hyracotherium* was believed a collateral rather than a direct ancestor, and *Hypohippus* was not yet distinguished in the Old World. Except for a considerable time gap between *Paleotherium* and *Anchitherium* the temporal sequence is almost complete, but there are no transitional forms between the sharply different genera. Thus, the phylogeny seems to have proceeded in a series of abrupt steps, with

[8] Although almost all the phylogenetic details were later found to be wrong, the achievement was real, and its conclusion fundamentally correct. Kowalevsky deciphered many of the essentials of the structural development of the modern horse. He did not identify the precise phyletic lines, because they were not in the materials then available.

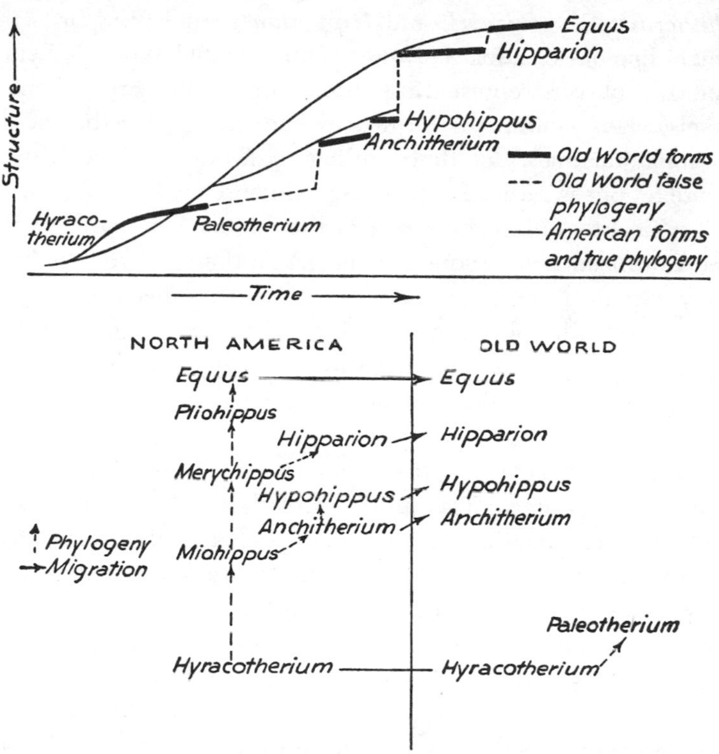

FIGURE 46. APPARENT SALTATORY AND TRUE CONTINUOUS PHYLOGENY OF THE EQUIDAE. Old World series, a structural sequence of immigrant forms, is saltatory; American series, a true phylogeny, is continuous. American genera not immediately related to the migrating forms are omitted.

little evolutionary progress within any one genus and instantaneous jumps of various sizes from one genus to the next. Discoveries in North America during the last seventy years have revealed the true phylogeny and explained the saltations as a peculiarity of the record, not of evolution (Fig. 46). *Hyracotherium* was common to Europe and North America, and in Europe it (or some close ally) gave rise to *Paleotherium* and other Eocene genera really well removed from the ancestry of the horse. *Equus* arose in North America by continuous progressive evolution in which the transitions between genera are not more abrupt than between species within the genera. The four Old World genera subsequent to *Paleotherium* represent as many migrations from America.

Anchitherium, Hypohippus, and *Hipparion* are all from branches off the main line of evolution to *Equus.* Thus the Old World "phylogeny" is doubly indirect, representing discontinuous migration from what were elsewhere continuous sequences and arising, for the most part, from collateral lines, not that leading to *Equus.* It is a fairly good Stufenreihe but is rather far from the Ahnenreihe (see Chapter VII).

It is typical that in the examples of the Puerco-Torrejon lineages and of the European horse sequence the discontinuities were reduced or entirely removed by later discovery. This has been the universal tendency as fossil collecting continues and becomes more intensive. Every year apparent discontinuities in phylogeny are filled in, and each year it becomes less logical to conclude that remaining discontinuities represent saltation.

There remains, however, the point that for still higher categories discontinuity of appearance in the record is not only frequent but also systematic. Some break in continuity always occurs in categories from orders upwards, at least, although the break may not be large or appear significant to most students. It was estimated above (Table 26) that the mammalian orders had unknown origin sequences (assuming, for the moment, that they did not arise by saltation) that may have averaged as much as one-third as long as the recorded sequences, although probably considerably shorter. As further detail for this example, the present records of all the mammalian orders are summarized in Table 28. The magnitudes of the morphological and adaptive discontinuities are quite different in different orders and tend, significantly, to be correlated with the indicated lapse between occurrence of possible ancestry and appearance of the order. Thus the earliest Condylarthra occur in association with some forms that would be suitable ancestral types and do, in fact, intergrade almost completely with these, as shown earlier in this chapter. On the other hand, the earliest Proboscidea appear with large adaptive discontinuity from any probable ancestors, but probable ancestors are not known after the Paleocene and there is a lapse at least on the order of 20 million years between them and the first Proboscidea.

That is the first of a number of points that, taken together, seem to me to explain the systematic discontinuities of records of the origin of higher categories and to make their interpretation as due to saltation entirely untenable. These points, each of which will be commented on somewhat further, are as follows:

1. There is time corresponding with the unrecorded sequences; the discontinuities follow time gaps.

2. The discontinuities are not absolute; they may be very slight or almost nominal, and isolated finds of fossils of intermediate type may occur in them.

3. Gaps comparable in length are known which certainly had occupants even though these do not appear in the fossil record.

4. The origin-sequences are often of such length that under conditions of paleontological sampling it is almost impossible for their record to be continuous.

5. Chances of discovery of transitional types are often reduced by the fact that they moved between quite different ecological conditions.

6. Relatively small populations were often involved.

7. The rate of evolution was often relatively fast and fewer individuals occurred per structural stage than in average well-recorded sequences.

8. The transitions were in restricted geographic regions and not always in the same region throughout any one transition.

These points bear on the present subject as follows:

1. It is perhaps impossible to prove that time for transition *always* precedes the discontinuous appearance of a higher category, and there is some danger of circular argument. It is, however, true that in some instances there certainly was a time gap and that in no instance do any known facts rule out this possibility. The only difficulties arise from the fact that bradytelic and horotelic lines may arise from the same ancestry and that late stages of a bradytelic line may be mistaken for that ancestry. Nothing in the structure of Recent Didelphidae absolutely rules them out as ancestors for the Recent Australian marsupials. Other considerations, however, make it quite certain that the real common ancestry was much older, probably not more recent than about the end of the Cretaceous, and that there is a gap of perhaps 65 million years between the didelphid-like ancestral stage and the oldest known Australian marsupials. When bradytelic or, at least, slowly evolving lines do not occur, the presence of a gap is usually evident. For instance the last known reptiles that could possibly be ancestral to birds ap-

TABLE 28

AVAILABLE RECORDS OF THE MAMMALIAN ORDERS

Order	*Known Occurrence of Possible Ancestry*	*Recorded Span*	*Continuity of Record (for any considerable part of order)*	*Remarks*
Monotremata	Late Triassic	Pleistocene—Recent	Very short	Probably Australian for much of history, where few fossils are known
Multituberculata *	Late Triassic	Late Jurassic—early Eocene	Fair	
Triconodonta *	Late Triassic	Middle Jurassic—early Cretaceous	Poor	Known from only three ages
Symmetrodonta *	Late Triassic	Late Jurassic	None	All known are of one age
Pantotheria *	Late Triassic	Middle—late Jurassic	Poor	Only two ages
Marsupialia †	Late Jurassic	Late Cretaceous—Recent	Good and Poor	Good record in America; very defective in Australia
Insectivora †	Late Jurassic	Late Cretaceous—Recent	Good	
Dermoptera	Late Cretaceous—Paleocene	Late Paleocene—Recent	Poor	Enormous gap, early Eocene—Recent, probably because of small size and Asiatic occurrence
Chiroptera	Late Cretaceous—Paleocene	(Late Paleocene?) middle Eocene—Recent	Poor	Generally rare fossils, small, fragile, and not likely to occur in normal sediments
Primates	Late Cretaceous—Paleocene	Middle Paleocene—Recent	Fair	Uncommon as fossils, but many scattered occurrences
Tillodontia *	Late Cretaceous—Paleocene	Late Paleocene—middle Eocene	Good	
Taeniodonta * ‡	Late Cretaceous	Early Paleocene—late Eocene	Good	

Edentata †	Late Cretaceous—Paleocene	Late Paleocene—Recent	Good	
Pholidota	Paleocene	(Oligocene?) Pleistocene—Recent	Poor	Fossil record valueless
Lagomorpha †	Late Cretaceous—Paleocene	Late Paleocene—Recent	Good	Good later than Eocene; practically lacking before
Rodentia †	Late Cretaceous—Paleocene	Late Paleocene—Recent	Good	
Cetacea †	Paleocene	Middle Eocene—Recent	Good	
Carnivora †	Late Cretaceous	Early Paleocene—Recent	Good	
Condylarthra * †	Late Cretaceous—Paleocene	Early Paleocene—late Eocene (Miocene?)	Good	
Litopterna * †	Paleocene	Late Paleocene—Pleistocene	Good	
Notoungulata * †	Paleocene	Late Paleocene—Pleistocene	Good	
Astrapotheria *	Paleocene	Early Eocene—middle Miocene	Fair	
Tubulidentata	Paleocene—Eocene	(Early Eocene?) early Pliocene—Recent	Poor	Eocene occurrence dubious and other fossil record unimportant
Pantodonta * †	Paleocene	Middle Paleocene—middle Oligocene	Good	
Dinocerata *	Paleocene	Late Paleocene—late Eocene	Good	
Xenungulata *	Paleocene	Late Paleocene	None	Two probably contemporaneous occurrences
Pyrotheria *	Paleocene	Early Eocene—early Oligocene	Fair	
Proboscidea †	Paleocene	Late Eocene—Recent	Good	
Embrithopoda *	Paleocene	Early Oligocene	None	Only one occurrence
Hyracoidea	Paleocene—Eocene	Early Oligocene—Recent	Fair	
Sirenia †	Paleocene	Middle Eocene—Recent	Good	
Perissodactyla †	Paleocene	Early Eocene—Recent	Good	
Artiodactyla †	Paleocene	Early Eocene—Recent	Good	

* Extinct orders.

† Orders with long, good records.

pear in the Triassic and there is a gap of at least 15 million years before appearance of a (very reptilian) bird in the record.

It may be mentioned in passing that some paleontologists have leaned the other way and have assumed that appearance of a higher category implies its origin a geological period or two earlier. For instance, the possible occurrence of mammals in the late Triassic (occurrence that now seems to be quite dubious although possibly correct) used to be taken as indication that the Class Mammalia arose in the Permian, which is surely much too early on present evidence. Even now it is more usual than not to assume that the animal phyla, all of which appear in the Cambrian and Ordovician as far as they have fossil records, "must have" arisen far back in the pre-Cambrian. Pre-Cambrian life must, indeed, have been more varied than the few known fossils of that age, but there is no real reason to assume that all animal phyla arose in the pre-Cambrian or that any except the protistans arose long (geologically speaking) before the Cambrian.

2. Some of the discontinuities in question are large, some almost insignificantly small. It may be quite misleading to state simply that all higher categories do have some degree of discontinuity when they appear. A great many reptiles were evolving at moderate rates and more or less steadily in the direction of mammals throughout the Permian and Triassic. There is really very little difference between the last well-known therapsids and true mammals. The earliest possible mammals among known fossils are represented only by scraps of teeth and could well prove actual intergradation if well known. More or less 10 million years later what are certainly true mammals appear. A gap occurs, but it is not really profound and could readily be bridged by simple continuation of the sort of evolution actually recorded *before* the gap.

Nor are the gaps absolute. Isolated finds have been made in what would otherwise be large gaps, and such finds always are intermediate in structure just as if transitional evolution had been going on all the time. *Archaeopteryx* is the most famous case in point. Schindewolf (1950a) disposes of it by saying that it is "a true bird" and so cannot close the discontinuity between reptiles and birds. But if we did not know that *Archaeopteryx* had feathers, or if we found its last featherless ancestors, then of course we would have "a true reptile." The break can be maintained in words even when it is closed by specimens. The indisputable fact is that *Archaeopteryx* is as reptilian as avian throughout (e.g., Heilmann, 1926) and that it is precisely the sort of animal we would expect from the middle Jurassic if the Aves were transitionally developing their new adaptation from toward the end of the Triassic into the late Cretaceous. It is obvious that isolated discoveries do not fill the gap, and it is always possible to insist that the gap is still there on one side or the other (there really remains a *smaller* gap on each side, of course), but these discoveries are evidence as to what was going on during the break in the fossil record.

An even more extraordinary example is provided by the late Devonian [9] amphibians from Greenland, known from complete skeletons (but not yet fully described by Jarvik, who has them in hand), and the perhaps earlier *Elpistostege* from Canada (see Westoll, 1943), known only from skull fragments. In these creatures we really do have an essentially continuous transition between two very high categories: classes. The Greenland forms are so completely and amazingly intermediate as to be animals that could not possibly exist if higher categories arose as such and by saltation.

3. If transitional lines did occur in the gaps in the record, then groups capable of preservation as fossils sometimes existed, occasionally for as much as tens of millions of years, without being found as fossils. It is therefore highly pertinent that we know, as indubitable fact, a good many instances in which fossilizable animals have left no known record over spans even longer than any required to explain the most discontinuous appearances of higher categories. Coelacanth fishes are abundant fossils in some formations, but the last known fossil coelacanth is from the late Cretaceous, and now, more than 75 million years later, there are still living coelacanths (*Latimeria*). Dermoptera are known from the late Paleocene and early Eocene, then not again for more than 50 million years (*Cynocephalus* = "*Galeopithecus*," the colugo or "flying lemur"). The entire mammalian fauna of Australia must have had varied, abundant, distinctive ancestors back to the early Cenozoic, at least, perhaps back to the Triassic for the monotremes, but just one Tertiary specimen is known and before that for tens of millions of years nothing whatever. Sphenodonts are not extraordinarily rare in the late Triassic and Jurassic and they also occur in the early Cretaceous. Thereafter they are wholly absent from the record for well over

[9] There is some dispute as to the precise age, but that has no bearing on the present point.

100 million years, but they are still in existence today. There are many other examples of similar gaps. Of course circumstances can be conjectured to explain each example, but so do circumstances explain such gaps as exist in records of lines leading to higher categories.

4. A long origin-sequence into a new higher category is likely to have some gaps in the record just because it is long. Such sequences apparently extend quite commonly to 10^7 years or more. Really, completely continuous sequences of such length are relatively rare in any part of the record, although a few even longer are known. This brings up another point concerning paleontological sampling. Lineages leading to very high categories, such as those that do have systematic discontinuities, have certainly been far less common than specific and generic lines within such categories. It would be very conservative to put the ratio at 1:1,000. If chances of recovery were the same in both cases (and they seem really to average lower for the transitional lineages), then we should have at least 1,000 long, continuous phyletic sequences within higher categories for one equally long and good sequence leading to a very high category. We probably do not have 1,000 really long, really continuous sequences of any sort.

5. Paleontological sampling is necessarily spotty both as to area and as to facies. Some facies, e.g., lowlands, marshes, littoral zones, are quite well sampled. Others, e.g., alpine zones or the deep-sea, are hardly represented at all. The origin of a higher category frequently involves a marked change in ecology of the group concerned. Change of facies must often be involved, and when it is the chances of finding fossils of *both* ends of the sequence are greatly reduced. In the subsequent, expanding phase of the higher category numerous different facies are commonly occupied and the sampling chances of then finding some member of the group may be excellent (see Simpson, 1936a).

6. The greater part of the fossil record and most particularly those parts that do include long, good evolutionary sequences are made up of animals that not only had large populations in the sense of population genetics but also were even more extremely abundant. As discussed above, the lines leading to higher categories must often, or even usually, have had populations relatively quite small as compared with their descendants in the categories. This would markedly reduce the chances of their recovery as fossils.

7. Reasons have been given for believing that on an average the

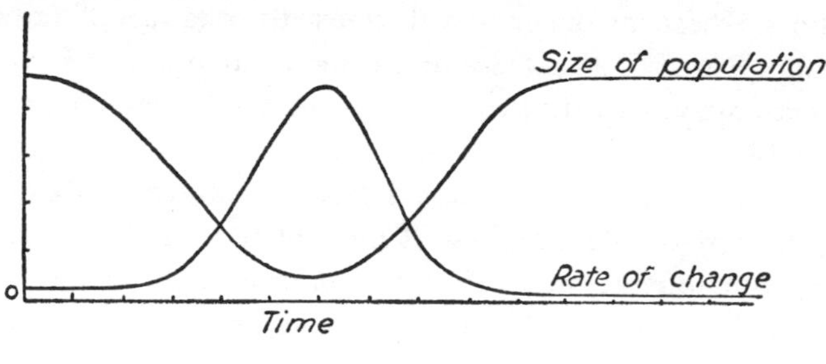

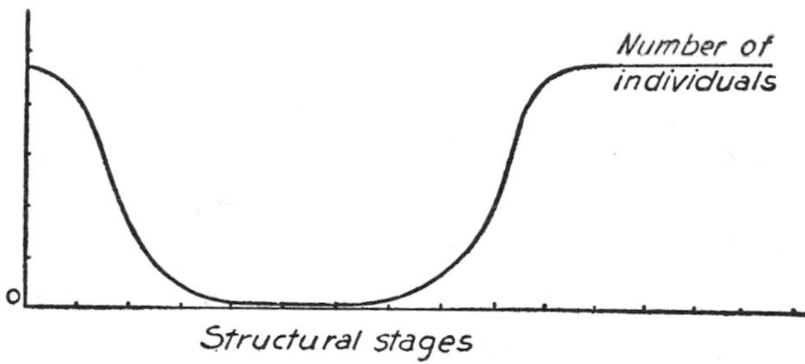

FIGURE 47. REDUCTION OF NUMBER OF INDIVIDUALS DEMONSTRATING MAJOR STRUCTURAL CHANGES. In such sequences increased rate of evolution coincides with decreased size of population (upper figure). There are, then, more distinguishable structural stages per unit of time and also fewer individuals per unit of time than in the normal evolution of large populations. The result (lower figure) is an extreme and long-continued reduction in the number of individuals exemplifying each structural stage.

lines leading to higher categories were evolving decidedly more rapidly than the intracategory groups that make up most of the fossil record. Even if chances of recovery were otherwise equal (and it is clear that they often were not) faster evolving lines have fewer individuals per given degree of change, and hence chances are less for the recovery of a sample of a particular stage in that change. This effect is greatly increased if, as seems often to be true, the faster lines in question also have smaller populations, a circumstance diagrammatically shown in Figure 47.

8. The practical certainty that the actually transitional line or lines will have restricted geographic distribution and the probability that this will not remain the same throughout the transition have been men-

tioned above. These factors greatly decrease the chances of finding transitional lines. Even if nothing else depressed chances, these factors alone make it extremely unlikely that the whole of a transition to a higher category will be found.

Even paleontologists do not always seem to realize how exceedingly spotty is the geographic sampling of most of the fossil record. It gives the impression, furthermore, of being especially poor for the times most crucial for origin of higher categories, a subjective impression doubtless heightened by frustration but also possibly indicating that there is, after all, some obscure correlation of general geographic (or physiographic) conditions and the rise of higher categories. For instance, early Paleocene is the most crucial single age for rise of higher categories in mammals and some other terrestrial groups. Fossils of that age are known from only four places, all in the same region: one each in New Mexico, Colorado, Utah, and Wyoming, and only two of these have yielded more than a few scraps. The whole of the rest of the world is still a complete blank for the terrestrial early Paleocene. The middle Paleocene is as restricted; larger faunas are known, but they also are all from the Rocky Mountain region of the United States. Even the late Paleocene collecting fields, in which forms transitional to several mammalian orders should occur, are still very restricted geographically: the Rocky Mountain region, again, two fair faunas from geographic pinpoints in Europe, one of strangely aberrant facies from a few square yards in Asia, and some fossils of uncertain exact correlation and already of island-continent type from two places in South America, a single limestone quarry in Brazil and a few square miles in Patagonia. For invertebrates, latest Permian and earliest Triassic exemplify a time evidently crucial for rise of many higher categories. Marine rocks of this age are known from widely scattered localities, but they are notoriously restricted in extent and often are of peculiar facies.

The challenge to specify just where the transition to a particular higher category occurred is one that can very rarely be met by fossil evidence that is available now or, indeed, ever likely to become available. It can always be shown that there is plenty of room, in areas that are and often that must forever remain paleontological blanks for the time in question. That, indeed, is why the problem is usually insoluble: there are too many possibilities that cannot be either investigated or ruled out. Aside from the fact that the record in continental areas is spotty as to geographic sampling and almost always full of large gaps as to sampling of facies and environments, there are vast areas quite likely to have harbored transitional lines and not even available for paleontological sampling. Important terrestrial higher categories are not likely to have arisen on oceanic islands, but they are likely to have arisen on continental islands or archipelagoes, for instance in the sub-Arctic (which had moderate climates during most of geological history), where such islands as remain above the sea have few or no sediments of the crucial ages. Oceanic islands and reeferies are likely places for marine invertebrate origins and they probably occurred throughout geological history, but none of them older than the Cenozoic (and generally late Cenozoic) are now available to us. (Numerous fossil reefs are known, but all are relatively young or were epicontinental.) The continental margins at full emergence are also likely places for invertebrate origins, and most of these are now submarine.

Paleontologists used to spend considerable time and paper trying to locate centers of origin for major groups. For instance, Osborn (e.g., 1910) and Matthew (e.g., 1915, 1939) devoted much attention to this problem for mammals and some other vertebrates and published maps showing centers of origin for various orders. Most of the centers for mammals were placed in central Asia, then paleontological *terra incognita,* but not one of these was confirmed—nor yet ruled out—by the subsequent brilliant discoveries of Granger (under Andrews as party leader) in central Asia.[10] For a few groups of peculiar geographic history a broad region of origin can, indeed, be specified. For instance, the Litopterna almost certainly arose in South America—but where on that large and varied continent, in what particular environment, and in what precise lineages are still unknown.

The subsequent geographic history of a group such as the Order Perissodactyla can often be worked out in considerable and well-documented detail, but most of us are now resigned to the fact that its place of origin is completely unknown, and likely to remain so. Neontologists, especially botanists, still seem to spend much time determining (as they believe) centers of origin of higher categories, but to a paleontologist some of this activity seems best classified as a fairly harmless pastime.

[10] Osborn almost persuaded himself that discovery of *any* member of a group in central Asia confirmed its origin there; this was mere human overenthusiasm.

All these factors combined make the systematic deficiencies of record of the origin of higher categories exactly what should be expected if, as seems surely to be the case, they did arise by transitional lines. Origin by saltation is highly improbable on other grounds. The fossil record provides evidence against it and no valid or impelling evidence for it. The surprising thing is not that there are discontinuities but that these are not absolute, that the fossil record does provide considerable information even regarding many of the major transitions.

This general survey of higher categories shows that their evolution has special features tending to distinguish it from the evolution of lower categories. It may be in some cases but is not typically merely a multiple of the evolution of species and other lower categories. The distinguishing features relate mainly to the scale and the adaptive relationships of the evolution of higher categories. They involve certain durations, intensities, and combinations of factors. There is no reason to believe that any different factors are involved than those seen in lower categories or in "microevolution." On the contrary, those factors are fully consistent with what we know of higher category evolution and quite capable of explaining it.

CHAPTER XII

Patterns or Modes of Evolution

IT HAS BEEN REPEATEDLY EMPHASIZED in this book that evolution is an incredibly complex but at the same time an integrated and unitary process. Violence is done whenever we pick out some one factor, process, or element of the pattern and attempt to consider this apart from the whole. Some of the great controversies about evolution have had little more basis than this: that different theories were based on one factor or another when the essential point was really the relationship between the two. Yet analysis is necessary. The complex cannot be understood without separate consideration of its parts. The thing to remember is that analysis does destroy the very thing that is being studied. It is essential that the original relationships of the parts be considered also, that synthesis accompany analysis.

Most of the analysis and much of the synthesis set as the aims of this study have now been presented. What remains is to try to see what sort of thing the history of life is. It is still impossible to see it whole; analysis is still necessary. It is however possible now to attempt a different analytical approach, and to synthesize in a somewhat bolder way.

The history of life as it has existed in nature is a vast succession of ontogenies of organisms, all the ontogenies being connected by the fact that each arises with material continuity from one or from two others. This is the objective pattern of evolution, not objective in being now present and verifiable to our senses, which it unfortunately is not, but in being what we otherwise verify as having been the real and physical embodiment of life through the ages. The continuum of ontogenies is phylogeny, and phylogeny has a material pattern traced by the descent of living matter and its development in individuals.

Looking more closely into the pattern, we see that it involves also the organic changes that have occurred in the sequence and the rates at which these changes have occurred. Trying to see how these arose, were transmitted, and became what they did, we find ourselves grappling finally with every factor and element that is in life or that affects life.

Thus the pattern of phylogeny brings together everything in the complex, integrated phenomenon of evolution. It may, however, first be considered simply as organic descent. Whether examined in its finest mesh, at the level of individuals, or more broadly, the whole tremendous intricacy of phylogeny involves only three simple sorts of events or processes, or we may say that the pattern has only three elements. The lines of organic descent may join, separate, or continue. Among individuals at the basically objective level of phylogeny lines join in sexual reproduction; they separate in the production of offspring; they continue through each of the offspring. At a higher level are populations within which the processes of individual joining, separation, and continuation form an anastomosing pattern in time. Within a subdivided species, the populations themselves anastomose and form a web. The web is a line, and such lines form a larger pattern, which is also very complex as a whole but which is even simpler as regards the elements involved. These lines or phyletic lineages separate and continue; they do nothing else. Some populations that are beginning to be separate lines may have individual anastomosis with others, or may even join others, but this is only a slight complication before decisive lineage status is reached, after which lines never join. The relationships involved are summarized diagrammatically in Figure 48.

This book is devoted to major features of evolution, so that joining and anastomosis in phylogeny concern it only to the extent that they are part of the intimate mechanism of evolution which does rise to levels where joining no longer occurs. That aspect has been discussed, especially in Chapters III and V. The phylogenetic patterns here to be considered are those of lineages, and their elements are separation and continuation only. In order to bring in somewhat broader aspects than those commonly associated with speciation, which is the process by which separation of lineages occurs, this element in the broad phylogenetic pattern will be called "splitting." It is discussed below under that heading.

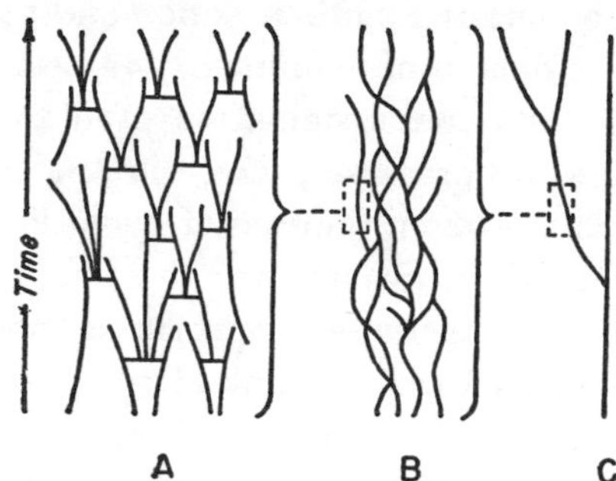

FIGURE 48. DIAGRAM OF PHYLOGENETIC PATTERNS. A, greatly simplified suggestion of individual descent; ascending lines are ontogenies and cross-bars represent sexual reproduction. B, simplified diagram of descent in a species; lines are demes or other local subdivisions, each of which has an included individual pattern as in A. C, diagram of descent in lineages; lines are lineages, each of which has an included pattern like B.

Of course the simple fact of splitting or continuation has little interest unless the sorts of changes involved and some of the circumstances are also considered. In the continuation of lineages, it seems that two limiting cases have special interest: (1) change does not occur, the lineage simply goes on as is, or (2) quite radical and sudden change occurs. The former case is arrested evolution, and it was sufficiently considered in Chapter X in connection with bradytely, the rate distribution especially associated with it. The latter limiting case has also been involved at various points in preceding chapters, but still needs some special consideration. It is discussed below as "quantum evolution." The continuation of lineages between these limiting cases, neither changeless nor changing in sudden shifts, will be discussed under the more general term of "phyletic evolution."

SPLITTING

In the predecessor of this book, the whole of the mode of evolution here designated as "splitting" was called "speciation." This led to some misunderstanding. Usages vary considerably, but most students understand by the word "speciation" one or both of two processes: (1) the origin of genetic isolation between two populations (or groups of populations), and (2) the origin of new characters and their distribution or differentiation among and within populations (a process that begins far below the level of genetic isolation and continues above it). These usages and the duality of definition have been stressed by Mayr (1942). Experimentalists and systematists used to be so exclusively concerned with these two processes, which are indeed of fundamental importance, that many of them thought of evolution as being, simply, cumulative speciation. (The Society for the Study of Evolution was preceded by a Society for the Study of Speciation, intended to be similar in scope.) It now seems clearer to think of speciation as, literally, the origin of species and particularly the origin of genetical or horizontal species. On this basis, speciation is the process of differentiation within populations and of the rise of genetic isolation between populations formerly part of the same species. Subsequent divergence is not, strictly speaking, speciation but an aspect of phyletic evolution. Of course this frequently has its roots in differentiation within populations and hence is not absolutely distinct from speciation. This merely again exemplifies the fact that what we are analyzing is integral and that its elements do not have absolute distinctions.

Speciation is the basic mechanism of the splitting of lineages,[1] although it does not always or usually give rise to separate, significantly progressing and diverging lineages.

At the lowest level, the process starts with differences between individuals, which in the most local population groups are usually very minor and fluctuate from generation to generation. In a subdivided population, the smallest collective units, the demes, usually differ so little and so temporarily that it is not practical to give them taxonomic recognition or to name them.[2] It is, nevertheless, from such differences that arise the somewhat more clear-cut and enduring distinctions commonly recognized as subspecific. A deme may simply become more distinctive and more populous, rising to subspecific status, but this is not the usual process. Demes are not by nature units that are evolving into subspecies, nor are subspecies by nature units evolving into species. These units are still in a genetic continuum, and a subspecies normally arises by limited exchange of genetic factors among demes so that a

[1] Rensch's (1947) term "cladogenesis" ("Kladogenese") is more elegant and is defined as "splitting of lineages" ("Stammverzweigung") but Rensch, himself, applied his term especially to progressive specialization and to the rise of new structures and structural plans, processes in which splitting is irrelevant.

[2] C. Hart Merriam's classification of North American bears was essentially an unsuccessful attempt to distinguish and name demes, although he called them "species." This now seems like an example of what not to do in taxonomy.

number of them come to have a distinctive facies, generally adaptive in nature, sufficiently defined and lasting for useful taxonomic recognition.

Species apparently do frequently arise from subspecies, but few subspecies become species, and it often happens that species are not, strictly, derived from a single subspecies. Subspecies are still impermanent by nature. Even without extinction as such, they can disappear by genetic fusion with other subspecies or with the general specific population. In this shifting blend, one species can arise from two or more subspecies, and two or more species can also separate from one subspecies or from a species without subspecies. These events lead to splitting, but they are still occurring in the reticulate part of the phylogenetic pattern, where joining is an element and splitting only becomes definitive at the last stage.

Most differentiations of either subspecies or species (all included in speciation in customary usage) represent occupation of subzones within an adaptive zone, or niches of a biotope or biochore in ecological terms, or of adjacent minor peaks on one part of the selection landscape in Sewall Wright's symbolism (Chapter V). Two processes are particularly involved, although, as usual, the two grade into each other and are by no means clear-cut alternatives. One is the differentiation of a more broadly adapted population into temporarily (subspecies) or permanently (species) separate populations each more narrowly adapted to part of the original adaptive range (Fig. 49A). The other is spread from a parental population, which may be either broadly or narrowly adapted, into adjacent different adaptive subzones with temporary or permanent differentiation of populations in those subzones (Fig. 49B).

The direction of evolution in such events, even though locally adaptive, is often shifting, erratic, and, until or unless definitive separation of species has occurred, readily and quickly reversible. The whole event depends largely, often exclusively, on variation already present in the parental population. The materials for evolution are segregated more than they are progressively changed.

Such is the speciation usually studied by experimentalists and neontologists. It has given rise to queries whether species are on their way to becoming higher categories, whether speciation is in fact involved in progressive change that does have more or less consistent direction. Some verbal confusion is involved in the argument, but apart

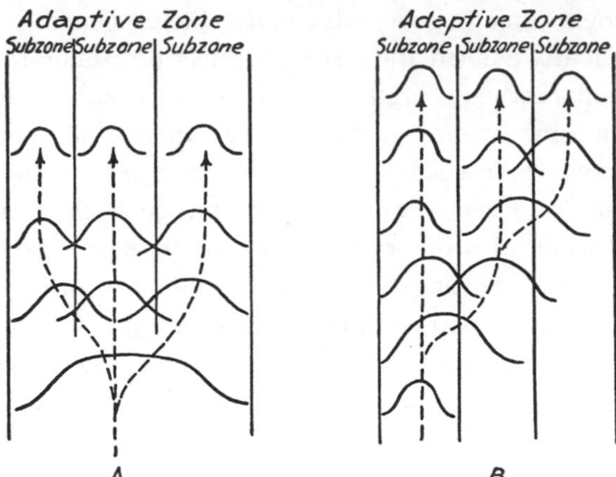

FIGURE 49. TWO PATTERNS OF SPECIATION. A, a single widespread population becoming differentiated into more specialized, locally adapted groups; B, a local population spreading into adjacent subzones, to each of which descendants become specially adapted. The curves suggest cross-sectional variation patterns in the populations.

from this the questions are pertinent and answerable. These particular, frequent aspects of speciation rarely eventuate in significant progression or in categories higher than genera. They do frequently give rise to genera in the largely formal sense that when an ancestral species has proliferated into a number of similar but more specialized or differently specialized descendant species, systematists may find it convenient to call the species-cluster a genus. When speciation is completed in the process, then there has literally been definitive splitting, but this is not a splitting of phyletic lineages that has much if any significance in a broader phylogenetic pattern.

More broadly significant splitting is not the usual result of speciation but is a special or limiting case. Intergradation with the more common cases is seen when successive occupation of adaptive subzones has a more or less consistent adaptive trend. If this trend is gradational, it will (at least at first) be an ecocline, which will correspond with a chronocline in the temporal sequence of changing populations. It may be equally or more frequent for progression to be more or less stepwise, but that makes no difference from the present point of view. The result is a population that is adaptively quite different from the original parental population and not merely a more or less casual adaptive ad-

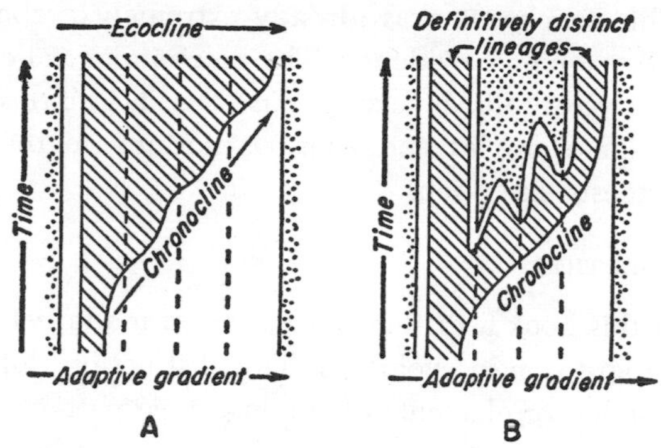

FIGURE 50. ONE PROCESS OF DEFINITIVE SPLITTING OF LINEAGES AND ITS RELATIONSHIP TO ONE OF THE PATTERNS OF SPECIATION. A, progressive speciational occupation of an adaptive gradient. B, same with unstable intermediate populations dying out. In both A and B, populations are represented by a cross-hatched area spreading on an adaptive grid. It is immaterial for the comparison whether the subzonal populations of A become species, subspecies, or are continuous populations across the cline. A compares with Figure 49B.

justment of the original type to special or local conditions.

The divergent population may reach a threshold and cross into a zone discontinuous from the parental zone, or populations along the cline or graded steps may die out—their position is likely to be somewhat unstable relative to that of the parental or of the most progressive divergent populations. In either case, the result is sharp and now discontinuous differentiation of two (or more; the process may of course be multiple) lineages of different adaptive type (Fig. 50). This is splitting of lineages in a fully definitive sense and in a way essentially significant in the broad picture of phylogeny. It is not the inevitable or the usual result of speciation, but is rooted in speciation and cannot occur without it. Except for the different adaptive circumstances, its factors and processes are exactly the same as in "ordinary" speciation. Progression outside the parental zone involves prospective adaptation; but, as already discussed (Chapter VI), this is not a special evolutionary factor but only an *ex post facto* labeling of characteristics that took on special significance because of the adaptive circumstances in which they happened to occur.

Phylogenetic splitting of lineages, including those from which higher categories up to the highest later develop, thus occurs by speciation at their bases. This conclusion was already extremely probable from all the genetic and taxonomic evidence. The paleontological evidence cannot exclude the possibility of exceptions, but it confirms the conclusion in particular examples and there is nothing in the record that requires or suggests exceptions.

PHYLETIC EVOLUTION

Throughout this book a distinction has been made between diversification and progression in evolution. This distinction rests on the two evolutionary modes, or elements of phylogenetic pattern, splitting and phyletic evolution.[3] In extreme cases, the two may be fully distinct. When an ancestral species splits into two or more descendant species by segregation of existing variation, phyletic evolution is not an essential part of the picture; and when a single population undergoes extensive, cumulative change, there is no splitting. Other cases show, again, that the distinction is not absolute, that we are simply looking at two intimately connected parts of the whole. It is an essential part of splitting that differences arise between the two or more groups thus separated. With respect to any one of these groups, its characters, including those that are differential, arose in a single ancestral-descendant sequence, hence by phyletic evolution. On the other hand, phyletic evolution seldom continues long without splitting also occurring by speciation in the populations involved, and the line of longest or greatest phyletic change may take now one, now another branch of a recurrently splitting sequence. It would thus be incorrect to say that splitting or that speciation is not involved in phyletic evolution. Nevertheless, these are different elements in the phylogenetic pattern and may be studied as such.

Phyletic evolution may involve random, fluctuating, and reversible change such as occurs within populations at the levels where all materials for evolution arise. Its typical aspect is, however, a broader one,

[3] Wright (1949b and elsewhere), one of the relatively few geneticists to make and stress a really clear distinction between speciation and phyletic evolution, has used the term "transformation" for the latter. Etymologically this is a good choice, but unfortunately that word has long been used in some of the evolutionary, especially paleontological, literature for one special aspect or result sometimes involved in phyletic evolution: radical functional and corresponding morphological change in a structure, such as from terrestrial limb to wing or paddle, or quadrate and articular to auditory ossicles. Rensch's (1947) "anagenesis" also applies to only one aspect of phyletic evolution: the tendency of some major groups to have a fairly consistent and definable evolutionary direction, often exhibited by different lineages in parallel.

in geological spans of time and among definitely separate lineages. On this scale, phyletic evolution is usually characterized by progressive and essentially nonrandom change. (It may be well to remind the reader that in this book "progressive" means "in a progression," "with cumulative change," and has no implication as to direction or desirability of change, or progress.) Its features that are cumulative and that have persistence of direction are evidently nonrandom and apparently are always adaptive.

Phyletic evolution is particularly evident to paleontologists and the most abundant examples from the fossil record illustrate this aspect of evolution more clearly than any other. It has been largely exemplified and many of its characteristics have been discussed in various connections throughout this book. Details previously given need not be repeated here.

In the present context, as a major element in phylogenetic patterns, phyletic evolution has as its definitive characteristic the continuous maintenance of adaptation in ancestral and descendant populations. This maintenance also has different aspects or occurs in different ways, among them the following:

(1). Adequate adaptation requiring little or no progressive change in the organisms (and with inhibition of change in them) to meet such changes as occur in a relatively stable environment: arrested evolution (Chapter X).

(2). Progressive, more or less long-continued, cumulative change corresponding with secular environmental change (including changes in the populations, themselves, as in self-braking trends): trends (Chapter VIII).

(3). Short-range adaptation to occasional or nonrecurrent environmental changes: casual and episodic change (involved in discussion in Chapters VI, VII, and elsewhere).

(4). Short-range and relatively rapid shift from the ancestral adaptive zone into another that happens to be available, which may in some cases and not in others involve maintenance of adaptation in the face of radical environmental change: quantum evolution (below).

It is evident that these are not four separate things or different sorts of evolution. They are all simply the maintenance of adaptation in various circumstances, and these circumstances are not of four kinds or combinations but are along different parts of a continuous scale.

Arrested evolution may be indistinguishable from a very slow trend; a short trend may be indistinguishable from casual or episodic change; and a relatively rapid and striking episodic change may be indistinguishable from quantum evolution.

On this scale, the great majority of documented examples fall along the range of (2) and (3). This is a correct weighting in that phyletic evolution is in this range for longer periods and in more different lineages. Examples near the bottom end of the scale, (1), are also rather well documented in many cases, but they are less common and their over-all importance in the history of life has not been very great, although they do cast some special light on factors with a wider bearing. Examples near the top end of the scale, (4), on the other hand, have had an extremely important role in the whole of evolution, even though they are less common than (2) or (3), but they are rarely well documented.

Because such cases are really the most common and also are usually the best documented, most paleontological generalizations as to what is "typical" in evolution have been based on instances of phyletic evolution around (2) or (3) on this scale. The false generalization of "orthogenesis" was, for instance, paleontologically supported by phyletic evolution around (2) on the scale, ignoring or denying other cases. Generalizations as to horotelic rates and their distributions are based essentially on instances around (2) and (3), but this is believed to be logically valid because these are statistical generalizations about the most frequent cases and the occurrence of the less common cases is explicitly recognized. The phenomenon of phyletic evolution has led to some lack of mutual comprehension between neontologists, who do not deal with it, and paleontologists, who have to deal with it constantly. It also introduces some complications and difficulties into the theory and practice of systematics. If a single lineage is being considered and if it changes so that the descendant and ancestral populations differ about as much as is usual between two related contemporaneous species, we say that one species has given rise to another. This, the essence of change in phyletic evolution, is at least as fundamental for evolution in general as is the separation of two species from what was originally one. Both processes produce nominally new species, but they are distinctly different and only the latter is speciation in the usual sense of the word.

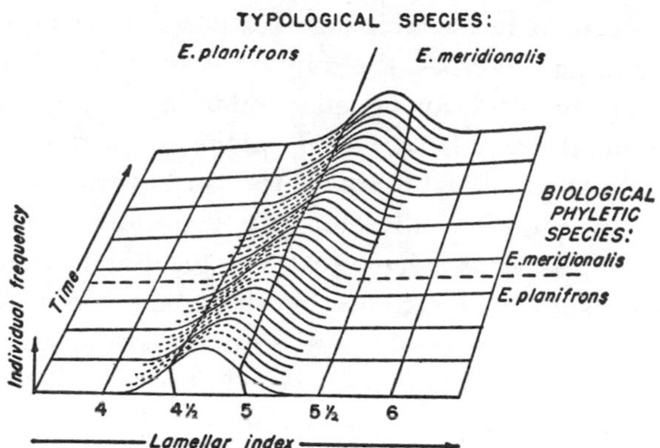

FIGURE 51. TRANSITIONAL LINEAGE BETWEEN TWO SPECIES OF EUROPEAN FOSSIL ELEPHANTS. See text for explanation. (After Trevisan, 1949, with modifications.)

An example, one of many available, may be drawn from the extinct European elephants as summarized by Trevisan (1949). In the lineage of *Elephas planifrons–E. meridionalis* there is rather rapid progression in molar structure indicated by change in lamellar index (number of enamel plates per 10 centimeters of tooth length) from about 4½ to about 5½. As would be expected, the change occurs as a gradual shift of the mode in variable populations (Fig. 51). This whole sequence is a single *evolutionary species,* one lineage with a unitary role in evolution, and at any one time it is a single species by any current neontological definition, genetical or morphological. Nevertheless the early and late forms differ as much as species usually do, and if taxonomy is to be significant and useful they must be distinguished.

The solution of the old typological systematics (followed, e.g., by Osborn, 1942, for fossil elephants), which classified individuals by comparison with types, was to put all specimens more like the type of *planifrons,* i.e., with lamellar index nearer 4½, in that species and all more like the type of *meridionalis,* index nearer 5½, in that one. This makes it appear that two species coexisted during the transition shown in Figure 51, but that *planifrons* steadily became less and *meridionalis* more abundant. This seriously misrepresents the real situation, in which there is only one interbreeding population throughout, and these "species" have no biological significance.

In the new systematics, which classifies populations as a whole, not individuals, and has no types, the solution is to draw a line in time at the point when the mode for lamellar index in the population was 5. Before this time, the populations are *E. planifrons* and after it they are *E. meridionalis*. Admittedly this procedure is just as arbitrary as the other —both separate species by division of a continuum—but it has the merit of making species that have real biological and evolutionary significance. All members of actually interbreeding populations (necessarily contemporaneous) are referred to one species, and the species are arbitrary but objectively real segments of an evolutionary lineage. It also happens that the procedure is more useful because it permits fine stratigraphic subdivision from the paleontological data and the typological procedure does not.[4]

This technical problem of classification is worth mentioning, but the important point here is a consequence of the fact that phyletic evolution is a continuum. When in this way one group gives rise to another, a species to a new species, one genus to another, etc., the relationship between the two species, genera, etc., is radically different from that between two species as defined and studied by neontologists and experimentalists. The latter are contemporaneous, genetically isolated groups; they are different phyletic lineages. The successive taxonomic units of phyletic evolution are segments of single lineages and of larger continuous, intergrading groups. They are not the same sorts of things at all, even though the same taxonomic terms are currently applied to them. That important point suffices for present purposes, but there are numerous other theoretical and practical problems involved in such situations (see Simpson, 1951b, and references there).[5]

[4] Parenthetically it may be noted that the typological procedure places some siblings in different species but that the biologically sounder method theoretically does even worse: it not only places some parents and their offspring in different species, but also places some single individuals in different species at different times in their lives. However, anomalies cannot be avoided when a line must be drawn in a continuum, and the latter inconsistency affects only one generation while the typological inconsistency affects all those in the transitional sequence. Moreover, in most cases in practice there is some hiatus in the record and the time line between species is drawn there.

[5] One other parenthetical point may be made, however: the time-line division of successive species makes it appear that one species arose from another over night, or literally instantaneously. In somewhat more veiled form, this semasiological artifact has been used to support the view that taxonomic units arise by saltation. The fallacy is obvious from this illustration.

QUANTUM EVOLUTION

The point emphasized by applying the term "quantum" to a particular set of evolutionary events is that these involve an all-or-none reaction. They are changes of adaptive zone such that transitional forms between the old zone and the new cannot, or at any rate do not, persist. Populations tending toward the new zone are carried fully into it. It has been emphasized above that this is not a different sort of evolution from phyletic evolution, or even a distinctly different element of the total phylogenetic pattern. It is a special, more or less extreme and limiting case of phyletic evolution.

That such evolutionary events do occur is well attested. Most of the evidence for them has been summarized in preceding chapters. Major incidents of quantum evolution have systematically poor records, for reasons discussed in Chapter XI. Nevertheless, we do have many partial records of quantum evolution, even at high levels (e.g., origin of classes), which can be completed by sound and unequivocal inference. We also do have fully complete and continuous records for some quantum shifts at intermediate and lower levels. The changes from brachydonty to hypsodonty and between different foot mechanisms in the Equidae are such examples, sufficiently summarized on earlier pages. Less complete records which nevertheless seem clearly to involve quantum evolution and have been well studied from this point of view include the rise of the Stylinodontinae (Patterson, 1949), also previously mentioned, and of the Artiodactyla (Schaeffer, 1947, 1948).

Quantum evolution may lead to a new group at any taxonomic level. It is probable that species, either genetic or phyletic, often arise in this way. Certainly genera and all higher categories may do so. The phenomenon naturally becomes clearer and more readily definable when the change in adaptation and structure is relatively large, and such changes commonly eventuate in the development of higher categories. There is no level at which clear-cut quantum evolution is the only mode of origin of new groups, but at high levels some element of quantum evolution is usually involved. That is the most important point about this mode of evolution and one of the reasons for its separate designation and special study.

Most of the factors involved in quantum evolution were discussed in

Chapters VI–VII and now need only brief review for reconsideration in a somewhat different context. The all-or-none element in quantum evolution arises from discontinuity between adaptive zones, discontinuity effectively existing from the start or developing pari passu with transition to the zone. As noted in Chapter VII, the latter situation seems to be more common, but the former is probable in some of the greater and more rapid shifts. In either case there must be some prospective adaptation, a threshold, and postadaptation (Chapter VI).

The most obscure and controversial point here is whether prospective adaptation as prelude to quantum evolution arises adaptively or inadaptively. It was concluded above that it usually arises adaptively, and this seems to be true in such examples of quantum evolution as have been carefully studied. At least, these examples can be interpreted as involving adaptive change only, although an inadaptive phase or element is not ruled out. In the nature of things, it is practically impossible to determine certainly whether a key mutation, for instance, was inadaptive or adaptive when it arose. The fact that it is retrospectively labeled "key mutation" means that it rather rapidly became involved in adaptation, whatever was its status to begin with. The precise role of, say, genetic drift in this process thus is largely speculative at present. It may have an essential part or none. It surely is not involved in all cases of quantum evolution, but there is a strong possibility that it is often involved. If or when it is involved, it is an initiating mechanism. Drift can only rarely, and only for lower categories, have completed the transition to a new adaptive zone (Chapter XI). One other point to be added is that an inadaptive phase can arise by lag when change of environment occurs more rapidly than population change can follow (Chapter IX). Like genetic drift, this phenomenon has extinction as its usual result but it, too, is at least a possible initiating mechanism for quantum evolution.

If a new adaptive type arises by evolution of adaptation in a zone or by episodic phyletic change of a whole population, a threshold effect may not occur and in any case is not likely to be apparent. Clear-cut quantum evolution will not then occur, and the evolutionary event grades into (3) of the continuous scale of phyletic evolution given above. However, in most cases where the ancestral adaptive type continues, and sometimes when it does not, a threshold effect does occur and the occupation of a new zone may be more clearly identified as quantum evolution. When there is selection both for the ancestral adaptation and for the new adaptation, there must be a point where the two balance. That is necessarily a point of unstable equilibrium, and by definition it is the threshold.

The threshold may occur early, middle, or late in the actual transitional sequence. It is most likely to occur early, and it may actually be crossed with the rise of the first prospective adaptation. A key mutation may, for instance, be subject to selection away from ancestral type immediately when it arises, and if so, the threshold is already crossed by the change produced by that mutation. This crossing of the threshold forthwith, which I think must be relatively uncommon, eliminates much of the instability of the threshold effect, although not all, since a mutation sufficiently "large" to be a key to adaptive change must harmonize poorly with its original genetic background. I cannot fully agree with Patterson (1949) that a one-mutation crossing of the threshold, if and when it occurs, means that no threshold is involved.

The populations making a quantum shift do not lose adaptation altogether; to do so is to become extinct. It is also clear that the direction of change is adaptive, unless at the very beginning in such shifts as may be initiated by drift. Indeed the relatively rapid change in such a shift is more rigidly adaptive than are slower phases of phyletic change, for the direction and the rate of change result from strong selection pressure once the threshold is crossed. Yet the very fact that selection pressure is strong can only be a concomitant of movement from a more poorly to a better adapted status. Selection is not linear but centripetal when adaptation is perfected. It is then "stabilizing selection" (Schmalhausen). The quantum change is a break-through from one position of stabilizing selection to another. To this degree and in this sense, the status of the population during the post-threshold stage, at least, of the shift is inadaptive relative to the new stable status when postadaptation is completed.

This sort of effect seems to me clearly involved in the origin of the artiodactyl tarsus, which was the basic feature that led to expansion of artiodactyls into a high category (an order). This was an all-or-none reaction, not in the sense that there was a "stage that the tarsus had to attain to prevent the actual elimination of the trend," but in the sense that once the direction of evolution of the tarsus was established (and the point at which it was established was a threshold), selection continuously and strongly acted in that direction until completion of the quantum step. For that reason, the change did go on quickly (the first artiodactyls differ hardly at all from contemporaneous and slightly earlier condylarths except in the tarsus and functionally related characters) and to an "apparently absolute limit" of biomechanical efficiency (since the first appearance of this mechanism it has undergone no essential change in any of the otherwise extremely varied artiodactyls).[6] No intermediate stage persisted, because intermediate stages were less efficient (i.e., *relatively* inadaptive).

The taeniodonts illustrate one additional point, illuminating as regards the relationship between quantum evolution and trend evolution, both of which are special cases of phyletic evolution. In this small order there are two adaptive types, the development of which is such that taxonomists label the two groups as subfamilies. One subfamily, Stylinodontinae, arose from the other, Conoryctinae, by a quantum shift (Patterson, 1949, and Fig. 52, here). After the separation, some similar skull and jaw characters (associated with powerful canines) arose in both, but much more slowly in the Conoryctinae. Primary characters of the quantum shift, enlarged claws and powerful limbs, did not develop in the Conoryctinae. Thus selection in some respects acted similarly on the related lineages and produced similar phyletic evolution at different rates. But in only one lineage did rapid shift of adaptive type occur on the basis of a prospective adaptation absent (or eliminated) in the other.

Quantum evolution usually is and at some level it may always be involved in the opening or so-called "explosive" phase of adaptive radiation. The relative rapidity with which a variety of adaptive zones are then occupied seems quite inexplicable except by a series of divergent and also, often, successive quantum shifts into the varied zones. The rates thereafter slow down. Stable current adaptation has been achieved. Quieter phyletic evolution ensues, often with long trends, sometimes with episodic changes of direction, occasionally with virtual cessation of change and drop to bradytelic rates. The explosive phase is rooted in significant splitting of lineages, behind which is speciation. All the

[6] Quotations are from Schaeffer (1948) and my interpretation depends on his data and discussion. It differs from his own conclusions in some respects, but I think the difference is more verbal than fundamental. His discussion has, incidentally, been instrumental in modification of my formerly more extreme views on quantum evolution.

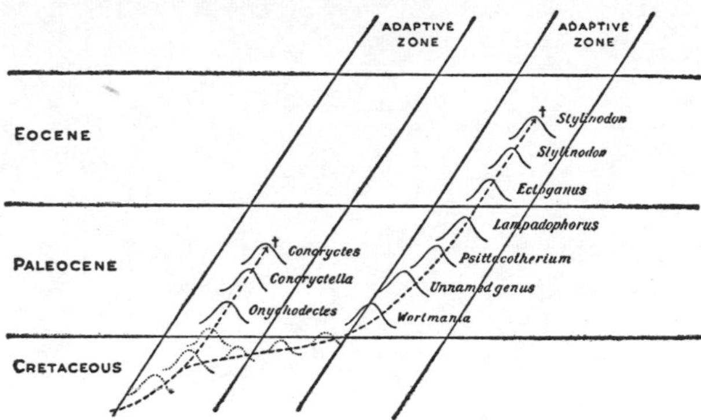

FIGURE 52. EVOLUTION OF THE TAENIODONTA. The Stylinodontinae (right) are shown as originating by quantum evolution from the Conoryctinae (left). (From Patterson, 1949.)

time, but especially during the longer, later phase of phyletic evolution, speciation of a less significant sort accompanies the process, filling the niches or local adaptive peaks of the constantly shifting environment.

In adaptive radiation and in every part of the whole, wonderful history of life, all the modes and all the factors of evolution are inextricably interwoven. The total process cannot be made simple, but it can be analyzed in part. It is not understood in all its appalling intricacy, but some understanding is in our grasp, and we may trust our own powers to obtain more.

Works Cited

Abel, O. 1911. Grundzüge der Palaeobiologie der Wirbeltiere. Stuttgart, Schweizerbart.

——— 1928. Das biologische Trägheitsgesetz. Biol. Gen., 4: 1–102.

——— 1929. Paläobiologie und Stammesgeschichte. Jena, Fischer.

——— 1935. Vorzeitliche Lebensspuren. Jena, Fischer.

Allee, W. C. 1943. Where angels fear to tread: a contribution from general sociology to human ethics. Science, 97: 514–525.

Allee, W. C., A. E. Emerson, O. Park, T. Park, and K. P. Schmidt. 1949. Principles of animal ecology. Philadelphia, Saunders.

Amadon, D. 1950. The Hawaiian honeycreepers (Aves, Drepaniidae). Bull. Amer. Mus. Nat. Hist., 95: 151–262.

Anderson, E. 1948. Hybridization of the environment. Evolution, 2: 1–9.

Arambourg, C. 1950. Le problème de l'extinction des espèces et des groupes. Colloques Internat. Centre Nat. Rech. Sci., 21: 89–111.

Ashton, E. H. and S. Zuckerman. 1950. The influence of geographic isolation on the skull of the green monkey (*Cercopithicus aethiops sabaeus*) I. A comparison between the teeth of the St. Kitts and the African green monkey. Proc. Roy. Soc. London, B, 137: 212–238.

Bailey, V. 1931. Mammals of New Mexico. U.S. Dept. Agric., Bur. Biol. Surv., North American Fauna, No. 53.

Baldwin, J.M. 1896. A new factor in evolution. Amer. Nat., 30: 441–451, 536–553.

Basse, E. 1938. Sur une nouvelle espèce de *Valenciennesia* (pulmoné thalassophile) du Cénomanien malgache: *V. madagascariensis* n. sp. Bull. Mus. Hist. Nat. (Paris), 10: 659–661.

Beadle, G. W. 1945. Biochemical genetics. Chem. Rev., 37: 15–96.

Beadle, G. W., and B. Ephrussi. 1936. The differentiation of eye pigments in *Drosophila* as studied by transplantation. Genetics, 21: 225–247.

Beirne, B. P. 1947. The history of the British land mammals. Ann. Mag. Nat. Hist., 11th ser., 14: 501–516.

Bertram, G. C. L. 1940. The biology of the Weddell and crabeater seals. Brit. Mus. (Nat. Hist.), Brit. Graham Land Exped., Sci. Repts., I (No. 1): 1–139.

Beurlen, K. 1929. Parallelentwicklung und Iterationen bei Decapoden. Palaeont. Zeitschr., 11: 50–52.

Blair, W. F. 1941. Techniques for study of mammal populations. Jour. Mammalogy, 22: 148–157.

Blum, H. F. 1951. Time's arrow and evolution. Princeton, Princeton Univ. Press.

Bohlin, B. 1940. Food habits of the machaerodonts, with special regard to *Smilodon*. Bull. Geol. Inst. Upsala, 28: 156–174.

Brinkmann, R. 1929. Statistisch-biostratigraphische Untersuchungen an mitteljurassischen Ammoniten über Artbegriff und Stammesentwicklung. Abh. Ges. Wiss. Göttingen, Math.-Phys. Kl., N.F., Vol. 8, Part 3.

Broom, R. 1932. The mammal-like reptiles of South Africa and the origin of mammals. London, H. F. and G. Witherby.

——— 1933. Evolution—Is there intelligence behind it? South Afr. Jour. Sci., 30: 1–19.

Bulman, O. M. B. 1933. Programme evolution in the graptolites. Biol. Rev., 8: 311–334.

Bumpus, H. G. 1898. The elimination of the unfit as illustrated by the introduced sparrow, *Passer domesticus*. Biol. Lectures Woods Hole Marine Biol. Lab., 6: 209–226.

Butler, P. M. 1939. Studies of the mammalian dentition; differentiation of the post-canine dentition. Proc. Zool. Soc. London, Ser. B, 109 (Pt. 1): 1–36.

——— 1946. The evolution of carnassial dentitions in the Mammalia. Proc. Zool. Soc. London, 116: 198–220.

Bystrow, A. P. 1938. Dvinosaurus als neotenische Form der Stegocephalen. Acta Zool., 19: 209–295.

Cain, A. J. and P. M. Sheppard. 1950. Selection in the polymorphic land snail *Cepaea nemoralis*. Heredity, 4: 275–294.

Cain, S. A. 1944. Foundations of plant geography. New York, Harper.

Camp, Charles L., and Natasha Smith, 1942. Phylogeny and functions of the digital ligaments of the horse. Mem. Univ. California, 13: 69–124.

Carter, G. S. 1951. Animal evolution, a study of recent views of its causes. London, Sidgwick and Jackson.

Caspari, E. 1948. Cytoplasmic inheritance. Advances in Genetics, 2: 1–66.

Chaney, R. 1948. The bearing of the living *Metasequoia* on problems of Tertiary paleobotany. Proc. Nat. Acad. Sci., 34: 503–515.

Charles, D. R., and R. H. Goodwin. 1943. An estimate of the minimum number of genes differentiating two species of golden-rod with respect to their morphological characters. Amer. Nat., 77: 53–69.

Clark, A. H. 1930. The new evolution. Zoogenesis. Baltimore, Williams and Wilkins.

Clark, W. E. Le Gros, and P. B. Medawar. 1945. Essays on growth and form presented to D'Arcy Wentworth Thompson. Oxford, Clarendon Press.

Cloud, P. E., Jr. 1949. Some problems and patterns of evolution exemplified by fossil invertebrates. Evolution, 2: 322–350.

Colbert, E. H. 1946. *Sebecus,* representative of a peculiar suborder of fossil Crocodilia from Patagonia. Bull. Amer. Mus. Nat. Hist., 87: 217–270.

——— 1946. Temperature tolerances in the American alligator and their bearing on the habits, evolution, and extinction of dinosaurs. Bull. Amer. Mus. Nat. Hist., 86: 327–374.

——— 1949a. Some paleontological principles significant in human evolution. Studies Phys. Anthrop., 1: 103–149.

——— 1949b. Progressive adaptations as seen in the fossil record. *In* Jepsen, Mayr, and Simpson, 1949: 390–402.

Cope, E. D. 1886. The origin of the fittest. New York, Appleton.

Corset, J. 1931. Les coaptations chez les insectes. Bull. Biol. France Belgique, Suppl., 13: 1–337.

Cowles, R. B. 1945. Heat-induced sterility and its possible bearing on evolution. Amer. Nat., 79: 160–175.

Crampton, H. E. 1932. Studies on the variation, distribution, and evolution of the genus *Partula;* the species inhabiting Moorea. Pub. Carnegie Inst. Washington, No. 410, pp. 1–335.

Crombie, A. C. 1947. Interspecific competition. Jour. Animal Ecol., 16: 44–73.

Cuénot, L. 1921. La Genèse des espèces animales. Paris, Félix Alcan.

——— 1925. L'Adaptation. Paris, G. Doin.

——— 1941. Invention et finalité en biologie. Paris, Flammarion.

Cuénot, L. (with the collaboration of A. Tétry). 1951. L'évolution biologique, les faits, les incertitudes. Paris, Masson.

Dacqué, E. 1935. Organische Morphologie und Paläontologie. Berlin, Borntraeger.

Darlington, C. D. 1939. The evolution of genetic systems. Cambridge, Cambridge University Press.

Darlington, C. D., and K. Mather. 1950. The elements of genetics. New York, Macmillan.

Darwin, C. 1859. On the origin of species by means of natural selection, or the preservation of favored races in the struggle for life. London, Murray.

Davis, D. D. 1949. Comparative anatomy and the evolution of vertebrates. *In* Jepsen, Mayr, and Simpson, 1949: 64–89.

De Beer, G. R. 1951. Embryos and ancestors. Revised edition. Oxford, Oxford Univ. Press.

Degerbol, M. 1939. The field mouse of Iceland, its systematic position (*Apodemus sylvaticus grandiculus* subsp. nov.) and biology. *In* The zoology of Iceland, vol. 4, pt. 76: 39–51.

Dementiev, G. P., and V. F. Larionov. 1945. The development of geographical color variations with special reference to birds. Proc. Zool. Soc. London, 115: 85–96.

De Vries, H. 1901. Die Mutationstheorie. Leipzig, Veit.

Dice, L. R. 1947. Effectiveness of selection by owls of deer-mice (*Peromyscus maniculatus*) which contrast in color with their background. Cont. Lab. Vert. Biol. Univ. Michigan, No. 34: 1–20.

——— 1949. The selection index and its test of significance. Evolution, 3: 262–265.

Dietrich, W. O. 1949. Stetigkeit und Unstetigkeit in der Pferdegeschichte. Neues Jahrb. Mineral. etc., Abh. (B), 91: 121–148.

Dobzhansky, T. 1937. Genetics and the origin of species. New York, Columbia University Press.

——— 1941. Same. 2d ed.

——— 1951. Same. 3d ed.

——— 1947. Adaptive changes induced by natural selection in wild populations of *Drosophila*. Evolution, 1: 1–16.

Dobzhansky, T., and H. Levene. 1948. Genetics of natural populations. XVII. Proof of operation of natural selection in wild populations of *Drosophila pseudoobscura*. Genetics, 33: 537–547.

Dobzhansky, T., and B. Spassky. 1947. Evolutionary changes in laboratory cultures of *Drosophila pseudoobscura*. Evolution, 1: 191–216.

Dobzhansky, T., and C. C. Tan, 1936. Studies on hybrid sterility. III. A comparison of the gene arrangement in two species, *Drosophila pseudoobscura* and *Drosophila miranda*. Zeit. indukt. Abstamm.-u. Vererblehre., 72: 88–114.

Dollo, L. 1896. Sur la phylogénie des dipneustes. Bull. Soc. Belge Géol. Pal. Hydr., 9: 79–128.

Doutt, J. Kenneth. 1942. A review of the genus *Phoca*. Ann. Carnegie Mus., 29: 61–125.

Dubinin, N. P., and others. 1934. Experimental study of the ecogenotypes of *Drosophila melanogaster*. Biol. Zhurn. (Moscow), 3: 166–216.

Dunn, Emmett Reid. 1935. The survival value of specific characters. Copeia, 1935, No. 2, pp. 85–98.

——— 1942. Survival value of varietal characters in snakes. Amer. Nat., 76: 104–109.

Dunn, L. C. (editor). 1951. Genetics in the 20th century. New York, Macmillan.

Du Noüy, L. 1947. Human destiny. New York, Longmans.

Edinger, T. 1942. The pituitary body in giant animals fossil and living: a survey and a suggestion. Quart. Rev. Biol., 17: 31–45.

——— 1948: Evolution of the horse brain. Mem. Geol. Soc. Amer., No. 25.

Eigenmann, C. H. 1909. Adaptation. *In* Fifty years of Darwinism. Henry Holt & Co. New York. Pp. 182–208.

Elias, M. K. 1942. Tertiary prairie grasses and other herbs from the high plains. Spec. Papers Geol. Soc. Amer., No. 41: 1–176.

Elton, C. S. 1924. Periodic fluctuations in the number of animals; their causes and effects. Brit. Jour. Exper. Biol., 3: 119–163.

——— 1933. The ecology of animals. London, Methuen.

——— 1942. Voles, mice and lemmings. Problems in population dynamics. Oxford, Clarendon Press.

Emerson, A. E. 1949. Ecology and evolution. *In* Allee, Emerson, Park, Park, and Schmidt, 1949: 598–729. (This section of the book is not separately signed by Emerson, but his responsibility for it is noted, and citation is accordingly simplified here.)

Eyster, W. H. 1931. Vivipary in maize. Genetics, 16: 574–590.

Fenton, C. L. 1935. Factors of evolution in fossil series. Amer. Nat., 69: 139–173.

Fisher, R. A. 1930. The genetical theory of natural selection. Oxford, Clarendon Press.

Florkin, M. 1949. Biochemical evolution. (Edited, translated, and augmented by S. Morgulis.) New York, Academic Press.

Ford, E. B. 1945a. Butterflies. London, Collins.

——— 1945b. Polymorphism. Biol. Rev., 20: 73–88.

Ford, H. D., and E. B. Ford. 1930. Fluctuation in numbers and its influence on variation in *Melitaea aurinia*. Trans. Ent. Soc. London, 78: 345–351.

Gause, G. F. 1934. The struggle for existence. Baltimore, Williams and Wilkins.

Gazin, C. Lewis. 1941. The mammalian faunas of the Paleocene of central Utah, with notes on the geology. Proc. U.S. Nat. Mus., 91: 1–53.

George, T. N. 1948. Evolution in fossil communities. Proc. Roy. Phil. Soc. Glasgow, 73: 23–42.

Gilluly, J. 1949. Distribution of mountain building in geologic time. Bull. Geol. Soc. Amer., 60: 561–590.

Goldschmidt, R. 1938. Physiological genetics. New York and London, McGraw-Hill.

——— 1940. The material basis of evolution. New Haven, Yale Univ. Press.

——— 1945. Mimetic polymorphism. A controversial chapter of Darwinism. Quart. Rev. Biol., 20: 147–164, 205–230.

——— 1946. "An empirical evolutionary generalization" viewed from the standpoint of phenogenetics. Amer. Nat., 80: 305–317.

Gorjanovic-Kramberger, K. 1901. Ueber die Gattung *Valenciennesia* und einige unterpontische Limnaeen. Beitr. Paläont. Geol. Öst.-Ung., 13: 121–140.

——— 1923. Ueber die Bedeutung der Valenciennesiiden in stratigraphischer und genetischer Hinsicht. Paläont. Zeit., 5: 339–344.

Gray, S. W. 1946. Relative growth in a phylogenetic series and in an ontogenetic series of one of its members. Amer. Jour. Sci., 244: 792–807.

Green, E. L., and M. C. Green. 1946. Effect of the short ear gene on number of ribs and presacral vertebrae in the house mouse. Amer. Nat., 80: 619–625.

Gregory, W. K. 1935a. "Williston's Law" relating to the evolution of skull bones in the vertebrates. Amer. Jour. Phys. Anthrop., 20: 123–152.

——— 1935b. Reduplication in evolution. Quart. Rev. Biol., 10: 272–290.

——— 1935c. The roles of undeviating evolution and transformation in the origin of man. Amer. Nat., 69: 285–404.

——— 1936. On the meaning and limits of irreversibility of evolution. Amer. Nat., 70: 517–528.

Gregory, W. K. 1951. Evolution emerging. 2 vols. New York, Macmillan.

Grüneberg, H. 1947. Animal genetics and medicine. London, Paul B. Hoeber, Inc.

——— 1948. Genes and pathological development in mammals. *In* Symposia of the Soc. for Exper. Biol., No. 2, Growth in relation to differentiation and morphogenesis: 155–176.

Günther, K. 1949. Über Evolutionsfaktoren und die Bedeutung des Begriffs der "ökologischen Lizenz" für die Erklarung von Formenerscheinungen im Tierreich. *In* Ornithologie als biologische Wissenschaft (Festsch. zu . . . Stresemann): Pp. 23–54.

Haas, O., and G. G. Simpson. 1946. Analysis of some phylogenetic terms with attempts at redefinition. Proc. Amer. Phil. Soc., 90: 319–349.

Haldane, J. B. S. 1932. The causes of evolution. New York and London, Harper.

——— 1942. New paths in genetics. New York and London, Harper.

——— 1949. Suggestions as to quantitative measurement of rates of evolution. Evolution, 3: 51–56.

Hamilton, W. J., Jr. 1939. American mammals. New York and London, McGraw-Hill.

Heberer, G. (editor). 1943a. Die Evolution der Organismen, Ergebnisse und Probleme der Abstammungslehre. Jena, Fischer.

——— 1943b. Das Typenproblem in der Stammesgeschichte. *In* Heberer, 1934a: 545–585.

Hegner, R. W., and K. A. Stiles. 1951. College zoology. New York, Macmillan.

Heilmann. G. 1936. The origin of birds. London, Witherby.

Heim de Balsac, H. 1936. Biogéographie des mammifères et des oiseaux de l'Afrique du Nord. Bull. Biol. France et Belgique, Suppl., 21: 1–447.

Heitz, E. 1944. Über einige Frage der Artbildung. Arch. Julius Klaus-Stift. Vererbungsf., 19: 510–528.

Henderson, L. J. 1913. The fitness of the environment. New York, Macmillan.

Hersh, A. H. 1934. Evolutionary relative growth in the titanotheres. Amer. Nat., 68: 537–561.

Hesse, R., W. C. Allee, and K. P. Schmidt. 1951. Ecological animal geography. 2d ed. New York, Wiley.

Heuts, M. J. 1947. Experimental studies on adaptive evolution in *Gasterosteus aculeatus* L. Evolution, 1: 89–102.

——— 1949. On the mechanism and the nature of adaptive evolution. La Ricerca Scientifica, Suppl., 19: 35–42.

Hogben, L. 1946. An introduction to mathematical genetics. New York, W. W. Norton & Co.

Hough, J. 1950. The habits and adaptation of the Oligocene saber tooth carnivore, *Hoplophoneus*. Prof. Paper, U.S. Geol. Surv., No. 221-H: 125–137.

Hovasse, R. 1950. Adaptation et évolution. Paris, Hermann.

Hubbs, C. L. 1934. Racial and individual variation in animals, especially fishes. Amer. Nat., 68: 115–128.

——— 1938. Fishes from the caves of Yucatan. Carnegie Inst. Washington, Pub. No. 491: 261–295.

Hutchinson, G. E. 1945. [Review of Tempo and mode in evolution.] Amer. Jour. Sci., 243: 356–358.

Huxley, J. S. 1932. Problems of relative growth. London, Methuen.

——— 1939. Clines: an auxiliary method in taxonomy. Bijdr. Dierk., 27: 491–520.

——— (editor) 1940. The new systematics. Oxford, Clarendon Press.

——— 1942. Evolution, the modern synthesis. New York, Harper.

Huxley, J. S., J. Needham, and I. M. Lerner. 1941. Terminology of relative growth-rates. Nature, 148–225.

Huxley, T. H. 1870. Lay sermons, addresses, and reviews. New York, D. Appleton & Company.

Ivanow, A. N. 1945. À propos de la phase soit-disant "prophétique" dans l'évolution des Kosmoceratidae. Bull. Soc. Nat. Moscou, n.s., 50: 31–32.

Ives, P. T. 1950. The importance of mutation rate genes in evolution. Evolution, 4: 236–252.

Jacot, A. P. 1932. The status of the genus and the species. Amer. Nat., 66: 346–364.

Jepsen, G. L. 1949. Selection, "orthogenesis," and the fossil record. Proc. Amer. Phil. Soc., 93: 479–500.

Jepsen, G. L., E. Mayr, and G. G. Simpson (editors). 1949. Genetics, paleontology, and evolution. Princeton, Princeton Univ. Press.

Kachkarov, D. N., and E. P. Korovine, translated, revised, and expanded by T. Monod. 1942. La vie dans les déserts. Paris, Payot.

Kaufmann, R. 1933. Variationsstatistische Untersuchungen über die "Artabwandlung" und "Artumbildung" an der oberkambrischen Trilobitengattung *Olenus* Dalm. Abh. Geol.-Pal Inst. Univ. Greifswald, 10: 1–54.

——— 1935. Exakt-statistische Biostratigraphie der Olenus-Arten von Sudöland. Geol. Foren. Stokholm Förhandl., 1935: 19–28.

Kerkis, J. 1936. Chromosome conjugation in hybrids between *Drosophila melanogaster* and *Drosophila simulans*. Amer. Nat., 70: 81–86.

Kirikov, S. V. 1934. Sur la distribution géographique du hamster noir et ses relations avec la forme normale de *Cricetus cricetus*. Zool. Zhurnal, 13: 361–368.

Knopf, A. 1949. Time in earth history. *In* Jepsen, Mayr, and Simpson. 1949: 1–9.

Kowalevsky, W. 1874. Monographie der Gattung *Anthracotherium* Cuv. und Versuch einer naturlichen Classification der fossilen Hufthiere. Paleontographica, N.F., 2 (No. 3): i–iv, 133–285.

Kozlowski, R. 1947. Les affinités des graptolithes. Biol. Rev., 22: 93–108.

Krumbein, W. C., and L. L. Sloss. 1951. Stratigraphy and sedimentation. San Francisco, Freeman.

Lack, D. 1947. Darwin's finches. Cambridge, Cambridge Univ. Press.

Lerner, I. M., and E. R. Dempster. 1948. Some aspects of evolutionary theory in the light of recent work on animal breeding. Evolution, 2: 19–28.

Lotsy, J. P. 1916. Evolution by means of hybridization. The Hague, Martinus Nijhoff.

Lull, R. S., and S. W. Gray. 1949. Growth patterns in the Ceratopsia. Amer. Jour. Sci., 247: 492–503.

Lutz, B. 1947. Trends towards non-aquatic and direct evolution in frogs. Copeia, 1947: 242–252.

——— 1948. Ontogenetic evolution in frogs. Evolution, 2: 29–39.

MacArthur, J. W. 1949. Selection for small and large body size in the house mouse. Genetics, 34: 194–209.

MacLulich, D. A. 1937. Fluctuations in the numbers of the varying hare (*Lepus americanus*). Univ. Toronto Studies, Biol. Ser., No. 43: 1–136.

Mampell, K. 1945. Analysis of a mutator. Genetics, 30: 496–505.

Manton, I. 1950. Problems of cytology and evolution in the Pteridophyta. New York and London, Cambridge Univ. Press.

Mather, K. 1941. Variation and selection of polygenic characters. Jour. Genetics, 41: 159–193.

——— 1949. Biometrical genetics, the study of continuous variation. Dover Publications.

Mather, K., and B. J. Harrison, 1949. The manifold effect of selection. Heredity, 3: 1–52, 131–162.

Matthew, W. D. 1910. The phylogeny of the Felidae. Bull. Amer. Mus. Nat. Hist., 28: 289–316.

——— 1914. Time ratios in the evolution of mammalian phyla; a contribution to the problem of the age of the earth. Science, N.S. 40: 232–235.

——— 1915. Climate and evolution. Ann. N.Y. Acad. Sci., 24: 171–318.

——— 1926. The evolution of the horse; a record and its interpretation. Quart. Rev. Biol., 1: 139–185.

——— 1929. On the phylogeny of horses, dogs, and cats. Science, 69: 494–496.

——— 1937. Paleocene faunas of the San Juan Basin, New Mexico. Trans. Amer. Phil. Soc., N.S., 30: i–viii, 1–510.

——— 1939. Climate and evolution. 2d edition, rev. and enl. Spec. Pub. N.Y. Acad. Sci., 1: i–xii, I–223.

Matthew, W. D., and S. H. Chubb. 1913. Evolution of the horse. Amer. Mus. Guide Leaflet, No. 36.

Mayr, E. 1942. Systematics and the origin of species. New York, Columbia Univ. Press.

——— 1949a. Speciation and selection. Proc. Amer. Phil. Soc., 93: 514–519.

——— 1949b. Speciation and systematics. *In* Jepsen, Mayr, and Simpson, 1949: 281–298.

Miller, A. H. 1947. Panmixia and population size with reference to birds. Evolution, 1: 186–190.

Miller, G. S., Jr., and J. W. Gidley. 1918. Synopsis of the supergeneric groups of rodents. Jour. Washington Acad. Sci., 8: 431–448.

Mohr, O. L. 1932. On the potency of mutant genes and wild-type allelomorphs. Proc. 6th Int. Congr. Genet., 1: 190–212.

Morgan, C. L. 1896. Habit and instinct. London and New York, Edward Arnold.

Mosauer, W. 1932. Adaptive convergence in the sand reptiles of the Sahara and of California; a study in structure and behavior. Copeia, 2: 72–78.

Muller, H. J. 1930. Radiation and genetics. Amer. Nat., 64: 220–251.

——— 1939. Reversibility in evolution considered from the standpoint of genetics. Biol. Rev., 14: 261–280.

——— 1945. Age in relation to the frequency of spontaneous mutations in *Drosophila*. Year Book Amer. Phil. Soc., 1945: 150–153.

——— 1947. The gene. Proc. Roy. Soc., B, 134: 1–37.

——— 1949a. The Darwinian and modern conceptions of natural selection. Proc. Amer. Phil. Soc., 93: 459–470.

——— 1949b. Redintegration of the symposium on genetics, paleontology, and evolution. *In* Jepsen, Mayr, and Simpson, 1949: 421–445.

——— 1950. Evidence of the precision of genetic adaptation. *In* The Harvey Lectures, 43: 165–229. Springfield, Thomas.

Murphy, R. C. 1936. Oceanic birds of South America. 2 vols. New York, Amer. Mus. Nat. Hist.

Newbigin, M. I. 1948. Plant and animal geography. London, Methuen.

Newell, N. D. 1948. Infraspecific categories in invertebrate paleontology. Jour. Paleont., 22: 225–232.

——— 1949. Phyletic size increase—an important trend illustrated by fossil invertebrates. Evolution, 3: 103–124.

——— 1952. Periodicity in invertebrate evolution. Jour. Paleont., 26: 371–385.

Nopcsa, F. 1923. Vorläufige Notiz über Pachyostose und Osteosklerose einiger marinen Wirbeltiere. Anat. Anz., 56: 353–359.

Olson, E. C. 1944. Origin of mammals based upon cranial morphology of the therapsid suborders. Geol. Soc. Amer., Special Papers, No. 55.

Osborn, H. F. 1896. [Abstr.] [A mode of evolution requiring neither natural selection nor the inheritance of acquired characters.] Trans. N.Y. Acad. Sci., vol. 15: 141–142, 148.

——— 1910. The Age of Mammals in Europe, Asia and North America. New York, Macmillan.

——— 1915. Origin of single characters as observed in fossil and living animals and plants. Amer. Nat., 49: 193–240.

——— 1925a. The origin of species as revealed by paleontology. Pp. 1–12. Reprinted from Nature, 115: 925, 926, 961–963.

——— 1925b. The origin of species. Part 2. Distinctions between rectigradations and allometrons. Proc. Nat. Acad. Sci., 11: 749–752.

——— 1926a. The origin of species, 1859–1925. Sci. Monthly, 22: 185–192.

Osborn, H. F. 1926b. The problem of the origin of species as it appeared to Darwin and as it appears to us today. Science. 64: 337–341.

——— 1927. The origin of species; Part 5: speciation and mutation. Amer. Nat., 61: 5–42. [Parts 1–4 of this series are Osborn, 1925a, 1925b, 1926a, and 1926b].

——— 1929. The Titanotheres of ancient Wyoming, Dakota and Nebraska. U.S. Geol. Surv., Mon. 55, 2 vols.

——— 1934. Aristogenesis, the creative principle in the origin of species. Amer. Nat., 68: 193–235.

——— 1936, 1942. Proboscidea. 2 vols. New York, American Museum Press.

Parr, A. E. 1926. Adaptiogenese und Phylogenese: zur Analyse der Anpassungerscheinungen und ihre Entstehung. Abh. Theor. organ. Ent. 1: 1–60.

Patterson, B. 1949. Rates of evolution in taeniodonts. *In* Jepsen, Mayr, and Simpson, 1949: 243–278.

Pearl, Raymond. 1940. Introduction to medical biometry and statistics. 3d ed. Philadelphia and London, Saunders.

Phleger, Fred B., Jr. 1940. Relative growth and vertebrate phylogeny. Amer. Jour. Sci., 238: 643–662.

Phleger, Fred B., Jr., and W. S. Putnam. 1942. Analysis of *Merycoidodon* skulls. Amer. Jour. Sci., 240: 547–566.

Pitelka, F. A. 1951. Principles of animal ecology. Evolution, 5: 81–84. (Review of Allee, Emerson, Park, Park and Schmidt, 1949).

Pittendrigh, C. S. 1948. The bromeliad-anopheles-malaria complex in Trinidad. I—The bromeliad flora. Evolution, 2, 58–89.

Quayle, H. J. 1938. The development of resistance in certain scale insects to hydrocyanic acid. Hilgardia, 11: 183–225.

Reeve, E. C. R. 1950. Genetical aspects of size allometry. Proc. Roy. Soc., B, 137: 515–518.

Reeve, E. C. R., and J. S. Huxley. 1945. Some problems in the study of allometric growth. *In* Clark and Medawar, 1945: 121–156.

Reeve, E. C. R., and P. D. F. Murray. 1942. Evolution in the horse's skull. Nature, 150: 402–403.

Rensch, B. 1929. Das Prinzip geographischer Rassenkreise und das Problem der Artbildung. Berlin, Borntraeger.

——— 1947. Neuere Probleme der Abstammungslehre, die transspezifische Evolution. Stuttgart, Enke.

——— 1948. Histological changes correlated with evolutionary changes in body size. Evolution, 2: 218–230.

Rhoades, M. M. 1938. Effect of the Dt gene on the mutability of the a_1 allele in maize. Genetics, 23: 377–397.

Richards, O. W., and A. J. Kavanagh. 1945. The analysis of growing form. *In* Clark and Medawar, 1945: 188–230.

Riggs, E. S. 1934. A new marsupial saber-tooth from the Pliocene of Argentina and its relationship to other South American predaceous marsupials. Trans. Amer. Phil. Soc., N.S., 24: 1–31.

Robb, R. C. 1935a. A study of mutations in evolution. Part 1: Evolution in the equine skull. Jour. Genetics, 31: 39–46.

——— 1935b. A study of mutations in evolution. Part 2: Ontogeny in the equine skull. Jour. Genetics, 31: 47–52.

——— 1936. A study of mutations in evolution. Part 3: The evolution of the equine foot. Jour. Genetics, 33: 267–273.

——— 1937. A study of mutations in evolution. Part 4: The ontogeny of the equine foot. Jour. Genetics, 34: 477–486.

Robson, G. C., and O. W. Richards, 1936. The variations of animals in nature. London. New York, Toronto, Longmans Green.

Rode, P. 1946–1947. Petit atlas des mammifères. New ed. 4 vols. Paris, Boubée.

Romer, A. S. 1936 [Review of Säve-Söderbergh, 1935]. Jour. Geol, 44: 534–536.

——— 1941. Man and the vertebrates. 3d ed. Chicago, Univ. Chicago Press.

——— 1942. Cartilage an embryonic adaptation. Amer. Nat., 76: 394–404.

——— 1945. Vertebrate paleontology. 2d ed. Chicago, Univ. Chicago Press.

——— 1948. Relative growth in pelycosaurian reptiles. *In* Robert Broom Commemorative volume, special Publ., Roy. Soc. South Africa, Cape Town. Pp. 45–55.

——— 1949a. Time series and trends in animal evolution. *In* Jepsen, Mayr, and Simpson, 1949: 103–120.

——— 1949b. The vertebrate body. Philadelphia, Saunders.

Rosa, D. 1899. La riduzione progressiva della variabilita e i suoi rapporti coll'estinzione e coll'origine delle specie. Torino, Carlo Clausen.

——— 1931. L'Ologénèse; nouvelle théorie de l'évolution et de la distribution géographique des êtres vivants. Paris, Félix Alcan.

Ross, H. H. 1951. The origin and dispersal of a group of primitive caddisflies. Evolution, 5: 102–115.

Rowe, A. W. 1899. An analysis of the genus *Micraster*, as determined by rigid zonal collecting from the zone of *Rhynchonella Cuvieri* to that of *Micraster cor-anguinum*. Quart. Jour. Geol. Soc. London, 55: 494–547.

Ruedemann, R. 1918. The paleontology of arrested evolution. New York State Mus. Bull., No. 196, pp. 107–134.

——— 1922a. Additional studies of arrested evolution. Proc. Nat. Acad. Sci. 8: 54–55.

——— 1922b. Further notes on the paleontology of arrested evolution. Amer. Nat., 56: 256–272.

Samter, M. 1905. Die geographische Verbreitung von *Mysis relicta, Pallasiella quadrispinosa,* und *Pontoporeia affinis* in Deutschland also Erklärungsversuch ihrer Herkunft. Abh. Akad. Wiss. Berlin, 1905, Abh. 5: 1–34.

Säve-Söderbergh, G. 1934. Some points of view concerning the evolution of the vertebrates and the classification of this group. K. Vet. Akad. Stockholm, Ark. Zool., 26A (No. 17): 20.

Säve-Söderbergh, G. 1935. On the dermal bones of the head in Labyrinthodont Stegocephalians and primitive Reptilia. Medd. om Grønland, Kom. Vid. Unders. Grønland, 98 (No. 3): 1–211.

Schaeffer, B. 1947. Notes on the origin and function of the artiodactyl tarsus. Amer. Mus. Novitates, No. 1356: 1–24.

——— 1948. The origin of a mammalian ordinal character. Evolution, 2: 164–175.

——— 1952. The Triassic coelacanth fish *Diplurus,* with observations on the evolution of the Coelacanthini. Bull. Amer. Mus. Nat. Hist., 99: 25–78.

Schindewolf, O. H. 1936. Paläontologie, Entwicklungslehre und Genetik. Berlin, Borntraeger.

——— 1950a. Grundfragen der Paläontologie. Stuttgart, Schweizerbart.

——— 1950b. Der Zeitfaktor in Geologie und Paläontologie Stuttgart, Schweizerbart.

Schlaikjer, E. M. 1935. Contributions to the stratigraphy and paleontology of the Goshen Hole Area, Wyoming. Part 4; New vertebrates and the stratigraphy of the Oligocene and early Miocene. Bull. Mus. Comp. Zool. Harvard, 76: 97–189.

Schmalhausen, I. I. 1949. Factors of evolution, the theory of stabilizing selection. (Translated by I. Dordick, edited by Th. Dobzhansky). Philadelphia, Blakiston.

Scott, W. B. 1937. A history of land mammals in the Western Hemisphere. Revised ed., rewritten throughout. New York, Macmillan Co.

Shimer, H. W., and R. R. Shrock. 1944. Index fossils of North America. New York, Wiley.

Sholl, D. A. 1950. The theory of differential growth analysis. Proc. Roy. Soc. B, 137: 470–474.

Simpson, G. G. 1931a. Origin of mammalian faunas as illustrated by that of Florida. Amer. Nat., 65: 258–276.

——— 1931b. *Metacheiromys* and the relationships of the Edentata. Bull. Amer. Mus. Nat. Hist., 59: 295–381.

——— 1936a. Data on the relationships of local and continental mammalian faunas. Jour. Paleont., 10: 410–414.

——— 1936b. Census of Paleocene mammals. Amer. Mus. Novitates, No. 848: 1–15.

——— 1937a. Super-specific variation in nature from the viewpoint of paleontology. Amer. Nat., 71: 236–267.

——— 1937b. Additions to the Upper Paleocene fauna of the Crazy Mountain field. Amer. Mus. Novitates, No. 940: 1–15.

——— 1941. The function of saber-like canines in carnivorous mammals. Amer. Mus. Novitates, No. 1130: 1–12.

——— 1944a. Tempo and mode in evolution. (First edition.) New York, Columbia Univ. Press.

——— 1944b. The principles of classification and a classification of mammals. Bull. Amer. Mus. Nat. Hist., 85: i–xvi, 1–350.

——— 1946. Fossil penguins. Bull. Amer. Mus. Nat. Hist., 87: 1–100.

——— 1947a. A continental Tertiary time chart. Jour. Paleont., 21: 480–483.
——— 1947b. The problem of plan and purpose in nature. Sci. Monthly, 64: 481–495.
——— 1948. The beginning of the Age of Mammals in South America. Part I. Bull. Amer. Mus. Nat. Hist., 91: 1–232.
——— 1949a. The meaning of evolution. New Haven, Yale.
——— 1949b. Rates of evolution in animals. *In* Jepsen, Mayr, and Simpson, 1949: 205–228.
——— 1949c. Essay-review of recent works on evolutionary theory by Rensch, Zimmermann, and Schindewolf. Evolution, 3: 178–184.
——— 1950a. Evolutionary determinism and the fossil record. Sci. Monthly, 71: 262–267.
——— 1950b. History of the fauna of Latin America. Amer. Scientist, 38: 361–389.
——— 1950c. L'orthogenèse et la théorie synthétique de l'évolution. Colloques Internat. Centre Nat. Rech. Sci., 21: 123–163.
——— 1951a. Horses. New York, Oxford Univ. Press.
——— 1951b. The species concept. Evolution, 5: 285–298.
——— 1952. Periodicity in vertebrate evolution. Jour. Paleont., 26: 359–370.
Simpson, G. G., and A. Roe. 1939. Quantitative zoology. New York and London, McGraw-Hill.
Sinclair, W. J., and W. Granger. 1914. Paleocene deposits of the San Juan Basin, New Mexico. Bull. Amer. Mus. Nat. Hist., 33: 297–316.
Sinnott, E. W. 1946. Substance or system; the riddle of morphogenesis. Amer. Nat., 80: 497–505.
Sinnott, E. W., L. C. Dunn, and Th. Dobzhansky. 1950. Principles of genetics. 4th ed. [the first by these authors]. New York, McGraw-Hill.
Sloss, L. L. 1950. Rates of evolution. Jour. Paleont., 24: 131–139.
Small, J. 1946. Quantitative evolution VIII. Numerical analysis of tables to illustrate the geological history of species number in diatoms; an introductory summary. Proc. Roy Irish Acad., 51, B: 53–80.
Smith, H. M. 1946. Handbook of lizards. Ithaca, Comstock.
Spath, L. F. 1933. The evolution of the Cephalopoda. Biol. Rev., 8: 418–462.
Spencer, W. P. 1947a. Mutations in wild populations in *Drosophila*. Advances in Genetics. 1: 359–402.
——— 1947b. Genetic drift in a population of *Drosophila immigrans*. Evolution, 1: 103–110.
Sperber, I. 1944. Studies on the mammalian kidney. Zool. Bidrag Uppsala, 22: 249–432.
Stadler. L. C. 1932. On the genetic nature of induced mutations in plants. Proc. VI Int. Cong. Genet., 1: 274–294.
Stebbins, G. L., Jr. 1949. Reality and efficacy of selection in plants. Proc. Amer. Phil. Soc., 93: 501–513.
——— 1950. Variation and evolution in plants. New York, Columbia.
Stebbins, R. C. 1944. Some aspects of the ecology of the iguanid genus *Uma*. Ecol. Mon., 14: 311–332.

Stenzel, H. B. 1949. Successional speciation in paleontology: the case of the oysters of the *sellaeformis* stock. Evolution, 3: 34–50.

Stern, C. 1949a. Gene and character. *In* Jepsen, Mayr, and Simpson, 1949: 13–22.

——— 1949b. Human genetics. San Francisco, Freeman.

Stirton, R. A. 1940. Phylogeny of North American Equidae. Univ. California Pub., Bull. Dept. Geol. Sci., 25: 165–198.

——— 1947. Observations on evolutionary rates in hypsodonty. Evolution, 1: 32–41.

Stromer, E. 1940–1941. Kritische Betrachtungen. 3. Die Lückenhaftigkeit der Fossilüberlieferung und unserer derzeitigen Kenntnisse und die Folgerungen daraus. Zent.-Bl. Min. etc., 1940B: 262–276; 1941B: 1–23.

——— 1944. Gesicherte Ergebnisse der Paläozoologie. Abh. Bayer. Akad. Wissensch., n.F., 54: 1–114.

Stubbe, H., and F. von Wettstein. 1941. Über die Bedeutung von Klein- und Grossmutationen in der Evolution. Biol. Zentralb., 61: 264–297.

Sturtevant, A. H. 1929. The genetics of *Drosophila simulans*. Pub. Carnegie Inst. Washington, No. 399, pp. 1–62.

——— 1937. Essays on evolution. 1. On the effects of selection on mutation rate. Quart. Rev. Biol., 12: 464–467.

Sturtevant, A. H., and T. Dobzhansky. 1936. Geographical distribution and cytology of "sex ratio" in *Drosophila pseudoobscura*. Genetics, 21: 473–490.

Svärdson, G. 1945. Chromosome studies on Salmonidae. Repts. Swedish State Inst. Fresh-water Fishery Research, No. 23.

Swinnerton, H. H. 1923. Outlines of palaeontology. London, Arnold.

——— 1932. Unit characters in fossils. Biol. Rev., 7: 321–335.

——— 1947. Outlines of palaeontology. 3d ed. London, Arnold.

Teilhard de Chardin, P. 1942. New rodents of the Pliocene and lower Pleistocene of China. Pub. Inst. Geobiol. Pekin, No. 9.

——— 1950. Sur un cas remarquable d'orthogénèse de groupe: l'évolution des siphnéidés de Chine. Colloques Internat. Centre Nat. Rech. Sci., 21: 169–173.

Thomas, H. D. 1940. Fossils and fashions. Proc. Trans. South-East. Union Sci. Soc. for 1939: 52–54.

Thorpe, W. H. 1930. Biological races in insects and allied groups. Biol. Rev., 5: 177–212.

——— 1940. Ecology and the future of systematics. *In* Huxley, 1940: 349–364.

Timofeeff-Ressovsky, N. W. 1934. Über die Vitalität einiger Genmutationen und ihrer Kombination bei *Drosophila funebris* und ihre Abhängigkeit vom "genotypischen" und vom äusseren Mileau. Zeitschr. indukt. Abst.- u. Vererbungslehre, 66: 319–344.

——— 1935. Über geographische Temperaturrassen bei *Drosophila funebris*. Arch. Naturgesch., n.F., 4: 245–257.

——— 1940. Mutations and geographical variation. *In* Huxley 1940, pp. 73–136.

Trevisan, L. 1949. Lineamenti dell'evoluzione del ceppo di elefanti eurasiatici nel Quaternario. La Ricerca Scientifica, 19, Suppl.: 105–111.

Truemann, A. E. 1930. Results of recent statistical investigations of invertebrate fossils. Biol. Rev., 5: 296–308.

Umbgrove, J. H. F. 1942. The pulse of the earth. The Hague, Martinus Nijhoff.

——— 1946. Evolution of reef corals in East Indies since Miocene time. Bull. Amer. Assoc. Petrol. Geol. 30: 23–31.

Villee, Claude A. 1942. The phenomenon of homoeosis. Amer. Nat., 76: 494–506.

Waagen, W. 1868. Die Formenreihe des *Ammonites subradiatus.* Benecke geog.-pal. Beit., 2: 179–257.

Waddington, C. H. 1939. An introduction to modern genetics. New York, Macmillan.

——— 1950. The biological foundations of measurements of growth and form. Proc. Roy. Soc., B, 137: 509–515.

Watson, D. M. S. 1919. The structure, evolution and origin of the Amphibia. The "orders" Rachitomi and Stereospondyli. Phil. Trans. Roy. Soc. London, B, 209: 1–73.

——— 1926. The evolution and origin of the Amphibia. Phil. Trans. Roy. Soc. London, B, 214: 189–255.

——— 1940. The origin of frogs. Trans. Roy. Soc. Edinburgh, 60 (Pt. 1): 195–231.

——— 1949. The evidence afforded by fossil vertebrates on the nature of evolution. *In* Jepsen, Mayr, and Simpson: 45–63.

——— 1950. L'évolution des amphibiens et son mécanisme. Colloques Internat. Centre Nat. Rech. Sci., 21: 45–50.

Weber, M. 1928. Die Säugetiere. 2d ed., with the collaboration of O. Abel. Jena, Fischer.

Weidenreich, Franz. 1941. The brain and its role in the phylogenetic transformation of the human skull. Trans. Amer. Phil. Soc., n.s., 31: 321–442.

Weigelt, J. 1943. Paläontologie als stammesgeschichtliche Urkundenforschung. *In* Heberer, 1943a: 131–182.

Weldon, W. F. R. 1898. Opening address [to Zoological Section of the British Association]. Nature, 58: 499–506.

Westoll, T. S. 1943. The origin of the tetrapods. Biol. Rev., 18: 78–98.

——— 1949. On the evolution of the Dipnoi. *In* Jepsen, Mayr, and Simpson, 1949: 121–184.

——— 1950. Some aspects of growth studies in fossils. Proc. Roy. Soc., B, 137: 490–509.

Wheeler, J. F. G. 1942. The discovery of the nemertean *Gorgonorhynchus* and its bearing on evolutionary theory. Amer. Nat., 76: 470–493.

White, M. J. D. 1951. Evolution of cytogenetic mechanisms in animals. *In* Dunn, 1951: 333–367.

White, W. J. 1945. Animal cytology and evolution. Cambridge, Cambridge Univ. Press.

Williams, J. 1951. Fall of the sparrow. New York, Oxford Univ. Press.

Willis, J. C. 1940. The course of evolution. Cambridge, Cambridge University Press.

Wilson, R. W. 1951. Evolution of the early Tertiary rodents. Evolution, 5: 207–215.

Wood, H. E., II. 1949. Evolutionary rates and trends in rhinoceroses. *In* Jepsen, Mayr, and Simpson: 185–189.

Worthington, E. B. 1937. On the evolution of fish in the great lakes of Africa. Int. Rev. Ges. Hydrol., Hydrogr., 35: 304–317.

——— 1940. Geographical differentiation in fresh waters with special reference to fish. *In* Huxley, 1940: 287–302.

Wright, S. 1931. Evolution in Mendelian populations. Genetics, 16: 97–159.

——— 1932. The roles of mutation, inbreeding, crossbreeding, and selection in evolution. Proc. 6th Int. Cong. Genetics, 1: 356–366.

——— 1934. Genetics of abnormal growth in the guinea pig. Cold Spring Harbor Symp. on Quant. Biol., 2: 137–147.

——— 1935. Evolution in populations in approximate equilibrium. Jour. Genetics, 30: 257–266.

——— 1937. The distribution of gene frequencies in populations. Proc. Nat. Acad. Sci., 23: 307–320.

——— 1939. Statistical genetics in relation to evolution. Paris, Hermann.

——— 1940. The statistical consequences of Mendelian heredity in relation to speciation. In Huxley 1940, pp. 161–183.

——— 1941a. The material basis of evolution. Sci. Monthly, 53: 165–170.

——— 1941b. The "Age and area" concept extended. Ecology, 22: 345–347. (A review of Willis, 1940.)

——— 1942. Statistical genetics and evolution. Bull. Amer. Math. Soc., 48: 223–246.

——— 1945. Tempo and mode in evolution: a critical review. Ecology, 26: 415–419.

——— 1948. On the roles of directed and random changes in gene frequency in the genetics of populations. Evolution, 2: 279–294.

——— 1949a. Adaptation and selection. *In* Jepsen, Mayr, and Simpson, 1949: 365–389.

——— 1949b. Population structure in evolution. Proc. Amer. Phil. Soc., 93: 471–478.

——— 1950. Genetical structure of populations. Nature, 166: 247–249.

Wright, Sewall, Th. Dobzhansky, and W. Hovanitz, 1942. Genetics of natural populations. Part 7: The allelism of lethals in the third chromosome of *Drosophila pseudoobscura*. Genetics, 27: 363–394.

Young, J. Z. 1950. The life of vertebrates. Oxford, Clarendon Press.

Zeuner, F. E. 1931. Die Insektenfauna des Böttinger Marmors. Fortschr. Geol. Pal., (9) 28: 1–160.

——— 1944. The Pleistocene period. Its climate, chronology, and faunal successions. London, Ray Society.

——— 1946a. Dating the past, an introduction to geochronology. London, Methuen.

——— 1946b. Time and the biologist. Discovery, 7: 242–249.

Zimmerman, E. C. 1943. On Wheeler's paper concerning evolution and the nemertean *Gorgonorhynchus*. Amer. Nat., 77: 373–376.

——— 1948. Insects of Hawaii. Vol. I. Introduction. Honolulu, Univ. of Hawaii Press.

Zimmermann, W. 1948. Grundfragen der Evolution. Frankfurt, Klostermann.

Zittel, K. A. von. 1913. Text-book of paleontology. [English revision by C. R. Eastman.] Vol. I. London and New York, Macmillan.

——— 1924. Osnovi paleontologii (Paleozoologiya). Part 1: Bespozvonochnie. [Russian revision by A. N. Ryabinin.] Leningrad. ONTI, NKTP, SSSR.

Zuckerman, S. (Leader of symposium) 1950. A discussion on the measurement of growth and form. Proc. Roy. Soc., B, 137: 433–523.

Index